The Way the Earth Works:

An Introduction to the New Global Geology and Its Revolutionary Development

Peter J. Wyllie

University of Chicago

JOHN WILEY & SONS

New York • Chichester • Brisbane • Toronto • Singapore

dedication

Dedicated to those I neglected during nights and weekends of writing, to old friends who understood, to new friends who wondered, and to acquaintances who might have become friends, given time.

Library of Congress Cataloging in Publication Data

Wyllie, Peter J.   1930–
       The way the earth works.

Bibliography: p.
Includes index.
1. Geology.   I. Title.

QE26.2.W93   550   75-23197
ISBN 0-471-96902-8
ISBN 0-471-96896-X pbk.

Printed in the United States of America

20 19 18 17 16 15 14

Cover Photo and Book Design By Angie Lee

preface

This textbook is based on the "Rocks" part of "Rocks and Stars," which is one of the courses offered for the physical science requirement in the liberal arts degree program at the University of Chicago. The central theme is plate tectonics, the new paradigm for the revolution in earth sciences. This format follows the recommendation of J. Tuzo Wilson that education in the earth sciences begin with this unified theory. The traditional geological topics are introduced and outlined only where they are required to develop the framework of plate tectonics.

The theme is placed in the historical setting of the continental drift debate, beginning early this century, and the rapid and remarkable developments of the revolutionary 1960s are recounted in some detail. The theory of plate tectonics is presented first, followed by the supporting evidence from earthquakes. Next comes the evidence from the earth's magnetic field and magnetized rocks that permit sea-floor spreading to be dated through tens of millions of years and the drifting continents to be tracked through hundreds of millions of years. Selected topics, including the geological cycle, the lost continent of Atlantis, and the extinction of dinosaurs are reviewed in the context of plate tectonics. The revolution in earth sciences was accompanied by an immense increase in our

understanding of planetary sciences, and this development is followed in the history of the lunar exploration program, with an introduction to some of the scientific conclusions. The final chapter returns again to earthquakes in connection with environmental geology and power plants.

This textbook was designed for nonscience majors, but science majors will find that the detailed explanations and the history of the revolution will improve their comprehension of many topics covered briefly in conventional courses that adopt the traditional approach to earth sciences through geology. I hope that it will prove useful as a supplement for such courses. Beginning graduate students whose undergraduate training did not stress plate tectonics need an overview of the topic and its development, and they will find it in this book.

Conventional college introductions to the earth sciences attempt to cover as many topics and basic principles as possible. This leads either to an encyclopedic textbook with far more pages than a student can reasonably be expected to read, or to a dictionarylike version in which each topic receives passing mention and little explanation. Most of them contain enough geological jargon to require a glossary. In today's technological society, scientific literacy is just as important as familiarity with the "great books," but I see no need for geologists to expect future humanists and social scientists to remember words such as epeirogeny, monadnock, pedalfer, phenocryst, or taconite.

The elegance and simplicity of the basic ideas in the new global geology provides an opportunity to present students with a grand, revolutionary theory using a minimum of geological and geophysical jargon. The theory is presently the main focus of research activity for many earth scientists; this is where the excitement is. I agree with Tuzo Wilson that this should be the topic for the first course. After the global framework in the first semester or term, a teacher can follow with almost anything: minerals and rocks, geomorphology, historical geology, oceans, atmospheres, planetary sciences, or environmental geology and natural resources. Teachers who prefer not to delay the specialized details of familiar topics could run a conventional laboratory sequence in physical geology parallel with lectures covering the subject of this textbook.

I have selected the main lines of evidence for plate tectonics and explained them in more detail than is customary in textbooks, devoting most attention to the topics that my students have found difficult to understand. Instead of borrowing diagrams directly from the research literature, I have taken them apart, simplified them, and put them together again in an attempt to help students visualize objects and space in three dimensions. This is important, for example, in the distinction between oceanic magnetic anomaly stripes and magnetized ocean-floor strips. I justify the lengthy explanations with the belief that students gain more satisfaction from really understanding a few significant topics than from learning large numbers of small facts. The narrative is carefully constructed so that each chapter builds on preceding chapters, assumes no other prerequisites, and prepares the way for following chapters. I have tried to alternate the explanatory sections with the more lively topics and historical debates and anecdotes.

Parts of chapters 2, 13, 17 and 18 are similar to parts of my article "Revolution in the Earth Sciences" published in *The Great Ideas Today, 1971*. I thank Encyclopaedia Britannica, Inc., for permission to use this material.

Several people were kind enough to review parts of the manuscript for me, and their critical comments have improved it. I thank G. Forney, J. R. Goldsmith, J. Goldsmith, L. Grossman, Y. Herman, W. S. McKerrow, W. D. Means, A. A. Meyerhoff, T. J. M. Schopf, J. V. Smith, H. Spall, L. R. Sykes, D. S. Wood, and F. R. Wyllie. Tom Schopf was especially critical and helpful. Only Mrs. Glenda York was more helpful, with her speedy, accurate, and cheerful typing.

Peter J. Wyllie

University of Chicago
Chicago, Illinois
April 1975

contents

chapter 1

Introduction: Time for a New Theory

This book presents, in simple language, the revolutionary theory of the earth, commonly known as the "new global geology." It explains the main lines of evidence on which it is based and outlines some of the implications and applications of the theory. This novel approach to the study of geology and the earth sciences differs from the classical pyramidal approach, in which groups of data and interpretations are gradually built up into a theory. Here we examine the grand global theory first, and the various subjects can then be comprehended at a more advanced level than in the conventional approach; each subject dovetails into its appropriate place in the framework of the global theory.

THE "NEW GLOBAL GEOLOGY," CONTINENTAL DRIFT, AND PLATE TECTONICS

The majority of geologists, geophysicists, and geochemists now believe that the continents have drifted across the surface of the earth through much of earth history, that is, for billions of years. Supercontinents have rifted apart; oceans have opened, expanded, and disappeared again as continents collided. Continental collisions thrust up great mountain ranges such as the Himalayas. The continents are still drifting: the Atlantic Ocean is increasing in size, and the Pacific Ocean is becoming smaller. Earlier positions of the drifting continents are shown in Figures 1-1, 12-3, and 12-4.

Evidence for and against the idea of continental drift has been debated since early in this century, but it was not generally accepted until the late 1960s as part of a new theory, called plate tectonics. "Tectonics" comes from the same stem as "architecture." It is the branch of geology concerned with the process of mountain building and, on a smaller scale, with the way in which rocks become buried within the earth, folded, and broken.

According to plate tectonic theory, the surface of the earth is covered by a series of relatively thin, shell-like plates. Figures 1-2 and 5-1 show the boundaries of these plates at the earth's surface, and Figure 2-3 shows a cross section through them. These rigid plates of rock slide over the earth's interior, grinding against each other, rather like ice floes drifting on a lake. The plate boundaries are sites of geological and tectonic activity; earthquakes and volcanoes are concentrated along them, and some boundaries are sites of mountain building.

Fig. 1-1 The continents have drifted. 200 million years ago the continental land masses formed a single supercontinent. This broke up and fragments drifted apart, with oceans forming between them. Compare Figs. 12-3 and 12-4. (a) 200 million years ago. (b) 65 million years ago. (c) Today.

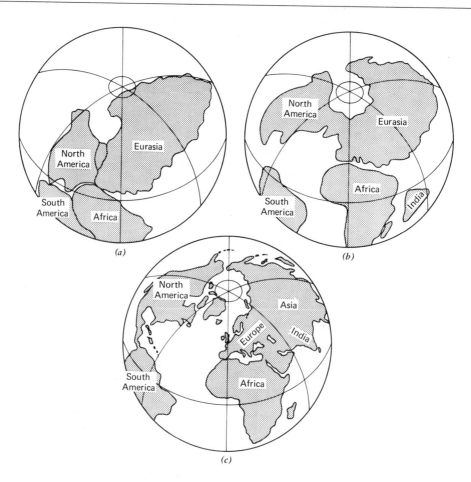

Acceptance of this theory involved such a significant reversal of scientific opinion that the theory has been proclaimed as a scientific revolution. For the first time, earth scientists have a global scheme that appears to explain many aspects of the behavior of the earth's surface and that links together phenomena that previously seemed to be unrelated and difficult to explain.

Various aspects of the history and behavior of the whole earth and its surface have been studied by experts trained in the diverse fields of geology, biology, physics, and chemistry. Most of these investigations are embraced by the term "physical sci-

ences." We should begin in proper scientific fashion by defining our terms, and in a good dictionary we find:

physical science. *the natural sciences (as mineralogy, astronomy, meteorology, geology) that deal primarily with nonliving materials.*

So then we turn the pages back to:

natural science. *any of the sciences (as physics, chemistry, biology) that deal with matter, energy, and their interrelations and transformations or with objectively measurable phenomena.*

and then to:

geology: 1a. *a science that deals with the history of the earth and its life especially as recorded in rocks;* b. *a study of the solid matter of a celestial body (as the moon).*

Communication and exchange of ideas among the various experts has not always been easy, partly because they used different symbolic languages—different jargon—the same kind of symbolism that apparently turns many young students away from science before they have the opportunity to learn enough to see the awesome beauty of a great theory. Because of the revolution, however, and the coherency of the new concepts involved, we now find research workers from most of the classical disciplines in physical sciences, as well as many from the biological sciences, collaborating on earth problems that turn out to be of mutual interest. Who would have anticipated, just a few years ago, that the combination of evidence from the study of fossilized skeletons of minute marine animals in rocks beneath the ocean and the evidence from the earth's magnetic field could have produced an elegant proof that the Atlantic Ocean has been increasing in width at a constant rate of 4 cm per year for at least 75 million years (see Chapter 14)?

The revolutionary theory of plate tectonics provides a central focus for the research activity of many earth scientists. This is where new discoveries are being made. This is where the excitement is. What could be more mind bending than the realization that the slow, northward movement of the Pacific Ocean plate will bring Los Angeles alongside San Francisco 10 million years from now (Chapter 12)?

The 1960s was a decade of extraordinary scientific ferment. It was time for a great idea. Most of this book is an account and evaluation of the theory of plate tectonics, but it includes other contemporaneous developments in the earth sciences. A special deep-sea drilling ship began to gather data from the ocean floor, and the space program landed astronauts on the moon. Growing public awareness of ecology, pollution, and recycling led to the development of environmental geology as an active research area.

DEEP-SEA DRILLING

About 70% of the earth's surface is covered by ocean water, and the floor of the ocean basins submerged beneath 4 km of water

Fig. 1-2 The surface of the earth is covered by a number of shell-like plates, with boundaries independent of the coastline between continents and oceans. The rigid plates move relative to each other, carrying the continents with them, and thus causing continental drift. Compare Figs. 2-3 and 5-1.

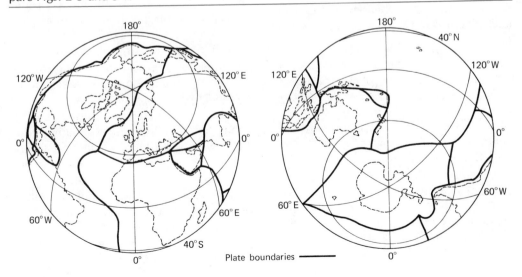

Plate boundaries ⎯⎯⎯

has long seemed as inaccessible as the surface of the moon. Most of the surface of the solid earth has therefore remained a mystery to geologists, whereas the near side of the moon could at least be examined through high-powered telescopes. A special drilling ship called the *Glomar Challenger* began a series of voyages in August 1968, and this has yielded an enormous quantity of data about the ocean floor. The *Glomar Challenger* is a strange-looking ship constructed specifically for the purpose of drilling and recovering long cores of sediments from the ocean floor (Figure 14-1).

When Charles Darwin was a young man of 22, he sailed in *H. M. S. Beagle* as a naturalist on a round-the-world expedition that lasted five years, from 1831 to 1836. This was the preparation for his major contribution to science, *On the Origin of Species by Means of Natural Selection,* which was published much later in 1859. The impact of this work was so great that, over a century later, the title *The Origin of Species* is probably more widely known than that of any other scientific book. The data obtained from the voyages of the *Glomar Challenger* will exceed in quantity that collected during the voyage of the *Beagle,* and it seems possible that the results and syntheses arising from the data may even match the thesis developed by Darwin in scientific significance.

In a December 1969 issue of *Science,* P. H. Abelson of the Geophysical Laboratory, Washington D. C., wrote in an editorial:

This has been a remarkable year for geology and associated disciplines. Two great developments have enriched these fields with new knowledge, new puzzles, and new objects for study. The most widely publicized of these developments is the exploration of the moon.... The second development is the success of an extensive program of drilling of the deep-sea bottom.... One cannot foresee ... the new opportunities that will arise from the two great recent developments. It is clear, however, that before another decade is over our understanding of the earth and the solar system will be substantially increased.

LUNAR EXPLORATION

The scientific revolution in earth sciences received its share of reports in the popular press during the late 1960s, but it was overshadowed by the publicity given to the exploration of the moon.

The story of the century was the first manned landing on the moon by Neil Armstrong and Edwin Aldrin in a space flight that began in Florida on July 16, 1969 and ended with a successful splashdown in the Pacific Ocean on July 24. The lift-off of Apollo 11 was watched by half a million visitors in Florida, including about 3500 journalists from the United States and 55 other countries. Television cameras brought the flight and the historic walk on the moon into the living rooms of the world. Geologists sitting in comfort at home watched the astronauts collect rock samples, in anticipation that these would solve some of the problems related to the origin of the earth.

Undoubtedly the greatest scientific event of 1970 was the Apollo 11 Lunar Science Conference held in Houston, January 5 to 8. The 22 kg of rock brought back from the moon by the crew of Apollo 11 was delivered into quarantine at the Lunar Receiving Laboratory in Houston on July 25, 1969 and held there for a month in case some kind of space plague should develop from alien pathogenic organisms. Intensive search for organisms by a variety of methods, during and after quarantine, showed that the moon rocks are devoid of reproductive life forms.

Results of research in Houston by the Lunar Sample Preliminary Examination Team were published in *Science* on September 19, at about the time that samples were being distributed to the 142 principal investigators who had earlier submitted proposals to study the lunar rocks. Most investigators had a number of professional associates involved in the research. For about three frenzied months, more than 500 scientists in nine countries devoted their efforts to the small samples they received from the 22 kg total. On January 5, 1970, they convened at the conference in Houston to present and compare their findings and to discuss their implications for the origin of the rocks, the origin of the moon, and the origin of the solar system.

The results were presented to an audience of about 1000 scientists and many journalists. Worldwide interest was shown in the results of this unique conference. The general public, waiting for some dramatic announcement about the origin of the moon, was probably rather disappointed by the lack of consensus among the experts about the broader applications of the research; the discovery of lava flows, with its confirmation that the moon had been very hot and chemically differentiated, was of less interest to the public than to earth scientists, who had debated this question for many years.

The intensive study of moon rocks by many specialists from many countries was continued with samples returned from other locations on the moon by the missions Apollo 12, 14, 15, 16, and 17. The Apollo 17 mission was completed in December 1972, about 3.5 years after the first landing in July 1969. It was the first of the missions to take off at night, and the spectacle of Apollo 17 rising on a pillar of fire was viewed by millions on television and by about 50,000 people on Florida highways and beaches. Public interest had waned since Apollo 11, which had been viewed by half a million Florida spectators.

The Fifth Lunar Science Conference was held in Houston in March 1974. By this time, 12 astronauts had explored the moon's surface, numerous scientific experiments had been conducted on the moon and from spacecraft orbiting around the moon, and 380 kg of rock had been returned to Houston. More than 1000 scientists in 19 countries had studied the lunar rock samples. Results were also available from about 150 g of lunar soil returned by Russia's unmanned spacecraft Luna 16 and Luna 20.

After nearly four years of exploration and intensive research, it was established that the origin of the solar system was more complex than previously imagined. There was no consensus regarding the origin of the moon, but a broad picture of the history of the moon was beginning to emerge. The first billion years of the moon's history provides information about the early history of the earth, which is useful because little trace has been found in the geological record of what was happening on earth during this period.

The significance of the moon's exploration was emphasized by B. Mason and W. G. Melson of the Smithsonian Institution in their book, *The Lunar Rocks*, which reviewed the results presented at the First Lunar Conference. Chapter 1 opened thus.

By any standards the decade of the 1960's was surely unique for science as a whole and for geology in particular. It began with a first view of the far side of the Moon and it ended with the first look at the material from which the Moon is made. In the intervening ten years the Moon was intensively mapped by remote-controlled spacecraft, both Russian and American.

This phase of scientific exploration of the moon is now over. It will be many years before the lunar rock specimens have revealed all of their secrets, but the initial excitement that captured headlines has abated. Developments arising from the revolution in earth sciences are of more direct interest to nonscientists than the lunar results.

ENVIRONMENTAL GEOLOGY

On May 31, 1970, about three months after the first Lunar Science Conference, a disastrous earthquake killed 50,000 people in Peru. Plate tectonics explains the distribution of active belts of the earth's crust, as expressed by earthquakes and volcanoes. If a phenomenon can be explained, then there is a hope that it can be controlled. The fear of a major earthquake disaster in the densely populated regions of California and

Japan has led to considerable research effort in earthquake prediction and control.

It is fashionable to classify earthquakes as geological hazards, and these constitute one part of the burgeoning subject of environmental geology. According to W. R. Dickinson of Stanford University:

one of the great ideas, perhaps the key idea, of human civilization . . . the realization that mankind must learn to live in balance with his physical environment . . . and I believe its time has come.

This quotation is from one of several papers published in the *Journal of Geological Education* in November 1970, from an April symposium on "Education in Environmental Geology," held in San Jose, California. In his introductory remarks, G. B. Oakeshott of the California Division of Mines and Geology claimed that *all* geology is environmental, and that there is no such thing as a geological hazard without people. Greatly intensified and unwise use of the environment in metropolitan areas has turned ordinary geological processes into "geological hazards."

In recent years, there has certainly been a tremendous increase in public awareness of the environment, of ecology, and of the possible effects of pollution of the environment and of ecological systems. Many scientists have suggested that the present sensitivity of man to his own environment is not due merely to tearful eyes, gasping lungs, and views of muck-laden streams, but that it might have been stimulated by the sight of what astronaut Aldrin called "Beautiful, beautiful—magnificent desolation." The desolate lunar landscape may have awakened people to the actual and potential disasters on earth, disasters of their own making. Many have questioned the desirability and the ethics of spending billions of dollars on the exploration of space, proposing instead that the money should be applied to the construction of sewage plants, the reduction of effluents from fuels, and the alleviation of poverty.

One thing is certain. If we are to survive as a species, we must tend more carefully the resources we draw from the environ-ment: air, water, fuel, and a host of minerals, rocks, and ore deposits. One way to test the theory of plate tectonics is by making long-range predictions; but this will work only if our descendants survive to check the predictions.

THE GEOLOGICAL TIME SCALE

The casual statement above about the first billion years of the moon's history was probably meaningless to most readers. It is very difficult for our senses to grasp the significance of the immense span of geological time. Yet the study of the earth is a historical subject, and many of the topics reviewed in subsequent chapters have to be dated. Much effort during the nineteenth and twentieth centuries was devoted to unraveling the sequences of geological events from the study of rocks and locating them on a relative time scale. But all efforts to calibrate the scale were failures until a method based on radioactive elements was developed.

Many rocks contain small quantities of radioactive elements that break down at constant rates to yield other elements, particles, and radiation. If the rate of breakdown of a radioactive element is known, the age of a rock measured from its time of formation (Chapter 6) can be determined by measuring the proportions of the radioactive element and the products of its breakdown. The principles of the method have been understood since 1904, but it is only since 1950 that the difficult experimental techniques were refined sufficiently to permit the dating of rocks on a routine basis. The geological time scale calibrated by this method is shown in the left-hand column of Figure 1-3.

The earth was formed about 4.6 billion (4.6×10^9) years ago, and 85% of this time is classified as Precambrian. Traces of primitive life forms have been detected in rocks as old as 3 billion years, but abundant fossils of more advanced forms are found only in rocks no more than 570 million years old, from the beginning of the Paleozoic era. At this time, creatures began

Fig. 1-3 The radiometric geological time scale compared with the geomagnetic time scale. The left-hand column shows the time since the earth was formed, 4.6 billion years ago. The scale for each successive column is increased by a factor of 10. The time scale is divided into named geological eras. The geomagnetic time scales are reviewed in Chapters 9 and 10. Compare Figs. 10-9*b* and 9-3.

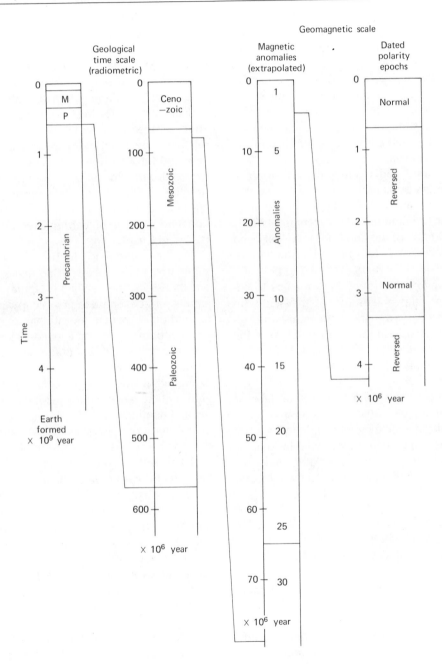

to build shells, which are more readily preserved as fossils in rocks than are the earlier soft-bodied animals.

The scale for each successive column in Figure 1-3 is increased by a factor of 10, and the second column is an expanded version of the time scale since the Precambrian, showing the durations of the Paleozoic, Mesozoic, and Cenozoic eras. These are subdivided into smaller units, each with its own name. In the geological literature, events tend to be described in terms of the named period during which they occur, but in this book we will avoid this long string of names by using the calibrated time scale and citing ages in terms of millions of years.

The third and fourth columns are each expanded by another factor of 10, so that the right-hand column, covering a span of just over 4 million years, represents only 1/1000 part of the duration of the left-hand column. Erect, two-legged primates developed only about 5 million (0.005×10^9) years ago, which may help readers to appreciate the magnitude of the scales in Figure 1-3. Now mark on the right-hand column the position for the destruction of the Cretan civilization, 1500 B.C. (Chapter 14), and see the supposed antiquity of ancient history!

The two right-hand columns are labeled "Geomagnetic scale." These give calibrated dates of magnetic polarity reversal epochs, which will be reviewed in Chapter 9, and the extrapolated dates of linear magnetic anomalies identified by numbers, which will be reviewed in Chapter 10.

APPROACH ADOPTED
IN FOLLOWING CHAPTERS

The central theme of this book is plate tectonics. The approach adopted with each new topic, wherever possible, is to present an overview, a complete framework, and then to examine selected aspects of the topic in more detail. Chapter 2 outlines the theory of plate tectonics and the history of its conceptual development, starting with continental drift in 1912. This serves as an introduction, or catalog, for many of the topics reviewed in detail in subsequent chapters. We first examine plate tectonics, then sea-floor spreading, and then continental drift.

In Chapter 3 we review the distribution of the large-scale features on the surface of the solid earth, on the continents, and beneath the oceans, and these are related to the distribution of earthquakes in Chapter 4. Earthquakes are signals of activity and, in Chapter 5, the interpretation of earthquakes and the surface features of the solid earth are combined to show the distribution and relative movements of lithosphere plates now and through historic time. Earthquake waves also provide an X-raylike view of the structure of the earth's interior, which is presented in Chapter 7. Methods for prediction and control of earthquakes are discussed in Chapter 18.

The geological cycle (in Chapter 6) summarizes the processes that make rocks and that shape the detailed sculpture of the earth's surface. These topics occupy a large proportion of many conventional introductory textbooks in earth sciences. Here we consider only the broad outline. Rocks become magnetized, and this process is examined in more detail because it provides critical evidence for sea-floor spreading and continental drift.

The study of anomalies (irregularities) in the earth's present magnetic field, associated with sea-floor spreading, defines the directions and rates of movement of plates through the past 4.5 million years and, by extrapolation, back to almost 100 million years. These periods are shown by the two right-hand columns in Figure 1-3. This is only a small fraction of the total time scale. The explanation of the evidence used is reviewed in Chapters 8, 9, 10, and 14.

Independent evidence for continental drift and plate movements comes from paleomagnetism, the study of fossil magnetism preserved in rocks. Chapters 11 and 12 explain how the relative movements of the continents are traced back for 500 million years or more. Even this takes us back only through about 1/10 of the earth's history (Figure 1-3).

The revolution in earth sciences has not occurred without controversy and dissent. Chapter 13 includes a detailed account of the development of the revolution and a critical evaluation by dissenters of some of the evidence considered in previous chapters. Similarly, in Chapter 17, we examine some of the scientific controversies about the history of the moon, both before and after the Apollo missions. Science is not always as cut and dried as it may seem in some textbooks.

Some applications and implications of plate tectonics for diverse topics such as the lost city of Atlantis, the extinction of dinosaurs, and the replacement of our depleted natural resources are discussed in Chapters 14, 15, and 18.

In Chapter 16 we consider how the earth works and review the proposed driving mechanisms for the moving plates.

Chapters 17 and 18 deal with the exploration of the moon and with environmental geology, two topics that developed more or less at the same time as the revolution in earth sciences.

SUMMARY

Conventional college introductions to the earth sciences and physical sciences attempt to cover as many of the basic principles as possible, building data and interpretations into theory. In most terminal one-year courses there is not enough time to cover the basics in sufficient detail for students to reach an appreciation of the beauty of science. The revolutionary theory of plate tectonics, the new global geology, provides an opportunity to present a simple, elegant theory of the earth, which is presently the central focus of research activity for many scientists. In this volume we start with the beauty of a conceptual model, and the various topics then reviewed can be appreciated at a more advanced level than in the conventional approach.

Most geologists, geophysicists, and geochemists are now convinced that the continents have drifted across the surface of the earth through much of earth history. Great mountain ranges were thrust up by continental collisions. Acceptance of this theory in the late 1960s involved such a significant reversal of opinion that it is considered to represent a scientific revolution. Associated with this revolution were three other developments. In 1968, a specially designed oceanographic vessel, *Glomar Challenger,* began a series of world voyages in an extensive program to drill into the deep-sea bottom. The first manned landing on the moon in 1969 began a series of lunar field trips that returned many kilograms of rock for study in unprecedented detail. Environmental geology became a burgeoning subject because of concern in the 1960s about expanding world population, increased exploitation of natural resources, and the pollution caused by industrial development.

The evidence for the new theory, based largely on earthquakes, the earth's magnetism, and magnetized rocks, is examined in terms of recent activity, of sea-floor spreading through the past 100 million years, and of continental drift for more than 500 million years. The earth was formed 4,600 million years ago.

SUGGESTED READINGS are at the end of the book.

chapter 2
Revolution in the Earth Sciences

When the revolution in earth sciences was proclaimed in 1967, many students were taking an active role in fomenting and formulating social revolutions on campus and in business districts. Their activities have significantly influenced many aspects of university life, including curricula and teaching methods.

Less well publicized than the social revolutions but probably with greater long-range influence on education and our lifestyles are the scientific revolutions that are reshaping the conceptual frameworks of one subject after another. Consider the following quotations from *The Great Ideas Today*, published in 1967 by Encyclopedia Britannica. According to Professor S. Toulmin:

In the world of science, as in the worlds of society and technology, the second half of the 20th century is proving to be a period of "perpetual revolution," and scientists are learning to live with the fact.

And, referring to the biological sciences, Professor T. T. Puck wrote that:

It has become increasingly clear during the past decade that biology is undergoing a new and revolutionary development. . . . Molecular biology has provided for the first time a master plan . . . provides

at least a generalized picture into which everything we know of the more complex systems seems to fit.

Similarly, the concept of plate tectonics in earth sciences has produced revolutionary developments. Whereas the revolution in biological sciences was based on microscale studies, that in the earth sciences was based on a macroscale scheme. Molecular biology seems to be capable of explaining large, more complex biological systems. Plate tectonics is the master scheme for the whole earth; it provides a framework for the explanation of localized geological phenomena and seems to accommodate and link together many aspects of earth sciences that previously had been treated as subjects for independent study.

Many writers have compared the potential effects of these revolutions in biological and earth sciences with those produced earlier by (1) the change from a belief in the separate creation of living things to a belief in Darwin's theory of evolution, and (2) the acceptance of quantum mechanics and Einstein's theory of relativity in physics.

A revolution involves a radical change in beliefs. But the seeds of this revolution in earth sciences were planted much earlier, and the sprouting ideas were nourished through half a century by the debate about continental drift before they eventually flourished, ripened, and scattered their own seeds into all corners of the earth in 1967.

Before discussing the grand theory of plate tectonics, we should view the developments leading up to the proclamation of revolution. These are summarized in Table 2-1. In a brief historical outline, some of the important topics will be introduced without adequate explanation. These topics are covered in suitable detail in later chapters, which are cited in Table 2-1 and in the following review.

CONTINENTAL DRIFT: THE GREAT DEBATE

The earth sciences have been shaped by a series of great and usually acrimonious controversies and a host of minor disputes. The arguments have been such as to give the impression that geologists enjoyed the excitement of debate more than the prospect of getting together to define the terms of their disagreements in an effort to resolve a dispute. The debate of this century has been about continental drift.

Continental drift is an old idea, formulated originally to explain the striking parallelism of the Atlantic coasts (see Figures 1-1 and 1-2). The idea of a mobile earth was familiar to geologists by 1900, but the concept of continental drift was not taken seriously until 1915, when Alfred Wegener published an expanded version of his 1912 paper in his book, *The Origin of Continents and Oceans*. Wegener and his followers compiled an impressive list of evidence supporting continental drift; the list was drawn from paleoclimatology (the reconstruction of ancient climatic zones), paleontology (the study of fossils, ancient life forms), the geometrical fit of continental margins on opposite sides of ocean basins, and the matching of rock successions and truncated geological structures across oceans.

Wegener proposed the name *Pangaea* for a single supercontinent presumed to exist before continental drift began about 180 million years ago. Pangaea then split up, with the southern continents moving west or toward the equator, or both. According to Wegener, South America and Africa began to drift apart about 70 million years ago, with the opening of the North Atlantic being accomplished mainly during the past few million years.

The theory of continental drift suffered from a lack of definitive evidence, and each line of evidence developed in favor of drift was soon opposed by a counterargument. Geophysicists were unconvinced that continental drift could have occurred, arguing that what was known of the physical properties of the earth simply would not permit the proposed movements. Proponents of drift, on the other hand, maintained that the geological facts cited as proof that continental drift had occurred should not be ignored, simply because the geophysicists were not yet smart enough to conceive a physical explanation for drifting continents. Then other experts disputed what were claimed to be geological facts, referring to them instead as inferences based on inadequate or inconclusive data. The arguments continued in this indeterminate vein, like some medieval philosophical controversy, for many years.

The American Association of Petroleum Geologists organized a famous meeting of prominent geologists in New York in 1926 to review the theory of continental drift. Some speakers at the symposium argued that there was good evidence for drift, but most argued that there were severe objections to the theory, and they argued with deadly sarcasm. Largely as a result of this symposium, the continental drift theory received little further attention in North America, although the debate continued among geologists elsewhere, as in Europe, Africa, and Australia. Eventually, however, just about everything that could be argued for and against continental drift had been written not once but many times, and the debate faded into a quiescent stalemate in the 1930s for lack of additional evidence. Continental drift became an article of faith among the more revolutionary geologists,

Table 2-1. Stages in the Development of a Revolution

Date	Topic or event	Scientists involved[a]	Chapter
1912-1915	Continental drift proposed	Wegener	11
1915-1930	Great debate		11
1930	Death on Greenland ice sheet	Wegener	2
1930-1950	Stalemate. A lost cause in United States of America; debate continued elsewhere	Du Toit, Holmes	11
1950-1960	Revival of interest	Carey, King	11
	Exploration of ocean floor	Bullard, Ewing, Heezen, Menard	3
	Fossil magnetism in rocks and paths of polar wandering.	Blackett, Runcorn	8, 11, 12
1960-1962	Sea-floor spreading—geopoetry	Dietz, Hess	10, 12, 14
1963	Oceanic magnetic anomalies associated with sea-floor spreading; the magnetic tape recorder	Matthews, Morley, Vine	8, 10, 13
1963-1966	Polarity reversals of earth's magnetic field; fossil magnetism and accurate rock dating for lavas on land and for deep-sea sediment cores	Cox, Dalrymple, Doell, Foster, McDougall, Opdyke,	9,15
1965-1966	Transform faults and earthquake studies	Sykes, Wilson	4, 5
1966-1967	Revolution proclaimed after it all came together at the Goddard Symposium		11, 13
1967-1968	Plate tectonics incorporated spreading and drift	Le Pichon, McKenzie, Morgan, Parker	4, 5, 7, 16
	Earthquake synthesis	Isacks, Oliver, Sykes	4, 5, 6
	Polarity reversal time scale extrapolated, giving sea-floor isochrons	Heirtzler, Pitman	10, 13
	The bandwagon rolled		
1968-1970	Deep-sea drilling by *Glomar Challenger*	Maxwell and others	14
1970	"Geopoetry becomes geofact," *Time* magazine		13
Present	Recent geological features are being related to present plates and plate boundaries; ancient geological features are being analyzed in attempts to reconstruct plate history	Everybody—well, nearly everybody. Exceptions include Beloussov, Jeffreys, and Meyerhoff	3, 5, 6, 13, 15, 16, 18

[a]A few names are listed in alphabetical order—hundreds have been omitted. Some of these scientists are not cited in the text. Their contribution may be found in *Continental Drift,* by U. B. Marvin, or in other books listed in Selected References at the end of this book.

and it was either ignored by geophysicists and most geologists or treated as another crackpot hypothesis.

REVIVAL

There was a revival of interest in continental drift during the 1950s as results were obtained from research efforts in two areas quite distinct from those involved in the earlier debate: rock magnetism and ocean-floor exploration.

Some rocks become weakly magnetized when they are formed, and the direction of magnetization preserves a fossil record of the direction of the earth's magnetic field at the time and place of formation (Chapters 8 and 11). Research on paleomagnetism (fossil magnetism in old rocks) by geophysicists such as P. M. S. Blackett in London and S. K. Runcorn in Cambridge yielded results indicating that the former positions of the earth's magnetic poles had changed relative to the continents. They mapped paths of polar wandering, which showed the position of the magnetic pole relative to each continent back through hundreds of millions of years. Because there was evidence that the direction of the magnetic axis of the earth remained approximately coincident with the rotational axis, the data could be interpreted in terms of relative movement of the continents.

The new evidence led many geophysicists to consider the theory of continental drift seriously, whereas many geologists remained unimpressed by and suspicious of this new approach. This is a reversal of the situation during previous years, when geophysicists denied the physical feasibility of the continental drifting proposed by geologists. The interpretation of the paleomagnetic results did depend on statistical analysis of sparse data at that time, and many geologists were as shy of mathematics as a college nonscience major.

The controversy concerned the drift of continents, which occupy barely 30% of the earth's surface, and most of the evidence was drawn from the continents. The development of new techniques and the adaptation of existing geological and geophysical techniques for use with oceanographic expeditions produced a great deal of new information about the ocean floors (Chapter 3). Marine geologists and geophysicists, in particular, M. Ewing and B. C. Heezen, charted great mountain ranges beneath the oceans, including a continuous system girdling the world.

This worldwide, submarine mountain range, approximately 65,000 km long and typically 1000 km wide, covers about one third of the ocean floor, or one quarter of the total world surface (see Figure 3-8). The crest of the mountain range beneath the Atlantic Ocean almost bisects the ocean floor between Europe and Africa and the Americas. Moreover, there is a central valley incising the mountains to depths of 1 to 3 km below the irregular peaks and ridges on either side of the valley. There is evidence suggesting that this remarkable feature is a rift valley, which implies that the central region is in a state of tension—the valley is opening up, parallel to the coastlines bordering the Atlantic Ocean. Could this deep valley represent a scar caused by the rupture and separation of the original supercontinent of Pangaea?

Other evidence from exploration of the ocean floors led to the conclusion that the ocean basins were relatively young features of the earth's surface. Their ages appeared to be measurable in hundreds of millions of years instead of the billions of years that have passed since the earth was formed. Many geologists began to think more kindly of the old idea of continental drift.

There is an enormous reservoir of heat within the earth, produced mainly by the slow decay of radioactive elements distributed very sparsely through the interior; because of this there is a continuous outflow of heat through the surface. The small rate of heat flow can be measured, with some difficulty, and it is known to vary considerably from place to place. In the 1950s, E. C. Bullard initiated the measurement of heat flow through the ocean floor, which is quite a trick from a boat that may be floating as much as 3 or 4 km above the ocean floor.

The rate of heat flow from the crests of the submarine mountain ranges was found to be unusually high.

Many geologists and geophysicists related the tensional character of the oceanic mountains and the high heat flow from the central rift valley regions to some kind of upwelling of the hot, solid material within the earth, as indicated schematically in Figure 2-1a. It was also suggested that the rising material could be the cause of the elevation of the submarine mountain range.

The movement of material illustrated in Figure 2-1a could represent one part of a convection cell. Convection cells in a pan of thick soup are shown in Figure 2-1b. The application of heat at the bottom of the pan warms the soup, causing it to expand and become less dense, or lighter, than the cooler soup around it, so it rises to the top where it is forced to move sideways. The motion is called convection, and heat is transported upward by the physical movement of the hot fluid. Near the top, the soup becomes cooler, so it contracts, becoming more dense or heavier, and it sinks to the bottom again to complete the convection cell.

In the earth the convective motions, such as those illustrated in Figure 2-1a, occur not in molten material, as had been suggested in some nineteenth-century models of the earth, but in solid material, as suggested first by A. Holmes of Edinburgh in 1928. We can imagine convection like that in Figure 2-1b occurring in a very viscous fluid, such as tar, but it is harder to conceive of similar flow in solid rock material.

We are familiar with the properties of hard, brittle rock at the surface of the earth. Within the earth, however, under extreme conditions of high pressures and temperatures, rock properties are very different. We know that temperatures are high within the earth from several lines of evidence, the most dramatic being the eruption of red hot lava, which is molten rock at a temperature often greater than $1100°C$.

When a blacksmith takes a cold bar of steel, he is unable to bend it. But when he heats it to a red glow and beats it with his hammer, he is able to bend the still solid bar and shape it precisely to fit a horse's hoof. Similarly, rocks in the earth's interior

Fig. 2-1 *(a)* Vertical cross section through a submarine mountain range illustrating schematically the slow convection of hot, solid rock material rising beneath the central rift valley. *(b)* Convection cells in a pan of soup, with floating scum on the surface.

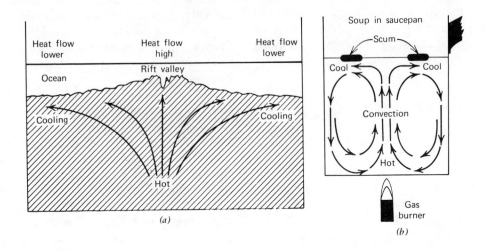

at high temperatures and pressures can be bent and deformed while still in the solid state.

The rates of movement envisaged in Figure 2-1a amount only to a centimeter or two a year, so the time factor is vastly different from that in Figure 2-1b. Given a force acting through the long periods of geological time, it is much easier to imagine hot rock deforming very slowly in such a way that there is net transport of material in one direction, which means that the rock flows, as indicated in Figure 2-1a. A force applied more rapidly, even under the extreme conditions of pressure and temperature within the earth, could cause solid rock to break or fracture instead of to flow.

Silly-putty is a familiar material whose properties are dependent on time. A ball of silly-putty bounced on the floor rebounds, as a ball should. In response to the brief impact with the floor, it behaves as an elastic body. On the other hand, if the ball of silly-putty is hit with a hammer, it breaks open like a rock. In response to the abrupt, forceful impact of the hammer, while confined between the hammer head and the surface on which it rests, the silly-putty behaves like a brittle material. If the ball is left to stand on a flat surface, it changes shape to a thin disc within an hour or two. Thus the solid flows slowly in response to the small force produced by its own weight. Silly-putty is a very different kind of material from a rock, but it illustrates the kind of flow that rocks are believed capable of at high pressures and temperatures within the earth in response to small forces persisting for very long periods of time.

SEA-FLOOR SPREADING

By 1960, it is notable that the attention of most investigators was directed toward the ocean basins instead of toward possibly drifting continents. The scene was set for formulation of the hypothesis of sea-floor spreading by the late H. H. Hess of Princeton. His paper entitled "History of the Ocean Basins" was presented orally in 1960 and published in 1962. Because the idea was fanciful and somewhat short on data, he called it "geopoetry."

Hess proposed that the major structures of the sea floor are direct expressions of a convection process within the earth's solid interior. Some oceanic mountain ranges mark the sites of rising convection cells, as depicted in Figure 2-1a. As the convecting material rises and then moves laterally, the sea floor spreads away from the rift valley produced above the rising convection currents. This crack in the ocean floor does not continue to grow as spreading proceeds because partial melting of the solid material below yields lava that rises to the surface, where it is erupted from volcanoes on the submarine mountain range and in the rift valley. The lava solidifies, forming new ocean floor, which is, in turn, spread laterally by the convecting material.

Continental drift was considered by Hess to be a direct consequence of the convection currents. Most earlier models for continental drift had the continents plowing through the material that forms the ocean floor. In Hess's model, the continents are carried passively by the internal convection currents as if they were on a moving conveyor belt. An analogy for this process is given by changing Figure 2-1b from a soup to a stew, one of those stews that develops a more or less solid scum floating on the surface. The action of the convection cells illustrated in Figure 2-1b would obviously carry the surface scum away from the center toward the cooler edges where the convection cell moves downward. The scum does not sink because it is too light, too buoyant.

The convection cells within the earth have rising and descending limbs, just like the soup in Figure 2-1b. Hess proposed that where convection cells converged at sites of downward movement, the older sea floor, which had been displaced laterally by the spreading process, was carried downward into the earth's interior along the descending limbs. These sites correspond to the deep oceanic trenches bordering island arcs and some continental margins (Chapter 3). In this way, the ocean floor was completely regenerated within 200 to 300 million years. Figure 2-3, a diagrammatic representation of plate tectonic theory, incorporates the essence of sea-floor spreading, but it differs in several respects from Hess' original scheme.

MAGNETIC ANOMALIES AND POLARITY REVERSALS

In 1963, a young graduate student at Cambridge, F. J. Vine, published a short article in collaboration with D. H. Matthews, a senior colleague, giving an explanation for anomalies (irregularities) in the earth's magnetic field that had been discovered over the Pacific Ocean only five years earlier. They suggested that these were caused by the effect of magnetized strips of the ocean floor, successively generated by sea-floor spreading, and carrying the imprint of periodic polarity reversals of the earth's magnetic field. This will be discussed in detail and illustrated in Chapter 10. It had not been established in 1963 that the earth's magnetic field had ever reversed its polarity, nor that the magnetic anomalies were associated with the submarine mountain ranges. It was the confirmation of these two items, and the conceptual model of Vine and Matthews, that transformed Hess's speculative geopoetry into a revolutionary movement.

From study of magnetized lava sequences on land from all over the world, it was established that the earth's magnetic field had reversed polarity in epochs with duration less than 1 million years. A geomagnetic polarity reversal time scale was calibrated back to about 4.5 million years in a series of publications after 1963 from two groups of scientists in friendly competition. A. Cox, R. Doell, and B. Dalrymple of the U.S. Geological Survey and I. McDougall of the Australian National University had been at Berkeley University together in the late 1950s. The results are reviewed in Chapters 8 and 9 (see Figure 1-3).

By 1966 it had been confirmed that a series of alternating magnetic anomalies in the earth's present magnetic field was present parallel to the crest of many submarine mountain ranges, and Vine demonstrated that these anomalies were consistent with the sea-floor spreading model (Chapter 10).

The significance of these developments became clear when papers published in 1966 and 1967 indicated that measurements of three different features of the earth all change in the same ratios. These are the ages of magnetic polarity reversals in lava sequences on land (Chapter 9), the depths to reversals of the direction of magnetization in rocks recovered from deep-sea cores (Chapter 9), and the widths of the magnetic anomalies parallel to the midoceanic ridges (Chapter 10). The fact that these ratios appeared to be the same in all parts of the world was regarded as convincing evidence that sea-floor spreading provided a valid global framework for the explanation and prediction of geological and geophysical phenomena. Refinement of the scheme, including precise estimates for the directions and rates of movement of large sectors of the earth's outer layer, led to formulation of the theory of plate tectonics, which incorporates sea-floor spreading and continental drift.

A detailed account of the sequence of events at this time and of the controversies that developed is given in the historical account of the revolution in Chapter 13.

TRANSFORM FAULTS AND EARTHQUAKE STUDIES

The ocean floor is scarred by large fracture zones, many of which are associated with the submarine mountain ranges. In 1965, J. T. Wilson of Toronto, a farsighted geologist whose ideas were influential in the development of the revolution, called these "transform faults." He explained previously puzzling features of them in the context of sea-floor spreading, and he demonstrated the consistency between his explanation for the fault zones and submarine mountains, and the distribution of earthquakes published in 1956 and 1963 by M. Ewing, B. Heezen, and L. R. Sykes of the Lamont Geological Observatory at Columbia University.

After Wilson published his transform fault concept, Sykes completed more detailed studies of earthquakes and provided convincing evidence for its validity in 1966 (Chapter 13). J. Oliver and B. Isacks, also at Lamont, studied the earthquakes that are concentrated at the other side of plates, where the surface is bent down into the earth's interior, according to Figure 2-3. Their results proved to be consistent with plate tectonics. In 1968, Oliver, Sykes, and Isacks published a paper reviewing worldwide earthquake data that tested the concepts of sea-floor spreading and plate tectonics and demonstrated a very satisfying consistency. The study of earthquakes now provides much of the basic evidence for plate tectonics (Chapters 4 and 5).

PLATE TECTONICS

Figure 1-2 illustrates the concept of plate tectonics. The surface of the earth is covered by a small number of relatively thin, shell-like plates of rigid rock that move over the earth's interior. The fact that the lithosphere plates are rigid means that if one part of a plate moves, the whole plate has to move as a unit. Consider a towel laid out on a table alongside a thin sheet of plywood with the same dimensions. If you push one side of the towel it crumples into folds—it is not rigid. If you push one side of the rigid plywood, the whole thing moves. Most plate boundaries are not associated with the boundaries between continent and ocean, but with active belts characterized by earthquakes and sometimes volcanic activity.

Figure 2-2 is a cross section through the earth illustrating the layers that are identified in the plate tectonics model of Figure 2-3. The outer plates, of relatively cool, rigid rock, are called the lithosphere. In most areas this is 100 km thick, but it is thinner beneath the crests of the oceanic mountain ranges. Note from Figure 2-3 that the continents are integral parts of the lithosphere.

Beneath the lithosphere is the asthenosphere layer, with variable thickness averaging about 200 km; below the asthenosphere is the mesosphere. The asthenosphere is a layer of weakness that is capable of flow. It is in this layer that convective

Fig. 2-2 Cross section through the outer layers of the earth.

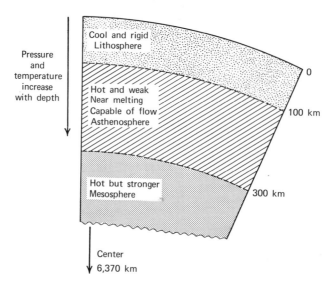

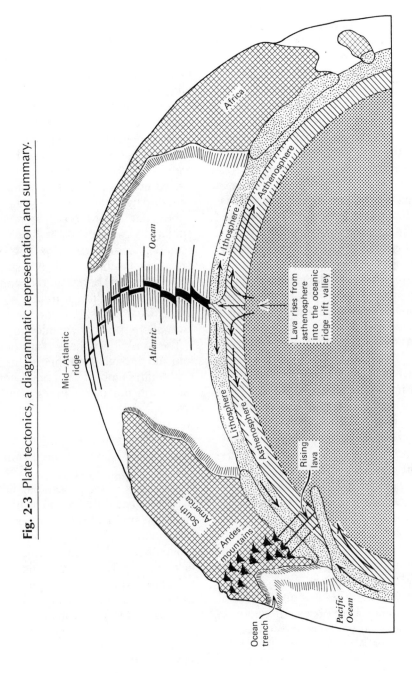

Fig. 2-3 Plate tectonics, a diagrammatic representation and summary.

motions are believed to be concentrated, according to most versions of plate tectonics. Contrast this with Hess's earlier model of sea-floor spreading, where the convection cell reaches upward almost to the ocean floor (Figure 2-1a).

The boundary between the lithosphere and the asthenosphere is caused by a change in the properties of the rock above and below. This is a function of increasing pressure and temperature with depth, and not one involving any change in chemical composition of the rocks. The cooler lithosphere has strength, whereas the asthenosphere has reached conditions near its melting temperature so that its strength is greatly diminished. It is even possible that within this layer, the temperature is high enough to produce just a trace of melting, so that a thin smear of molten rock between the mineral grain boundaries acts as a lubricant facilitating convection and the lateral migration of the overlying plates of lithosphere. There is evidence that the mesosphere has more strength than the asthenosphere.

The dynamic aspects of plate tectonics are illustrated in Figure 2-3. Sea-floor spreading in the region of the submarine mountain ranges is caused by (1) upward convection in the asthenosphere, (2) transportation of lithosphere away from the mountain crest as if on a conveyor belt, producing tension and possibly a rift valley, and (3) generation of new lithosphere between the older, spreading lithosphere. This is a complex process, including the eruption of lava in and near the rift valley, followed by solidification of the lava to produce new ocean floor.

The continental masses form an integral part of the lithosphere plates. Therefore, as the plates are transported by the convection currents in the asthenosphere, the continents move with them, similarly to logs frozen into ice floes on a lake surface.

The generation of new lithosphere above rising convection currents displaces the old lithosphere. Because the lithosphere plates are rigid, the whole ocean plate has to migrate away from the rising current. Obvi-

ously, this process cannot proceed very far unless the earth is expanding to allow for the increased area of surface, or unless the older lithosphere is being destroyed somewhere else. There is no good evidence to support an old hypothesis that the earth is expanding, and the preferred interpretation of lithosphere destruction is illustrated in Figure 2-3.

Where two plates moving in opposite directions collide, as illustrated along the western margin of South America, one of the plates is forced to descend into the earth's interior, where it becomes slowly heated and eventually assimilated into the surrounding material. Continental rocks are lighter than other lithosphere rocks, so the continents cannot sink, any more than the scum on a stew can sink (Figure 2-1b.) Therefore it is the oceanic lithosphere plate that descends beneath the continental plate. A deep valley or trench is produced in the ocean floor along the line where the lithosphere bends down into the interior.

The cool lithosphere plate descending into and through the asthenosphere forms a descending limb of the convection system balancing the upwelling associated with sea-floor spreading. We do not know how the convection cell is closed. The movement of material within the earth is probably far more complex than that of the simple cells in Figure 2-1b.

The edges of the descending lithosphere plate are heated by friction against the surrounding material, and temperatures rise high enough to cause partial melting. The molten material produced rises to the surface where it is erupted in the spectacular volcanoes typical of mountain ranges such as the Andes Mountains and of island arcs such as the Japanese islands.

REVOLUTION

The critical evidence from magnetism and earthquakes that became available in 1966 and 1967 convinced many geologists and geophysicists that sea-floor spreading was no longer a hypothesis, but a fact, and the fervor of revolution gripped the whole geological community. Since 1967, hundreds of papers dealing with various aspects of sea-floor spreading and plate tectonics have been presented at annual meetings of the American Geophysical Union, The Geological Society of America, and the American Association of Petroleum Geologists. The revolution was proclaimed and consolidated in 1968 by incorporation of the best features of sea-floor spreading and continental drift into the new model of plate tectonics.

Results published in 1970 from the deep-sea drilling program were hailed as proof. In an editorial for *Science* on December 5, 1969, P. H. Abelson wrote:

To date, examination of the cores have been conducted on shipboard, but major conclusions have already been announced . . . Results from the drilling strongly support hypotheses of sea-floor spreading and continental drift . . . the deep-sea drilling has changed speculation into something that must be regarded as established.

The theory of plate tectonics has enjoyed such success that it has already become a ruling theory, and critical comments tend to be disregarded by the scientific establishment. The advantages and disadvantages of this situation were stated by C. E. Wegmann in 1963.

The history of geology shows that a conceptual development in one sector is generally followed by a harvest of observations, since many geologists can only see what they are asked to record by their conceptual outfit.

The conceptual development of plate tectonics is, indeed, gathering a harvest of data and interpretations. Earth scientists are being forced to reevaluate almost everything that they thought they knew about geological processes. Geological observations that appeared to be isolated now seem to fall into place in the global scheme.

The second half of Wegmann's statement is also being borne out. Protagonists of plate tectonics tend to neglect the data and arguments not explained by the theory be-

cause they are confident that these can be ascribed to our lack of understanding instead of to inadequacies of the model and that time and improved understanding will bring forth explanations.

In Chapter 1, we noted the publicity given to the historic Apollo 11 Lunar Science Conference that convened in Houston on January 5, 1970. On the same date, *Time* magazine devoted a whole page to the revolution in earth sciences under the heading "Geopoetry Becomes Geofact." We all know that *Time* magazine publishes only the plain, unvarnished truth, so we really have to accept plate tectonics as the global scheme that actually works. It satisfies geophysicists who are accustomed to concepts based on theoretical models, and it satisfies the geologists who find various phenomena and puzzling fragments of geological history explained in delightfully consistent fashion by the model.

COUNTERREVOLUTION

Not all geologists consider sea-floor spreading and plate tectonics to be a panacea. There has been continuing opposition to continental drift through the developments of the past decade but, since the formulation of the plate tectonics model, the papers with antidrift arguments are rarely cited. A. A. Meyerhoff and associates have published several papers since 1970 marshalling the rather scattered antidrift arguments in counterrevolution. He presented geological and paleoclimatological evidence that appears to require that the Atlantic Ocean has remained in approximately its present position for the past 800 million years. According to plate tectonic theory, the Atlantic Ocean has closed and opened again during this period.

Continental drift was debated passionately for many years after 1912. The recent developments have also aroused passions, as shown by a letter from A. O. Kelly to *Science News* on January 3, 1970.

Now when some authority proposes a hypothesis like ocean floor spreading, every junior scientist in the country jumps on the bandwagon. Criticism, it seems, is rude, egotistical and out of style. . . . These me-too scientists are piling hypothesis upon hypothesis. . . . Reasons, it seems, are no longer necessary; one simply backs up one speculation with another.

V.V. Beloussov, who has a large following among Soviet geologists, has written several papers arguing against sea-floor spreading and plate tectonics. Beloussov emphasizes that although our knowledge of the structure of the ocean basins is still much more sketchy than our knowledge of continental geology, the new oceanic results

have cast a hypnotic spell and thrown a shadow over much that is old and familiar. The results of two centuries of data gathering from the continents should not be over-simplified in order to bring them down to the level of the schematic data available for the oceans.

Some Western geologists believe that Soviet scientists are following a scientific party line that denies the validity of plate tectonics. V. V. Beloussov is very influential in Russia. However, an interview with V. A. Magnitsky reported in *Science News* on July 11, 1970, after an International Symposium in Flagstaff, Arizona, suggests that the Russians are not rigidly dogmatic about the subject. Magnitsky is not an antagonist to the process; he thinks it is an interesting idea. He is quoted thus.

But I think the final solution to this problem will depend on the new material from the ocean floor which we expect to have in two or three years. Let's wait and see. . . . The history of the geotectonics of the whole century has been the introduction of new ideas and then their collapse. This is why so many of my colleagues are in this position.

Like other Soviet geologists, he is less impressed with the new evidence from the ocean floors than with geological data from the Eurasian continent. He advocated a little more patience.

There is no doubt that a revolution is in progress. Earth scientists are examining old evidence through new eyes. Proposals are being presented for the reorganization of geological curricula and the rewriting of textbooks. This volume is one outcome of those proposals.

If it is true that a new principle in the behavior of the earth has been discovered, then we are on the verge of understanding more clearly the origin and distribution of metallic ore bodies and the geological structures that trap oil. The practical and economic implications for the mining and petroleum industries are obvious. The distribution of earthquakes and volcanoes is directly related to the new global models, and earth scientists are exploring the implications for the prediction of earthquakes and eruptions and for the control of earthquakes. The social desirability of these explorations is equally obvious.

ALFRED WEGENER, DRIFTER AND EXPLORER

A revolution has occurred because earth scientists now believe that we live on a mobile rather than a static earth. The continents drift. Alfred Wegener is well known now for his advocacy of continental drift despite the severe criticisms from influential geologists and geophysicists, but he is less widely known for his other activities, which also involved adverse conditions. Wegener was one of the great arctic explorers. A brief account of this aspect of his life might remind readers that scientists are not stereotypes who stand around all day wearing long, white laboratory coats. They are people, too, just like humanists, historians, and truck drivers.

While he was a student in Berlin, earning a doctoral degree in astronomy, Wegener escaped to the mountains whenever possible to build up his endurance by climbing and skiing. He hoped that one day he would have the opportunity to explore in Greenland, a vast ice-covered island that was only peripherally known at that time. The opportunity came when he was 26 years old. He left his position as assistant at the Royal Aeronautical Observatory in Lindenberg and joined the Danmark Expedition as a meteorologist.

The Danmark Expedition (1906-1908) was one of the most successful in terms of accomplishments and exploration in northeastern Greenland. Conditions were severe. The leader of the expedition, Mylius-Ericksen, and two companions died of starvation and exposure during an overextended dog-sledge journey to the north.

This expedition was followed by a teaching appointment in meteorology and astronomy at the University of Marburg. In January 1912 Wegener presented his first statement about continental drift in two papers read at the Geological Association in Frankfurt-am-Main and the Society for the Advancement of Natural Science in Marburg. The paper was published in late spring, just before he left on his second expedition to Greenland.

J. P. Koch, another member of the Danmark Expedition, planned to cross the Greenland ice sheet at its widest part, traveling from east to west. There had been few crossings of the ice sheet, and the central part had not been explored at all. Koch planned the journey with ponies instead of dogs. During the summer and fall of 1912 the expedition struggled to drag its supplies from the east coast onto the edge of the ice sheet under very difficult conditions. They established a base and endured the winter of cold and three months of darkness. You cannot imagine what this means unless you have experienced it. I write with knowledge and feeling, because I spent 1952 to 1954 in the same place under much more comfortable conditions, as a member of the British North Greenland Expedition. Wegener broke a rib on September 16, but he recovered in a month. Koch fell 15 m into a crevasse and broke a leg, which laid him up for three months. On April 20, 1913, they began their journey across the ice sheet and reached the other side on July 4.

Wegener's first book, *The Origin of Continents and Oceans* (in German) was published in 1915, and the fourth edition, revised, was published in 1929, the year that he led a preliminary reconnaissance party onto the Greenland ice sheet to test equipment and prepare for a major expedition in

1930. German, British, and American groups were actively investigating weather conditions over the Greenland ice sheet, partly in anticipation of future airplane flights between Europe and America; the shortest route crosses Greenland. Wegener planned a large-scale expedition, with 20 members in three parties, one on the east coast, one on the west coast, and one in the center of the ice sheet. In order to establish station Eismitte (literally, "the middle of the ice"), the expedition was supported by ponies, 10 dog teams, and two air-propellor sledges.

In 1930 it took three journeys, each with 10 dog sledges, to equip station Eismitte at an elevation of 3000 m, nearly 400 km from both coasts. Two men were to stay there through the dark winter months but, after the third journey, supplies were still insufficient. They sent a message back with the last dog sledge stating that if additional petroleum did not reach them by October 20, the winter project would have to be abandoned, and they would leave Eismitte on foot. Wegener and two companions arrived on October 27 with additional supplies on two dog sledges to find that the wintering party had decided to stay after all. Temperatures had already fallen below −50°C, and a journey of 400 km over the ice with no supporting sledges could have been fatal under these conditions. They had transferred from their tent into a hut built from packing cases.

Wegener's fiftieth birthday was celebrated at Eismitte on November 1, just before he and one companion set off on the journey back to the coast. They never made it. The west coast station assumed that they had all remained at Eismitte during the winter, and it was not until the following spring when a relief party reached Eismitte in May that they learned of Wegener's disappearance. His body was subsequently found just about halfway between the two stations, where his Eskimo companion, Rasmus Villumsen, had buried him carefully in the snow. Villumsen attempted to complete the journey alone, but he and his sledge were never found.

Wegener apparently died from exhaustion in his tent. Trudging along on skis while a team of dogs pulls a lightly laden sledge is arduous. It must have been a grim race through the darkening winter, with unseen irregularities on the icy surface repeatedly snagging ski tips, and with the wind and drifting snow a constant icy lash exhorting both men and dogs to hurry to the safety of the mountains.

A major objective of Wegener's German Inland Ice Expedition was to determine the thickness of the ice by using a new technique involving echo-sounding methods. An explosion set off at the surface sends waves down through the ice, these are reflected from the rock below the ice sheet and recorded at the surface by special instruments. From the known speed of travel of the waves and the measured time of travel, the thickness can be calculated (see Figure 7-3). Although Wegener did not live to see the general acceptance of his theory of continental drift (Table 2-1), when he died he was pioneering a technique that later gave it strong support in its revised incarnation as plate tectonics. The development of echo-sounding methods from ships for exploration of the ocean floor was responsible for the revival of interest in continental drift in the 1950s. The explosion technique is now used to generate artificial earthquakes, and earthquake evidence provides persuasive evidence for plate tectonics.

SUMMARY

In 1912 Alfred Wegener proposed that the present continents had drifted apart from an original supercontinent. This initiated a great debate that stagnated in stalemate by the 1940s.

There was a revival of interest during the 1950s through research in rock magnetism and oceanography. Evidence from weakly magnetized rocks was interpreted in terms of paths of polar wandering, which indicated that continents had moved relative to each other. M. Ewing and B. C. Heezen charted an enormous submarine mountain range, girdling the world. In 1960 and 1961

H. H. Hess and R. S. Dietz proposed that this was the locus of sea-floor spreading, caused by convective uprise of hot, solid material from the earth's interior.

There were many developments between 1960 and 1968, when continental drift and sea-floor spreading were incorporated into the theory of plate tectonics. It was established that the earth's magnetic field had reversed polarity, and the polarity reversal time scale was calibrated back to 4.5 million years ago by A. Cox, R. Doell, B. Dalrymple, and I. McDougall. Between 1963 and 1966, F. J. Vine, with D. H. Matthews and J. T. Wilson, correlated these polarity reversals with irregularities (anomalies) in the intensity of the earth's present magnetic field that were symmetrically disposed across the crest of the submarine mountain range. They estimated the rates of sea-floor spreading from the correlated dates and distances.

Evidence from earthquakes provided strong support for sea-floor spreading and plate tectonics, as demonstrated between 1966 and 1968 by B. Isacks, J. Oliver, and L. R. Sykes. The surface of the earth is covered by a few shell-like plates of rigid rock that move over the earth's interior. These cool lithosphere plates are stable and relatively free of earthquakes. The plate boundaries are characterized by earthquakes and volcanoes. At divergent plate boundaries new lithosphere is generated, and at convergent boundaries one plate sinks down into the earth's interior.

Plate tectonics became the new paradigm for a revolution in earth sciences, and the theory has been remarkably successful in explaining and interrelating many diverse observations. Some scientists remain convinced that continents have not drifted. Alfred Wegener died an explorer's death in Greenland in 1930 while leading the German Inland Ice Expedition. He was pioneering an echo-sounding technique to measure the thickness of the ice cap, similar to techniques that later gave continental drift strong support in its revised incarnation as plate tectonics.

SUGGESTED READINGS are at the end of the book.

chapter 3

The Earth's Solid Surface

A major feature of Hess's sea-floor spreading hypothesis, and one that was carried over into the plate tectonic theory, is the conclusion that the major structures of the ocean floor are direct expressions of processes occurring within the earth's interior. The ocean basin floor is elevated above convection currents rising from the asthenosphere (Figures 2-1a and 2-3) and depressed where a descending lithosphere plate contributes to the convection pattern (Figure 2-3). The existence and distribution of the continents and ocean basins, of mountain ranges, and of volcanic chains are all controlled by internal processes.

The distribution of these surface features, therefore, is a primary clue to our understanding of the great scheme of plate tectonics. We have to become familiar with the static details of the three-dimensional geometry of the earth's surface before we can consider the more lively aspects of earthquakes and volcanoes. Great ideas in the sciences require a firm foundation of facts. We need to know what the major features are, their global distribution, and their shapes in three dimensions.

It is in great part due to the intensive curiosity of the late Maurice Ewing that we now know so much about the features of the ocean floor. In 1949 Ewing founded Lamont (now Lamont-Doherty) Geological Observatory of Columbia University, and this has become the world's leading center for marine geology and geophysics. Ewing's drive, energy, and innovative methods for investigation of the ocean floor provided a warehouse full of data that was influential in developing and testing the ideas associated with the revolution in earth sciences during the 1960s.

DISTRIBUTION OF CONTINENTS AND OCEANS

Figure 3-1 shows the primary division of the surface of the world into land and ocean. Areas on this map, as in many projections of a spherical surface onto two-dimensional paper, are considerably distorted. A good world globe is required to see the distribution of land and sea in proper perspective, but some rather strange features can be distinguished from Figures 3-1 and 3-2.

The continents are not distributed evenly over the surface of the earth. More than 65% of the total land is in the northern hemisphere, with a concentration near latitudes just south of the arctic circle. On

Fig. 3-1 Continents and oceans. The continental shelves are flooded parts of the continental masses extending out to the dotted lines. The areas are distorted in this projection. Compare Figure 3-2.

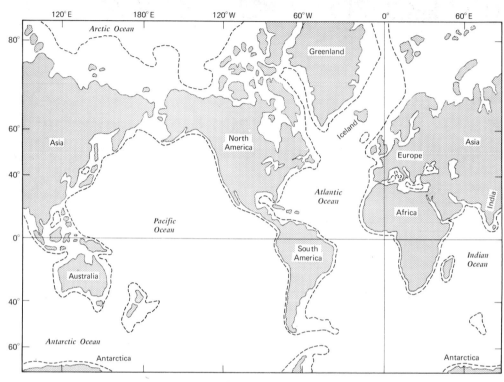

- - - - Edge of continental shelf

Fig. 3-2 (a) Continental hemisphere. (b) Oceanic hemisphere.

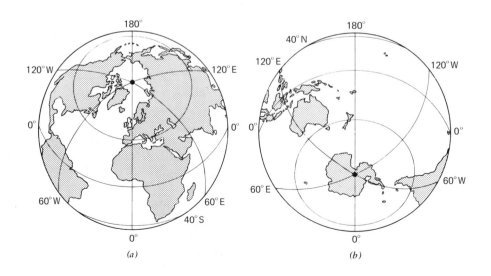

(a)   (b)

the northern hemisphere, 39% of the surface is land; on the southern hemisphere, only 19% of the surface is land. The north pole is centered on ocean, and the south pole is centered on land.

Actually, the continents are concentrated in one hemisphere of the surface, and the Pacific Ocean occupies most of the opposite hemisphere, as shown in Figure 3-2. If you rotate a globe so that you are looking at the Pacific Ocean, with New Zealand near the center of your field of view, you will see nearly all ocean and very little land. This view of the earth is known as the "oceanic hemisphere," with 89% ocean and only 11% land. The opposite hemisphere, centered around a point near Spain, is known as the "continental hemisphere," with 47% land and 53% sea. It is clear from Figure 3-2, if not from Figure 3-1, that the oceans cover more than one half of the world's surface.

The areas of continents and oceans have been measured with precison, and their relative areas are conveniently portrayed in Figures 3-3, 3-4, and 3-5.

The total area of the world surface is 510 million km² with oceans covering 362 x 10⁶ km², and with 148 x 10⁶ km² of land rising above sea level. Figure 3-3 a shows 71% of the total world area as ocean and 29% as land.

The present shoreline has no great significance except to the communities of marine animals living on or near the floor of the continental shelf and to the human societies occupying the ports, the lowland cities, and the farms. It is a transient boundary that has shifted backward and forward across the land throughout the history of the earth in response to many factors including, for example, the amount of water stored temporarily in ice sheets. If all of the ice covering Antarctica and Greenland were to melt, sea level would rise by about 70 m, and large areas of what is now defined as land would become covered by water. Figures 3-3 to 3-5 would have to be modified.

If the ocean water were removed from the surface of the earth, we would see a steep slope separating the continental and ocean-floor levels, but this slope would not correspond to the shoreline. The continental masses are partly covered by water at the present time, and the flooded portions include the continental shelf and the bordering continental slope, as shown in Fig-

Fig. 3-3 Percentages of world surface area covered by continents and oceans.

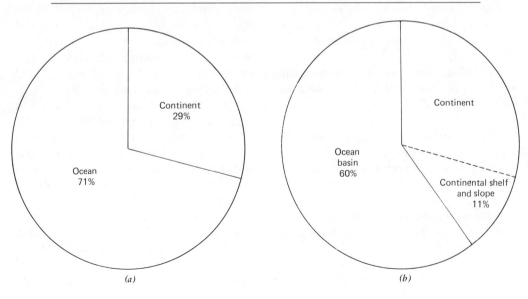

(a)   (b)

Fig. 3-4 Continents and oceans shown as percentage areas of total land and sea, respectively. *(a)* Continents. *(b)* Oceans.

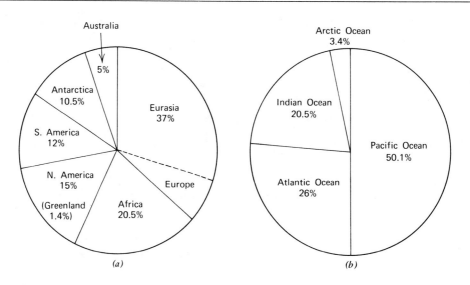

(a)   *(b)*

ure 3-1. Figure 3-3*b* shows that the area of continental slope and shelf is 11% of the total surface area, so that the continental masses, defined by their elevation above the ocean basin floor instead of by the transient shorelines, occupy 40% of the world's surface.

Now let us continue reviewing the statistical data for the present world map (Figure 3-1). The waters of the Pacific Ocean occupy just about one half of the total ocean surface, as shown in Figure 3-4*b*. The Arctic Ocean, which appears to be extensive in the projection of Figure 3-1, amounts to only 3.4% of the total ocean (compare Figure 3-2*a*.

Figure 3-4*a* shows the areas of the continents as a percentage of the total land area. Eurasia is almost twice as large as Africa, the next largest continent. Antarctica, which is barely shown at the bottom of Figure 3-1, is almost as large as each of the two American continents. Australia is barely half the size of these. The large island of Greenland is very distorted in Figure 3-1, because it represents only 1.4% of the total land area (included in North America in Figure 3-4*a*).

Figures 3-4*a* and 3-4*b* are combined in Figure 3-5, which shows the relative areas

of all continents and oceans. The Pacific Ocean covers more than one third of the world's surface, and it is larger than all of the continents combined (Figure 3-3). Each of the three major oceans is larger than Eurasia, the largest continent. Although the Arctic Ocean represents such a small percentage of total ocean surface, it is larger than Australia, and more than half the size of each of the continents of Antarctica, South America, and North America.

The surface of the ocean is the base level from which elevation or depression of the earth's solid surface is measured. We will review the surface relief of the solid earth in some detail in following sections but, while we are examining the distribution of land and sea, we should note the difference in elevation between land and ocean floor. Figure 3-6*a* shows that not only does the ocean dominate the land in terms of surface area, but that the average depth to the ocean floor, 3.7 km, is much greater than the average height of all land surfaces, 0.88 km.

The large difference between average elevation of continents and average depth of ocean floor, 4.6 km, is a significant feature. If the continental masses did not stand so high above the ocean floor there would

Fig. 3-5 Continents and oceans shown as percentages of total world surface area.

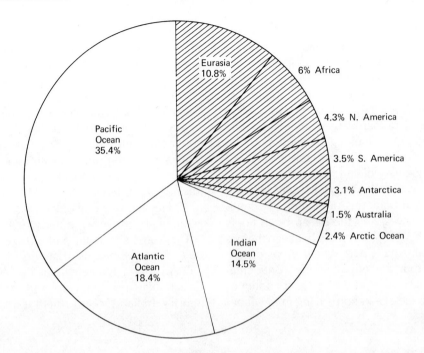

be no dry land. Figure 3-6*b* shows that if the surface of the solid earth were smoothed out so that the whole earth was covered uniformly by ocean water, the ocean would be 2.6 km in depth.

The continents are composed largely of rocks that differ from those underlying the ocean basin. Their density is lower. A fixed volume of continental rock weighs less than the same volume of suboceanic rock. In Chapter 2, we reviewed briefly the way that rocks are capable of flowing very slowly at high temperatures and pressures within the earth. It is this property that elevates the continental surface. The continents sit like enormous slabs of light rock "floating" on the denser rock, like a large slab of wood floating on water.

Fig. 3-6 (a) Average height of continents compared with average depth of oceans. (b) Depth of ocean if the surface of the solid earth were smoothed out so that the whole earth was covered uniformly by ocean water.

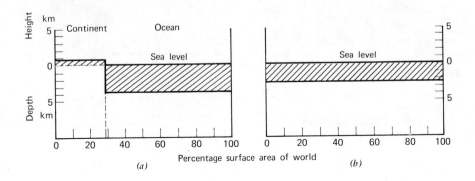

MAJOR FEATURES
OF THE SURFACE

Figure 3-7 shows the distribution of the major features of the surface of the solid earth as they would appear if the ocean water were removed from the 71% of the surface that it obscures. These features are listed in Table 3-1. In this section, we are concerned primarily with their areas and distribution but, looking ahead, the profiles shown in cross section in Figure 3-10 give a complementary picture of their shapes.

The two predominant features are the continental platforms, including the continental shelves, and the ocean basin floors (Figure 3-3). The boundary between them is the continental slope, which drops off abruptly from the edges of the continental shelves at angles of 3 to 6° and continues down to the ocean floors lying in the depth

range 3 to 6 km. Mountains rise from both of these platforms.

The most impressive mountain range is the worldwide system that extends for more than 65,000 km across the ocean basin floor. Its width varies considerably, but the average width is 1000 km. The term "midocean ridge," or "rise," was adopted because of the symmetrical position of the mid-Atlantic ridge between the bordering continents, but Figure 3-7 shows that this term is not appropriate for all parts of the ridge. The great extent and continuity of this ridge reported in 1965 by M. Ewing and B. C. Heezen was one of the most exciting geological discoveries in decades.

There are two main parts to the ocean ridge system. One extends across the Arctic Ocean, along the length of the Atlantic Ocean, between Africa and Antarctica, and into the Indian Ocean, where it meets the

Fig. 3-7 The major features of the solid earth as they would appear if the ocean water were removed. See Table 3-1.

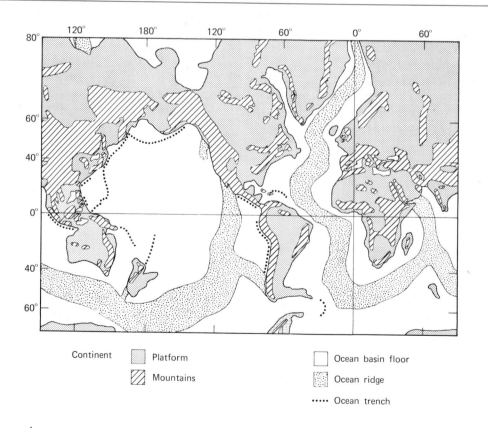

Continent   Platform   Ocean basin floor

Mountains   Ocean ridge

····· Ocean trench

Table 3-1 Major features of the Surface of the Solid Earth (Figure 3-8)

FEATURE	SURFACE AREA Percent of total
Major subdivisions	
Continent and continental shelf	} 40
Continental slope	
Ocean basin	60
Active belts, traversing ocean and continents	
Ocean ridge and rise	23.2
Mountain ranges	
Active, younger than 65 million years	2.8
Older than 65 million years	7.5
Volcanic island arcs	} 1.2
Ocean trenches	
Ocean-floor features	
Ocean-floor fracture zones (transform faults)	?
Oceanic volcanoes	?
Continental rise	3.8

other part. This one extends from the Red Sea, across the Indian Ocean, between Australia and Antarctica, and across the east Pacific Ocean to meet with North America near the Gulf of California.

Figure 3-7 shows that the elevated mountains and plateaus of the continents are not distributed in such an orderly fashion. If we consider only the distribution of those mountains that are young in geological terms, however, those formed within the last 65 million years, then a simple pattern emerges, as shown in Figure 3-12.

The volcanic island arcs represent extensions of these young, active mountain ranges across the ocean floor. Island arcs are elongated ridges with arcuate form, built largely of volcanic lavas, which are usually connected to a continent by one or both ends. They may consist of a series of islands representing the peaks of volcanoes rising from the submarine ridge, as in the Aleutian Islands strung between Asia and North America, or they may consist of larger islands with a line, or lines of volcanoes on the islands, as in Japan and Indonesia. The north and west sides of the Pacific Ocean are festooned with island arcs. There are just two arcs in the Atlantic Ocean, the West Indies and the Scotia arc,

looping from South America to Antarctica. The Indonesian island arc borders the Indian Ocean.

The volcanic island arcs and young continental mountain ranges delineate two belts, each of which follows approximately a great circle around the earth (Figure 3-12). The circum-Pacific belt extends through the Philippine Islands, Japan, Alaska, the Rocky Mountains, the Andes, and Antarctica. The Mediterranean-Asian belt follows the Alps, the Himalayas, Indonesia, New Guinea, and New Zealand.

Ocean trenches are relatively small but dramatically deep grooves in the ocean floor. They are located adjacent to island arcs or at the foot of the continental slope below young mountain ranges located along continental margins, such as the Andes.

In addition to the volcanoes of island arcs and those associated with the ocean ridge system, there are volcanoes scattered over the deep ocean floor. These are termed oceanic volcanoes in Table 3-1. Many of them appear to be randomly distributed, but others are spaced along specific volcanic lines. The peaks of many of the volcanoes rise above sea level, forming islands, but many more are completely submerged. An example is the Hawaiian island chain. Iceland is a different kind of volcanic island, rising from the mid-Atlantic ridge.

The ocean floor is scarred by a series of extensive fracture zones that produce a rugged surface with steep cliffs and deep valleys. These fracture zones typically trend perpendicular to the oceanic ridges, as indicated in Figure 3-12. The major fracture zones are listed in Table 3-1 as a specific feature of the surface, called transform faults.

The continental rise is a fairly gentle surface rising from the deep ocean basin floor to the continental slope. Its width varies from a hundred to several hundred kilometers.

The relative areas of some of the features of the solid surface of the earth are compared in Figure 3-8. The deep ocean basin, excluding the ocean ridges, occupies about one third of the total surface of the earth, which is almost equal to the area of the sta-

ble continental surface, excluding the young mountain ranges.

The ocean ridge system occupies almost one quarter of the earth's surface and more than one third of the ocean basins. The young mountain ranges traversing the continents, together with the volcanic island arcs that extend the system across the ocean floors, are much smaller in area, amounting to less than 5% of the total surface (Figure 3-12).

MAP CONTOURS
AND CROSS SECTIONS

The map in Figure 3-7 shows the distribution of the major features of the surface of the solid earth, but it tells nothing about the shapes of these features in three dimensions. All that we have established so far is that the two primary features, the continental masses and the ocean basins, occupy two distinct levels, separated by the steep

Fig. 3-8 Major features of the earth's solid surface shown as percentages of the total world surface.

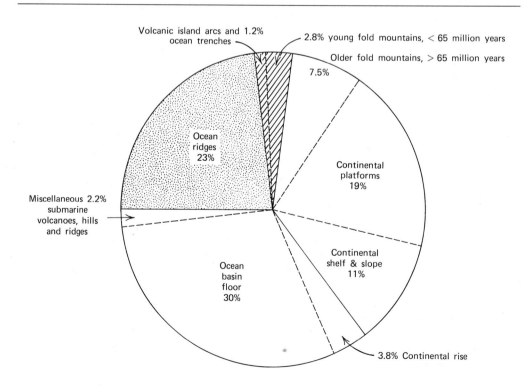

continental slope (see Figure 3-6a). In order to visualize in three dimensions the general shapes of the mountain ranges on land and ocean floor that rise above these two levels, we have to add contour lines to the map.

The position of each point on the surface of the solid earth can be defined in terms of its distance above or below mean sea level (the average between the levels of low and high tides). A contour line connects points on the surface at the same distance vertically above or below sea level. Contour lines for different levels are usually drawn at regular intervals of altitude. The contour interval is the vertical distance, or difference in altitude, between two adjacent contours. A map is produced by projecting these contour lines vertically onto the horizontal plane of the map, which already shows the shoreline as the zero-level contour.

Figure 3-9 is a simple contour map showing a coastline with cross sections constructed to show the profile of the slopes represented by the contours. The horizontal distances are given by the scale at the top of the map.

The contour interval in Figure 3-9a is 50 m, and the relief of the surface is depicted by the shoreline, at 0 m, by one contour line showing a depth of 50 m below sea level, and by two contour lines showing heights of 50 m along the sides of a ridge. Each contour line shows the position of the new shoreline if the sea level would fall or rise by 50 m.

A map with narrow contour intervals will show greater details for the shape of the solid surface than a map with wider contour intervals. Figure 3-9b maps the same area as Figure 3-9a, but the contour interval has been reduced to 10 m. Addition of four contour levels between the widely spaced contours of Figure 3-9a brings out a small ridge not even indicated in the first map.

For the practiced eye, the contours in Figure 3-9b are sufficient to give a three-dimensional picture of the sea floor and the ridges on land. For the unpracticed eye,

and for the purpose of exhibiting various aspects of a map, it is often useful to construct vertical cross sections through the features depicted by the contours. A cross section gives a profile of the surface in the selected direction.

Figures 3-9c to 3-9f are vertical cross sections through the map along the line XY. The distance along each horizontal axis is 750 m, according to the scale bar on the map. In Figures 3-9e and 3-9f, the vertical scale is the same as the horizontal scale, so these figures show profiles of the solid surface with no distortion. The angles between the various sloping surfaces and the horizontal are the same as those you would walk over if you visited the region. In Figures 3-9c and 3-9d, in contrast, the vertical axis has been stretched out compared with the horizontal scale. The horizontal distance in the sections that represent 100 m on the ground has been made equivalent to only 10 m on the vertical scale. The vertical scale has thus been expanded by a factor of 10. The sections are said to have vertical exaggeration of ×10.

The profile construction method is shown for Figure 3-9c. First, the horizontal and vertical scales are marked, and horizontal lines for the contour levels are drawn. The line of the section on the map, XY in Figure 3-9a, intersects each contour line at a point. Each point is defined in terms of horizontal distance in Figure 3-9c, and also in terms of elevation on the vertical axis. The intersection points are plotted on the cross section, as shown by the dashed lines connecting the map to the section. The curved line connecting the points is thus the profile of the solid surface along the line XY.

The points of intersection of the line XY with the contour lines in Figure 3-9b have been plotted in Figure 3-9d in the same way; only the points are shown. The line through the points gives the more detailed profile of the solid surface in the map, showing in addition to the general features illustrated in Figure 3-9c, the presence of a low ridge near the shoreline and of a shallow shelf extending some way out to sea

Fig. 3-9 Contour maps and vertical cross sections. (c) and (e) are cross-sections through map (a). (b) shows the same area with more closely spaced contours; corresponding cross sections (d) and (f) therefore show more surface detail.

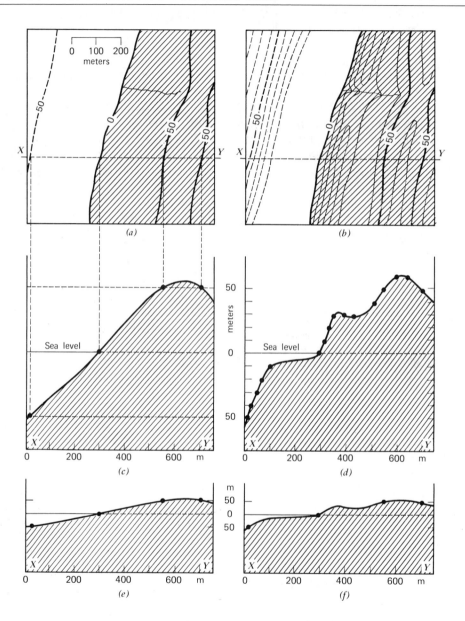

from the shore before the water depth increases abruptly.

The cross sections in Figures 3-9e and 3-9f were constructed in the same way, and the effect of vertical exaggeration in the profiles in very obvious. The apparent slopes in Figures 3-9c and 3-9d become much steeper than the true slopes in Figures

3-9e and 3-9f: The profile in Figure 3-9d illustrates more clearly than the true scale profile (Figure 3-9f) the details of the surface. In order to illustrate the surface shape of a feature having great horizontal dimensions compared with its vertical dimension, vertical exaggeration is required. This is clear from Figure 3-10.

SURFACE RELIEF
OF THE SOLID EARTH

Figure 3-10 shows a cross section through the solid surface of the earth that extends almost halfway around the world near the Tropic of Cancer, between latitudes 20°S and 25°S. The section is discontinuous across the Pacific Ocean, with a gap of about 11,500 km. The features intersected can be located on Figures 3-7 and 3-12.

The vertical exaggeration of ×100 is ten times as great as that in Figures 3-9c and 3-9d, so the angles of slopes are greatly distorted. The amount of distortion can be estimated by comparison with the section across the Tonga trench, which is drawn to true scale with no vertical exaggeration. In order to show the whole section of the figure to this scale, the horizontal axis would have to be extended by a factor of 100. It would be difficult to include such a figure in a textbook, even by means of using a large fold-out page designed by an expert in origami.

This cross section illustrates most of the major features of the solid earth's surface listed in Table 3-1. We see the continental platform and continental shelf, contrasted with the ocean basin floor. Continental slopes border Africa and South America. The section through the mid-Atlantic ridge shows the elevation of the ocean ridge system above the ocean basin floor and the rift valley at its crest. The active mountain ranges on land are represented by the Andes Mountains, rising from the South American continental platform along the Pacific coast. Volcanic island arcs are represented by the Tonga Islands, rising from the ocean floor and bordered by an ocean trench. Another ocean trench borders the Andes Mountain range. The other two features, oceanic volcanoes and ocean-floor fracture zones, are not represented specifically, although the rugged relief of the mid-Atlantic ridge is partly due to the transverse fracture zones (see Figure 3-12).

In addition to studying the elevations of specific features in specific cross sections, as in Figure 3-10, we can examine the statistical distribution of the elevations of the solid surface of the whole earth. This has been the subject of many studies since maps showing contours for land elevations and ocean depths became available.

The total vertical relief on the solid surface of the earth is about 20 km, the difference between the peak of Mount Everest, at 8.85 km, and the greatest known ocean depth of more than 11 km. For analysis of the altitudes of the earth's surface, a contour interval of 1 km is a suitable fraction of the total relief.

The first step is to take a world map, with 1 km contour intervals, and measure the total surface area within each pair of contours. For all land surfaces, the area is measured between sea level and the 1 km contour, between the 1 km contour and the 2 km contour, and so on, to the highest levels. Similarly, for the ocean floor, the area is measured between sea level and the 1 km depth contour, between the 1 km and the 2 km depth intervals, and so on, for the whole ocean floor.

This provides a list of areas for each contour interval, which adds up to the total area of the earth's surface, 510×10^6 km^2. The data can be analyzed either in the form of absolute areas measured or as percentages of the total area. We will use percentages, following the precedent set in other diagrams. The area determined for each contour interval is therefore calculated as a percentage of the total area of the earth's surface, and the results are given in Table 3-2. The areas of land within the contour intervals above the 5 km altitude amount to a very small percentage of the total surface, so these have been added together. So have the percentage areas of the ocean floor within the contour intervals deeper than 7 km.

Table 3-2 summarizes the statistical data for the percentage areas of the surface of the earth that occupy the specific height or depth intervals above and below sea level. This is called a frequency distribution, and there are standard graphs that display the data more conveniently than a table.

In general, the frequency of occurrence

Fig. 3-10 Major surface features of the earth shown in a cross section near the Tropic of Cancer, between latitudes 20°S and 25°S. Locate the line of the cross section in Figures 3-1 and 3-7. The distortion caused by using vertical exaggeration of ×100 can be seen from the

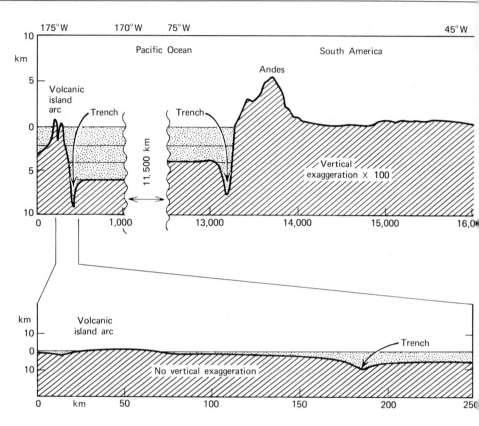

of a variable is measured by the numbers of items in a given class interval. In Table 3-2 the variable is the elevation, represented by the contour intervals, and the frequency of the occurrence of each selected contour intervals is measured by the area of solid surface within that interval. Each area has been recalculated as a percentage of the total area.

Figures 3-11a and 3-11b show two basic types of frequency distribution diagram. Figure 3-11a is a histogram, which shows the frequency distribution by means of a rectangle for each contour interval, with length representing the frequency, or percentage area, for that interval. In Figure 3-11b, the frequencies (percentage areas) are plotted as points in the centers of each contour interval. The line drawn connecting the points is called the frequency dis-

tribution curve. It shows the two dominant levels (Figure 3-6a).

Figure 3-11c shows a cumulative frequency curve, which is a third graph using the same data. Each point on the curve gives the percentage area of the earth's surface, which is at a higher level than the particular altitude for the point. For example, 29% of the earth's surface is at a higher level than altitude zero, sea level. This is the same as stating that 29% is dry land (Figure 3-3a). Figure 3-11c is called the hypsographic curve of the world's surface.

The curve is plotted using the data in Table 3-2. The third column gives the cumulative percentage area, which is higher than the lower limit of each contour interval, and this is obtained by adding successively the percentage areas within each contour interval. Only 0.1% of the surface

section through the Tonga Islands and Tonga trench (north of New Zealand) drawn with the same vertical and horizontal scales. The deeper layers of the earth along the same cross section are shown in Figure 7-6.

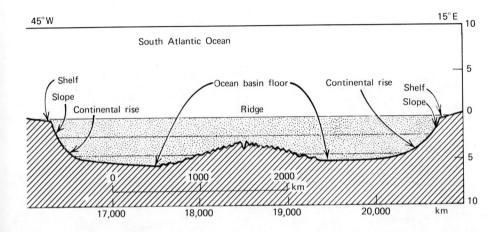

Table 3-2 The earth's solid surface: height and depth above and below sea level (Figure 3-11)

Height or depth interval, km	Percent of total area	Cumulative area (higher than lower limit of interval) as percent of total area
Above sea level: greatest height, Mt. Everest, 8.8 km; average height, 0.88 km		
>5	0.1	0.1
4-5	0.4	0.5
3-4	1.1	1.6
2-3	2.2	3.8
1-2	4.5	8.3
0-1	20.8	29.1
Below sea level: greatest depth, Marianas trench, >11 km; average depth, 3.7 km		
0-1	8.4	37.5
1-2	3.1	40.6
2-3	6.1	46.5
3-4	22.6	84.0
5-6	15.0	99.0
6-7	0.9	99.9
7-12	0.1	100.0

Fig. 3-11 Distribution of the areas of the solid earth between successive levels. See Table 3-2. *(a)* Histogram. *(b)* Frequency distribution curve. *(c)* Cumulative frequency distribution curve.

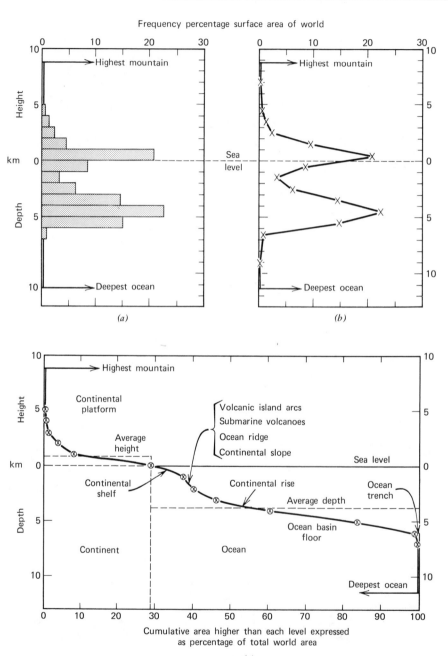

area is higher than 5 km. The area higher than 4 km above sea level is 0.1% plus 0.4%, the area within the 4 to 5 km contour interval. Similarly, the area higher than 3 km above sea level amounts to 0.1 + 0.4 + 1.1 = 1.6% of the total. And so the cumulative figures are added up to 100%.

Figure 3-11c illustrates in a general way some of the main features of the earth's solid surface, but other features are not distinguishable. The dominance of two principal levels corresponding to the continental plains and the ocean basin floor is clearly displayed (see Figure 3-10), and the curve also shows that the continental platform continues as the continental shelf at altitudes below sea level.

Only 1.6% of the world's surface rises above 3 km (compare Figure 3-8). At the other end of the curve, we see that only 1% of the world's surface is deeper than 6 km below sea level; yet there is a variation in depth of more than 5 km below the 6 km level within that 1%. This part of the diagram represents the ocean trenches.

The steep part of the curve between sea level and about 3 km depth bears a fortuitous resemblance to a cross section showing the continental slope connecting the continental shelf and the ocean basin floor, as in Figure 3-10, and it is labeled "continental slope." This part of the curve includes not only the altitudes of the continental slope, however, but also the altitudes of the continental rise, the ocean ridges, and the oceanic volcanic islands rising from the deep ocean floor (Figures 3-7, 3-10, and Table 3-1). These are physically distinct from the continents, but they have no independent representation in this statistical treatment of the whole world surface.

ACTIVE BELTS OF THE EARTH'S SURFACE

According to the theory of plate tectonics, the surface of the earth is covered by a series of shell-like, stable plates separated by active boundaries (Figure 1-2). One type of boundary is associated with the ocean ridge system, and another with the ocean trenches and nearby elevated belts, the volcanic island arcs, and the geologically young mountain ranges along continental margins (Figure 2-3).

The signals of activity are earthquakes and volcanoes in eruption. The distribution of these signals will be the topics of following chapters, but it is appropriate now to relate the active boundaries to the major features of the earth's surface that we have reviewed in the preceding pages. Figure 3-12 is a map showing the geologically active features of the world to be compared with Figure 3-7, which shows all major features.

Figures 3-7 and 3-8 show that the ocean ridge system itself is too large a feature to be considered as an active belt, but the rift valley at the center of many parts of the ridge does represent an active boundary, the site of sea-floor spreading. Figure 3-12 shows that the great rift system extends across the ocean basins and across some continents as well. It extends up the Red Sea and along the Jordan Valley. The African Rift valleys appear to be part of the globe-encircling system.

The rift is discontinuous because of displacement along fracture zones, as shown in Figure 3-12. Some of these fracture zones produce extremely rugged topography on the ocean ridges, with vertical relief of a few kilometers. High ridges and deep troughs associated with the zones may extend for hundreds of kilometers in a direction perpendicular to the rift valley or spreading center. These fracture zones certainly present evidence of activity, past or present, but their configuration is such that they cannot be boundaries between stable plates. We will see that, according to earthquake evidence, only the portion of a fracture zone between the offset parts of the rift valley is a plate boundary.

Figures 3-7 and 3-12 both show the distribution of ocean trenches, which delineate plate boundaries. These are associated with the volcanic island arcs and with parts of the geologically young mountain ranges on land. The young mountain

Fig. 3-12 Geologically active features of the world. Compare with Figures 3-7, 4-6, and 5-1. See Table 3-1.

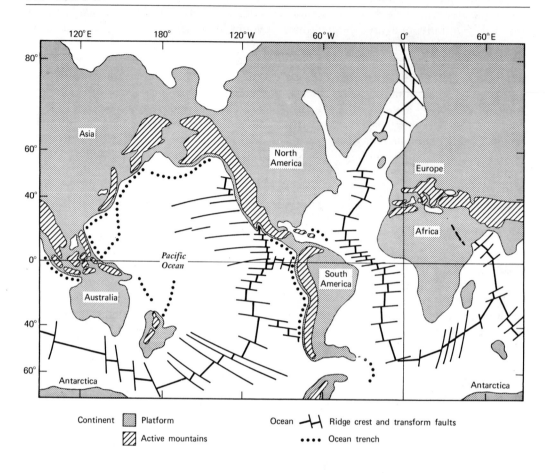

Continent ▓ Platform   Ocean ⊣⊢ Ridge crest and transform faults
                ▧ Active mountains   •••• Ocean trench

ranges are certainly active in a geological sense, and plate boundaries follow these across the continents. Evidence for this comes from the study of earthquakes, so let us now terminate our review of the distribution and shapes of surface features of the solid earth and go on to the next chapter, where the action is.

SUMMARY

The major subdivisions of the earth's solid surface are the ocean basins, covering 60% of the surface, and the continental platforms, which are separated from the ocean basin floor by the steep continental slope.

Ocean water covers 71% of the earth's surface, overlapping onto the continental shelves. The Pacific Ocean covers more than one third of the world's surface, and it is larger than all of the continents combined. More than 65% of the total land is in the northern hemisphere. The average depth to the ocean floor, 3.7 km, is much deeper than the average height of land surfaces, 0.88 km. The vertical distance between the highest mountain (8.8 km) and the deepest ocean trench (11 km) is about 20 km.

A worldwide submarine range of mountains rises from the ocean basin floor: the ocean ridge and rise system, covering nearly 25% of the total earth's surface. Continental mountain ranges cover only about 10% of the total earth's surface; another 1.2% is covered by their extensions across the ocean basins, as volcanic island arcs and associated deep ocean trenches. Ocean floor features include rugged fracture zones (transform faults), scattered oceanic volcanoes, and the continental rise at the foot of the continental slope.

The shapes of the features of the earth's surface are represented on maps by contour lines and are illustrated by vertical cross sections.

Plate boundaries coincide with the following major features of the surface of the solid earth: the crest of the ocean ridge system, the ocean trenches, the portions of transform faults that offset parts of the ridge crests and the trenches, and the geologically young continental mountain ranges.

SUGGESTED READINGS are at the end of the book.

chapter **4**

Earthquakes: Their Effects and Distribution

During an earthquake, the ground may vibrate up and down through fractions of centimeters or the land surface may undulate like the sea in a heavy swell. Land waves have been reported with heights of 0.5 m or more and with wave lengths (distance between wave crests) of 8 m. Individual waves may take 10 seconds to pass a point on the surface, and the process may continue for 3 or 4 minutes before the waves die down and calm is restored.

Calm is restored only to the tortured ground, but not necessarily to the human societies, because the passage of such vibrations through the solid earth wrecks buildings, buckles railway tracks, breaks water and gas mains, and collapses bridges and dams. In mountainous regions, landslides and avalanches are produced. Vibrations or other disturbances of the ocean floor may throw the surface of the ocean into giant waves, or *tsunamis,* which can traverse a whole ocean and pile up on unsuspecting shores as enormous "tidal" waves.

The waves of energy shaking the surface have their source at a focus within the earth where some kind of disturbance is initiated. In Chapter 1 we referred to rigid lithosphere plates sliding over the asthenosphere and

grinding against each other along their boundaries. Movement in the region of plate boundaries is the cause of most earthquakes.

Relative movement between a pair of plates is usually accomplished by a series of jerks. Each jerk follows a sudden fracture of rocks, and the fracture is a disturbance from which earthquake waves emanate in all directions. These vibrations in solid rocks are analogous to those produced by tossing a stone into a pool of water; concentric waves spread out in all directions from the point of disturbance where the stone splashed in and introduced energy into the system.

Figure 4-1*a* is a diagrammatic representation of the lithosphere with a plate boundary passing through it. The concepts of fracture and flow of rocks were introduced in Chapter 2. In response to very slow deforming forces, the rigid lithosphere rocks, although brittle, can bend in elastic fashion like steel, but only within very narrow limits.

Figure 4-1*b* shows that as the two adjacent plates move in opposite directions, the rocks are slowly bent. Elastic strain builds up along the locked plate boundary. When the elastic limit is reached, the rock fractures, and the strain energy is released as the two adjacent

Fig. 4-1 The relative movement of two lithosphere plates in opposite directions deforms the rocks, accumulating elastic strain that is suddenly released when the rock mass breaks along a fracture surface (fault). An earthquake results from the release of energy when fracture occurs. Compare Figure 5-2.

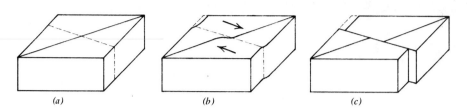

(a)   (b)   (c)

rock masses snap back into position through a few centimeters or meters. This is similar to what happens if you slowly bend a stick; suddenly, it breaks with a loud crack. The release of energy sends sound waves through the air.

The end result shown in Figure 4-1c is that the two blocks have moved relative to each other across a fracture plane. The fracture plane is called a fault. One kind of lithosphere plate boundary is actually a series of near-parallel faults of this type but, as we will see in Chapter 5 (Figure 5-2), there are other types, involving relative movements in different directions.

The focus of an earthquake is defined as the point of initial fracture. Once the break has started, however, relative movement can continue along the fracture plane for many kilometers.

INTENSITY AND MAGNITUDE OF EARTHQUAKES

An earthquake is a minor consequence of the global process of sea-floor spreading and plate tectonics, but the magnitude of the devastation that it can wreak in human terms is almost beyond belief. It has been estimated that between 1900 and 1968, earthquakes caused the death of about 800,000 people, with property damage of $10 billion. Most deaths are due to secondary causes such as building collapse, fires, landslides, and *tsunamis*. Underdeveloped areas with poorly constructed homes, often on loosely compacted earth, suffer most in loss of life.

The intensity of an earthquake in any region at the surface is defined on an arbitrary scale of 12 divisions, which is based largely on how the vibrations feel to people and on damage to various types of buildings. Table 4-1 lists some of the criteria used.

The intensity of an earthquake is determined by several factors. Of primary importance is the total amount of energy released at the focus of the earthquake. The highest intensity appears around the epicenter, which is the point on the surface directly above the focus (Figure 4-2). The fraction of the energy reaching the surface becomes less for points further away. Another variable factor is the nature of the ground below. The response of the surface to the earthquake waves when they arrive differs, depending on whether the ground is formed of solid rock or of potentially unstable material such as clay, sand, or gravel. A building on soil may collapse under conditions where the same structure built on solid rock might survive. Figure 4-2 shows the general distribution of intensity zones around the epicenter of an earthquake, with intensity defined according to the criteria listed in Table 4-1.

A more precise measure of earthquake strengths, based on measurements by instruments, is the Richter scale of earthquake magnitude shown in Table 4-2. The scale is based on direct measurement of the size (amplitude) of the earthquake vibration when it reaches a recording instrument at the surface. Allowance is made for the distance of the instrument from the epicenter.

An arbitrary zero point is for values near

Table 4-1 Scale of Earthquake Intensities

Intensity	Qualitative title	Description of effects
I	Negligible	Detected by instruments only
II	Feeble	Felt by sensitive people. Suspended objects swing
III	Slight	Vibration like passing truck. Standing cars may rock
IV	Moderate	Felt indoors. Some sleepers awakened. Sensation like a heavy truck striking building. Windows and dishes rattle. Standing cars rock
V	Rather strong	Felt by most people; many awakened. Some plaster falls. Dishes and windows broken. Pendulum clocks may stop
VI	Strong	Felt by all; many are frightened. Chimneys topple. Furniture moves
VII	Very strong	Alarm; most people run outdoors. Weak structures damaged moderately. Felt in moving cars
VIII	Destructive	General alarm; everyone runs outdoors. Weak structures severely damaged; slight damage to strong structures. Monuments toppled. Heavy furniture overturned
IX	Ruinous	Panic. Total destruction of weak structures; considerable damage to specially designed structures. Foundations damaged. Underground pipes broken. Ground fissured
X	Disastrous	Panic. Only the best buildings survive. Foundations ruined. Rails bent. Ground badly cracked. Large landslides
XI	Very disastrous	Panic. Few masonry structures remain standing. Broad fissures in ground
XII	Catastrophic	Superpanic. Total destruction. Waves are seen on the ground. Objects are thrown into the air

the limit of detection by instruments. For each step in the scale, the amplitude increases by a factor of 10. An earthquake with magnitude 2 therefore produces 10 times the amplitude of vibration as one with magnitude 1, and one with magnitude 8 causes 1 million (10^6) times the amplitude of one with magnitude 2.

The intensity of an earthquake, as measured by the amplitude of the earthquake waves, is closely related to the total energy released at the focus. One advantage of the Richter scale compared with the qualitative scale in Table 4-1 is that mathematical equations permit an estimate of total energy from the vibrations recorded by the instrument. The energy release increases by a factor of about 50 from one step to another in the Rich-

ter magnitude scale. Approximate energy release levels for some of the scale numbers are listed in Table 4-2. More than 90% of all earthquake energy is represented by earthquakes of magnitude 7 or more. A few large earthquakes thus account for most of the energy released.

The approximate relationship of the qualitative intensity scale of Table 4-1 to the Richter scale in Table 4-2 is shown by the roman numerals corresponding to the maximum intensity that would be experienced in the region of the epicenter (Figure 4-2).

Table 4-2 also shows the frequency of occurrence of earthquakes of various magnitudes. Approximately 1 million earthquakes are recorded by instruments each

Fig. 4-2 Earthquake focus within the earth compared with epicenter at the surface directly above the focus. Energy is transmitted in all directions, as shown by the arrows. The intensity of the earthquake decreases with distance from the epicenter, as shown by the zones of equal intensity. See Table 4-1.

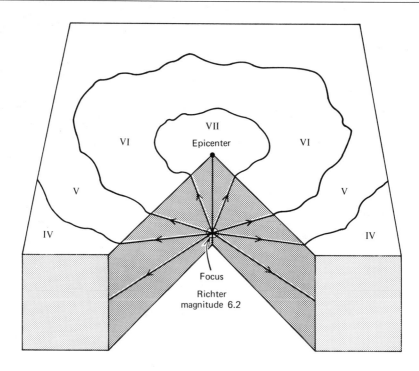

year, but of these only about 300,000 are felt by people. About 120 of these are destructive enough to cause some building collapse, 18 are major earthquakes causing widespread damage, and only 1 or 2 are great earthquakes strong enough to cause widespread devastation and to evoke newspaper headlines in another country for more than a day or two.

DEVASTATION IN PERU

Peru is one of the countries situated in the belt of compression between the Pacific and South American plates (Figures 1-2 and 4-3a). Figure 2-3 shows the thick slab of the Pacific plate extending down into the earth beneath the Andes. The Andes mountain ranges are slowly rising in response to the tremendous compressive pressure below, producing magnificent mountain peaks but great instability. The instability is shown by

the enormous range in vertical relief between the high peaks and the great depths in the ocean trench adjacent to the continental margin (Figure 3-10), and it is manifest in earthquakes and volcanic eruptions (Figure 4-6).

The deadliest earthquake in the recorded history of Latin America occurred in Peru on May 31, 1970. An estimated 50,000 people died in five minutes, another 100,000 were injured, and nearly 1 million out of the population of 13 million were left homeless. Estimates of property damage exceeded $250 million. The magnitude of the earthquake was 7.8 (Table 4-2).

The focus of the shock for the earthquake was located between the continent and the ocean trench, about 25 km west of the coast and 43 km below the surface (Figure 4-3). As the energy vibrations were transmitted to the continent, towns and villages collapsed. In an area covering 65,000 km² 80% of the adobe houses were destroyed or made uninhabitable. Many of the villages are situated on poorly consolidated alluvial soils. Additional damage was caused by secondary effects, including a massive avalanche of ice, water, mud, rocks, and other debris that swept down a valley in the Cordillera Blanca. This type of avalanche, a *huayco,* killed 41,000 in Ecuador and Peru in 1797, and another in 1939 killed 40,000 in Chile.

Vibration damage caused most of the destruction of buildings in 1970, and at least 60% of the fatalities. The second major cause of life loss was the *huayco* that buried about 90% of the resort town of Yungay, with a population of 20,000, and the neighboring village of Ranrahirca (Figures 4-3 and 4-4). The *huayco* had its source about 15 km east of Yungay and about 3700 m higher in the mountains. Apparently, the earthquake broke loose a mass of glacial ice and rock about 900 m wide and

1.6 km long. This moved downslope picking up earth and rocks and melting during descent. It was accompanied or preceded by a turbulent air blast. Survivors of the earthquake in Yungay were praying in the streets amid the wrecked buildings when they heard the thunderous roar of the *huayco* as it swept down the valley. In a few minutes the wreckage was covered by earth, mud, and boulders with thicknesses up to 14 m. This is represented diagrammatically in Figure 4-4.

The avalanche covered the 15 km from source to Yungay in two to four minutes. Thousands of boulders weighing up to 3000 kg were hurled more than 600 m across a valley in the middle part of its course, and calculations from the trajectories indicated a velocity of 400 km an hour. This high speed was due to the very steep slopes in the source region and the great vertical drop along the valley floor.

Table 4-2 Richter scale of earthquake magnitude, compared with intensities

Magnitude	Approximate maximum intensity	Number per year	Approximate energy released in explosive equivalents
0			
1		} 700,000	1 lb TNT
2			
2-2.9	II	300,000	
3			
3-3.9		49,000	
4	III		
4-4.9	Minor	6,200	
5	VI		Small atom bomb, 20,000 tons TNT (20 kilotons)
5-5.9	Damaging	800	
6	VII		Hydrogen bomb, 1 megaton
6-6.9	Destructive	120	
7	X		
7-7.9	Major	18	
8	XII		
8-8.6	Great	1 every few years	60,000 1-megaton bombs

Fig. 4-3 The Peru earthquake of May 31, 1970. Focus compared with the Andes and the ocean trench. Part *b* shows contours 2, 4, and 6 km both above and below sea level. Compare Figures 2-3 and 3-12.

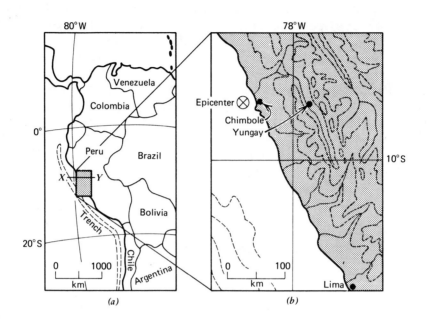

(a)

(b)

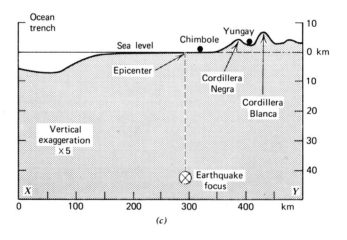

(c)

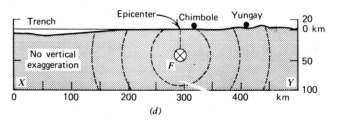

(d)

The mixture of melting ice and snow lubricated the mass of debris, reducing the frictional resistance, and there is evidence that locally the motion took place over a thin cushion of compressed air trapped between the debris and the valley floor.

It was estimated that the huge mass contained about 2.2 million m³ of water, mud, and rocks by the time it reached Yungay. Its momentum carried it across the Rio Santa near Yungay and 53 m uphill on the opposite bank. This sent a wave of water 14 m high rushing down through the narrow canyon occupied by Rio Santa. This wave of water was the third major cause of destruction arising from the earthquake.

Aerial photographs of the morass showed little evidence for the former existence of Yungay. The tops of four palm trees in the main square emerged above the debris, and a statue of Jesus was preserved because it stood on a small hill that divided the avalanche material. Many nations rushed aid to the stricken country, but it will probably take 10 to 15 years to repair the damage. Meanwhile, more strain builds up beneath the ocean trench and Andes mountains as

the Pacific plate is forced beneath the South American continent (Figure 2-3), making more earthquakes inevitable.

EARTHQUAKES IN CALIFORNIA

The San Andreas fault system in California is the boundary between the Pacific plate and the North American plate (Figures 1-2 and 5-1). As the Pacific plate moves north relative to the North American plate, occasional slippage along the fault system causes earthquakes (see Figure 4-1). Between 1906 and 1971, California experienced 18 destructive earthquakes with magnitude greater than 6, 4 of them being major, having magnitudes greater than 7. The most famous one, with estimated magnitude of 8.3, was the 1906 San Francisco earthquake. The property damage was dramatic, the estimated $400 million in 1906 being equivalent to about $1.6 billion today, allowing for inflation of the dollar. But the death of about 700 people was not excessive compared with the 50,000 killed in Peru in 1970 and with similar disastrous

Fig. 4-4 The Peru earthquake of May 31, 1970. Diagrammatic representation of the *huayco* that swept down from the Cordillera Blanca and destroyed the village of Yungay. Compare Figure 4-3c.

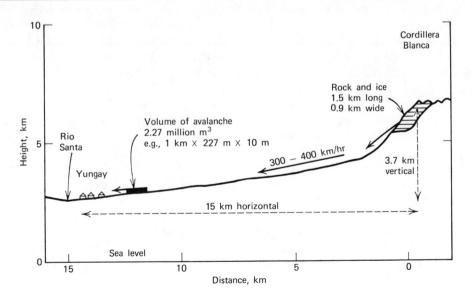

earthquakes. Much of the damage in San Francisco was caused by the great fire that raged out of control for three days because broken water mains left no water for fire fighting.

A report released in 1970 by the California legislature described what could happen if an earthquake equivalent to the 1906 San Francisco earthquake were to recur that year. An earthquake of this magnitude can be anticipated at intervals of about 100 years. Many buildings would collapse; the alluvial soils around the Bay Area would turn into quicksand, and buildings would sink. Communications would be disrupted as the freeways buckled and railway tracks twisted. At least one of the many dams in the area would fail, releasing flood water from the reservoir to drown large areas. Water mains would break. Estimated.property damage would amount to about $30 billion, and estimated deaths would number tens or even hundreds of thousands. The total of all previous losses caused by earthquakes in the United States is small in comparison with these estimates.

The population of California is often warned by the news media that sooner or later the state will be shaken by another major earthquake. This is fact, due to the passage of a plate boundary through the state (Figures 1-2 and 5-1). Prediction of precisely when an earthquake will occur is not yet possible, but this does not prevent some bold individuals from guessing. These individuals were supplied with useful information when the concept of plate tectonics was formulated in 1967 and 1968, because it was then announced that the earthquakes were caused by the slow movement of one part of California independently of the main American continent.

Some rather distorted versions of the plate tectonic theory became current among the population. Clairvoyants and soothsayers began to predict that California would soon split asunder and that one half would drift away and founder in the Pacific Ocean. This separation would, of course, be accompanied by earthquakes and tidal waves, causing additional devastation. A number of religious sects announced that Doomsday was imminent, and they prepared themselves for the end of the world.

For normal, educated citizens the topic provided entertaining material for cocktail party banter, but they knew perfectly well that California could not possibly sail off into the Pacific Ocean. Nevertheless, the occurrence of an earthquake certainly was an ever-present prospect, and the prophesied schism of the state did appear to have the respectable scientific backing of the highly touted new global theory of geology. Many articles in the popular press informed readers that most geologists now believed in the reality of continental drift.

By 1969, when the prophets of doom specified dates for the great schism, the telephone switchboards in geological survey offices and university earth science departments began to hum with enquiries from anxious people. The growing concern of the California public made newspaper headlines across the country. For example, an issue of the *Chicago Tribune* on March 30, 1969, reported:

Disaster talk runs rampant in California: quake will destroy state, seers say.

Monday, April 14, 1969, the day most frequently cited for the great event, was awaited with interest, with concern, or with disdain by various groups of people. The day came and went. Another Monday had started another workweek, and California remained in one piece. No earthquakes rumbled from the faults to test the stability of the fine high-rise buildings in the urban centers. The people laughed and told each other that they had known it was a hoax all along. It was reported that a number of prominent politicians whose business had forced them to be out of state on April 14 returned to their offices in California. Some disappointed religious sects prayed that next time Armageddon was announced it would really come.

The predictions of the seers and soothsayers actually erred only in one respect: they used the wrong time scale. As we will see in Chapter 12, one section of California is, indeed, destined to slide away from the rest of the state and to disappear 60 million years hence beneath Alaska. Where the seers erred, in their farsighted visions, was to compress the anticipated events of a few million years into a few hours.

The disaster talk abated, but a reminder came as the sun rose on February 9, 1971. An earthquake of magnitude 6.6 with focus near San Fernando, at a depth of 13 km, shook the residents of Los Angeles from their sleep. This was not a major earthquake, but it was one of the most devastating in the history of California because it occurred in a densely populated northern suburb of Los Angeles. This provided the first comprehensive test of earthquake provisions included in building codes. Engineers concluded that more stringent codes are required.

Widespread damage occurred. Thirty-eight patients died beneath the rubble of two collapsed hospitals. Freeways were closed by fallen overpasses, damaged bridges, and buckled roadways. Water pipes, gas lines, and electrical cables were broken, and several fires started. Thousands of people and homes were menaced by a cracked reservoir dam in the San Fernando Valley, and about 100,000 people were ordered evacuated by the police. Many of them did spend a few days in emergency refugee camps, while the water was drained from the reservoir.

Sixty-four people died, and estimated property loss was $1 billion. Many buildings in the Los Angeles area were so seriously damaged that they had to be abandoned. These included 450 homes, 60 apartment buildings, and 400 commercial buildings. Seventy-five schools were permanently vacated because of damage, and others were deemed in need of strengthening; but in October 1971, voters rejected a multimillion dollar bond issue to repair or replace the hazardous school buildings.

At 6 a.m., when the earthquake struck, the freeways, schools, offices, and streets were almost deserted; otherwise, the loss of life would have been greater. None of Los Angeles' tall buildings suffered serious earthquake damage, but the earthquake focus was too far away to test their ultimate strength. The damage would have been greater had the focus been located nearer the center of Los Angeles.

THE NEW MADRID EARTHQUAKES IN THE MIDWESTERN UNITED STATES

According to the theory of plate tectonics, earthquakes are concentrated in belts, and they are infrequent in the large stable plates (Figures 1-2 and 4-6). The most violent earthquakes known to have occurred in the United States, however, were located well within the North American stable plate.

From 1811 to 1812, a series of hundreds of earthquake shocks devastated the central part of the Mississippi Valley near the Missouri-Tennessee border, in area *B* of Figure 4-5. Three very large shocks had estimated intensities of XII, and they were felt over two thirds of the United States. In Washington, D.C., 1300 km away, sleepers were awakened, dishes and windows were rattled, and walls were cracked; even at this distance, the intensity was VII. The vibrations rang church bells in Boston. The earthquakes caused major changes in topography over 130,000 km². Large areas sank from 1 to 3 m, and depressions 30 to 50 km long and 6.5 to 8 km wide were produced. The course of the Mississippi River was changed. The area was sparsely populated at the time, and the loss of life was small, but not precisely known. According to estimates made from surface changes, the largest of the earthquakes may have had a magnitude of 8.5. Only few earthquakes have been recorded anywhere with larger magnitudes. The same area, *B*, experienced large earthquakes in 1843, 1865, 1895, 1923, 1927, and 1934.

The history of large earthquakes in the eastern United States is summarized in Figure 4-5. This does not give a useful indication of where future earthquakes might occur, except for the region *E* along the St. Lawrence River Valley, Lake Erie, and Lake

Fig. 4-5 Earthquake activity in central and eastern North America. (Data from B. F. Howell in *Earth and Mineral Sciences,* March 1973, Vol. 42, The Pennsylvania State University.)

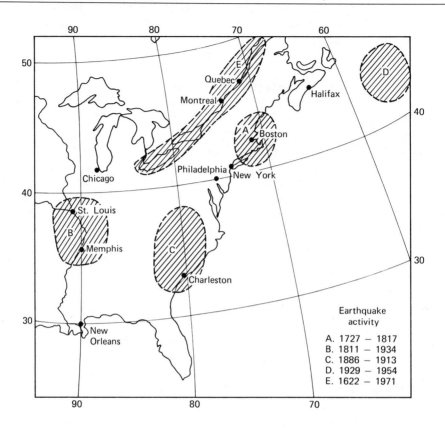

Ontario, where there has been at least one earthquake with intensity greater than VII every 14 years since 1638.

The center of earthquake activity has moved around within a large area. There was a burst of earthquake activity around Boston in area *A,* starting in 1727, with four additional large earthquakes between then and 1817. The activity in area *B* began with the New Madrid earthquakes in 1811, with the last one occurring in 1934. The Charleston, South Carolina earthquake was in 1886, and there were three later earthquakes in area *C* between 1897 and 1913. The most recent activity was on the continental shelf off Newfoundland, area *D,* with three earthquakes in 1929 and two in 1954.

In California and Japan, where earthquake activity is accepted as a fact of life, building codes include some provisions for earthquake-resistant construction. No such provisions are deemed necessary in the central United States, where the land is considered to be stable. If the New Madrid earthquake had occurred last week, the wreckage could have been catastrophic. At the present time, we have no way of predicting the likelihood of such an occurrence in the supposedly stable plates.

DISTRIBUTION OF EARTHQUAKE EPICENTERS

Every earthquake is caused by an event occurring within the earth, and each event can be defined in terms of place, time, and magnitude. The earthquake waves emanating from the event, the fracture, travel around and through the earth, and the waves are received at earthquake recording stations in many places. Their time of arrival at different stations can be measured precisely, and it is a fairly simple matter to obtain a fix on the position of the focus of the energy, of the place where the event occurred.

Each year, nearly 1 million earthquakes strong enough to be recorded by sensitive instruments occur. Of these, just a few major earthquakes account for most of the total energy released in the process of lithosphere plate readjustment. But the many minor tremors provide a record of the places where readjustment is occurring.

Figures 4-6 and 4-7 show the world distribution of earthquakes.

Figure 4-6 shows the position of the epicenter of every recorded earthquake from 1961 to 1967. These are not randomly distributed over the surface of the earth; they are obviously concentrated in certain belts. Comparison of the earthquake belts in Figure 4-6 with the map of active belts of the earth's surface shown in Figure 3-12 demonstrates that the active belts delineated on the basis of geological features of the surface of the solid earth do, indeed, correspond to the belts where the action is. The earthquakes follow very closely (1) the crest of the submarine mountain ranges, the world rift system, and (2) the chains of geologically young mountain ranges and their extensions along volcanic island arcs.

It was discovered by M. Ewing and B. C. Heezen in 1956 that the earthquake belt following the oceanic ridges is remarkably narrow. This contrasts with the belt following the young mountain ranges and island arcs, which expands into a rather broad band along some island arcs and across the Himalayas. In fact, although we can distinguish a concentration of epicenters following the Himalayas, this band extends north into an extensive area stretching through Tibet and into China. There is a similar broadening of the belt of earthquake epicenters in the western United States. Note also the scatter of epicenters in the region of the African rift valleys.

Figure 3-12 shows that the crest of the ocean ridges is not a continuous feature; it is repeatedly displaced by major fracture zones. The shallow earthquake epicenters in Figure 4-6 give little indication of these offsets. Figure 4-8, drawn to a larger scale, shows how earthquake epicenters are distributed along ridge crests and fracture zones.

Figure 4-8 is a diagrammatic map of a portion of the mid-Atlantic ridge. The rift valley along the crest of the ridge is represented by the pair of heavy lines, and the rugged relief along the fracture zones is represented by the pairs of finer lines dividing the rift valley into separate, displaced sections. The distribution of earthquake epicenters, represented by the circles, is restricted almost exclusively to the rift valleys and to the fracture zones. The rugged relief of the fracture zones extends well beyond the ridge crests, as shown in Figures 4-8 and 3-12, but the earthquakes occur only along those portions of the fracture zones between ridge crests. This was discovered by L. R. Sykes in 1966.

DISTRIBUTION OF EARTHQUAKE FOCI

The map of epicenters in Figure 4-6 shows the distribution of earthquakes with respect to surface area. But the focus of an earthquake is situated within the earth, as illustrated in Figure 4-2. Thus, for complete representation of the distribution in space of earthquakes, we need a third dimension, the depth of each focus.

Earthquakes are classified according to their depth of focus. Shallow-focus earthquakes occur between the surface and 70 km depth, intermediate-focus earthquakes in the depth interval 70 to 300 km, and deep-focus earthquakes between 300 and 700 km depth. There are very few earthquakes releasing energy at depths greater than 700 km.

Figure 4-7 shows the distribution of earthquake epicenters for all of the intermediate and deep-focus earthquakes in Figure 4-6; the shallow-focus earthquake

Fig. 4-6 Distribution of all earthquake epicenters recorded by U. S. Coast and Geodetic Survey, 1961-1967. (From M. Barazangi and J. Dorman, *Bull. Seismol. Soc. Amer., 59,* 369, 1969.)

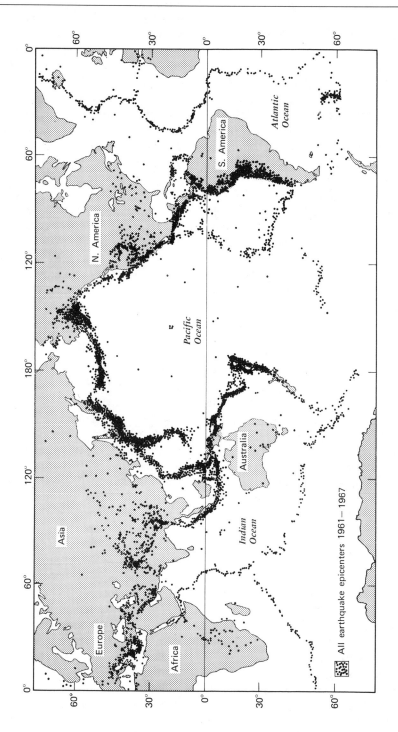

Fig. 4-7 Distribution of intermediate-focus and deep-focus earthquakes recorded by U. S. Coast and Geodetic Survey, 1961-1967. (From M. Barazangi and J. Dorman, 1969, see Figure 4-6.)

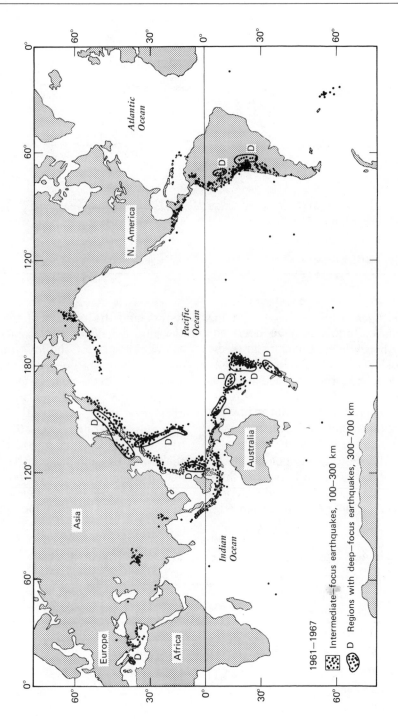

Fig. 4-8 Schematic map of portion of the mid-Atlantic ridge near the equator, illustrating the distribution of earthquake epicenters with respect to the central rift valley and the transverse fracture zones. Compare Figures 3-12 and 4-6. (Based on data of L. R. Sykes, 1967, *Jour. Geophys. Res., 72,* 2131-2153.)

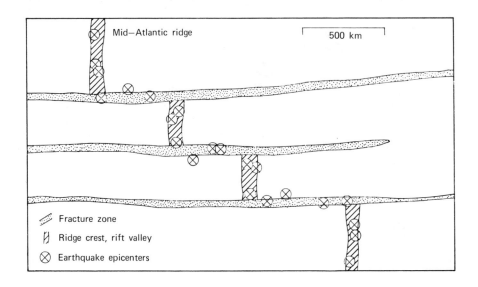

epicenters have been removed, and the deep-focus earthquakes have been enclosed by heavy lines. Comparison of the two maps shows the distribution of shallow focus earthquakes. The intermediate-focus and deep-focus earthquakes are more restricted in their distribution, with an obvious concentration around the Pacific margin.

Now, if we refer back to the world map of Figure 3-12, showing the geologically active belts of the earth, and compare these with Figures 4-6 and 4-7, we find the correlations listed in Table 4-3. The ocean ridge system and the portion of the fractures offsetting the ridges are characterized by the occurrence of shallow-focus earthquakes. There are no deeper earthquakes associated with the ridge system.

Deep-focus earthquakes are almost exclusively associated with the ocean trenches, in zones separated from the trenches by belts of intermediate-focus and shallow-focus earthquakes. The young, active mountain ranges that are not associated with ocean trenches experience shallow-focus earthquakes, with local concentra-

Table 4-3 Types of earthquakes associated with active belts at the earth's surface

Active feature	Earthquake depth of focus
Ocean ridge	Shallow
Transform fault	Shallow
Volcanic island arc with ocean trench	Shallow, intermediate ± deep
Young mountain ranges	
With associated trench	Shallow, intermediate, deep
Without trench	Shallow ± intermediate

tions of intermediate-focus, but no deep-focus earthquakes. The continental platforms experience scattered shallow-focus earthquakes. The ocean basin floors are clear except for local concentrations of shallow-focus probably associated with the oceanic volcanoes.

We have reviewed the damage caused by several specific earthquakes, and we can now relate these to the maps in Figures 4-6, 4-7, and 3-12. The earthquakes in Peru and Chile are associated with a concentrated zone of shallow-focus, intermediate-focus, and deep-focus earthquakes. California and the western United States experience only shallow-focus earthquakes. There is a minor concentration of shallow-focus earthquakes in the midwestern United States.

Figure 4-7 shows the concentration of intermediate-focus and deep-focus earthquakes around the Pacific Ocean. More than 90% of intermediate-focus earthquakes, and nearly all of the deep-focus earthquakes, occur in the circum-Pacific belt. Although it is not obvious from Figures 4-6 and 4-7, it is also true that more than 80% of the shallow-focus earthquakes occur in the same belt. In addition, 80% of all earthquake energy is liberated from this belt.

The spatial distribution of earthquakes in three dimensions is well displayed in the earthquake belt along the Tonga trench. We have already examined a cross section through the Tonga trench and volcanic island arc in Figure 3-10. The concentration of earthquake epicenters associated with this feature is shown in Figures 4-6 and 4-7, extending north from New Zealand. Not only is this one of the most active regions in the world for shallow earthquakes, but about 70% of the world's deep-focus earthquakes also occur between the Tonga and Fiji islands.

An enlarged map of this region is given in Figure 4-9. The Tonga trench is a prominent feature, but the volcanoes constituting the island arc barely reach above sea level. The circles represent the epicenters of deep-focus earthquakes that have occurred

through a period of years. Epicenters of intermediate-focus and shallow-focus earthquakes have been omitted because they would obscure the map of the trench and island arc if they were plotted. Figures 4-6 and 4-7 indicate the total width of the earthquake zone.

Figure 4-10 is a cross section through the earth's interior along line XY in Figure 4-9. The section is drawn perpendicular to the axis of the trench. There is no vertical exaggeration and, to this scale, the difference in elevation of the surface between island arc and trench floor is barely discernible; compare Figure 3-10. The circles represent the positions of earthquake foci for earthquakes that occurred within 125 km of the section. The depths and distances of each earthquake focus occurring within 125 km were plotted directly on the figure; this is the same as projecting each earthquake focus horizontally onto the section.

The deep-focus earthquakes mapped in Figure 4-9 within 125 km of section XY are shown as circles between 300 and 700 km depth in Figure 4-10. They are restricted to a remarkably narrow band. This band is extended upward toward the island arc and Tonga trench by the projected intermediate-focus and shallow-focus earthquakes. Most of the earthquake foci lie within a zone less than about 20 km thick, dipping west beneath the arc.

To a first approximation, this zone of earthquake foci may be treated as a nearly plane surface dipping into the earth's interior, a surface that is intersected by line ZW in the cross section of Figure 4-10. The surface is 250 km wide, extending to a depth of 700 km. This is more than one tenth of the radius of the earth. Similar examination of the positions of all the other earthquake foci show that this is just a small part of a much larger surface, extending the whole length of the Tonga trench in Figure 4-9 and dipping down beneath the volcanic island arc. In any cross section drawn perpendicular to the trench axis, the earthquake surface is intersected in a line similar to ZW in Figure 4-10.

Fig. 4-9 Schematic map of distribution of earthquake epicenters for deep-focus earthquakes between the Tonga trench and the Fiji Islands, north of New Zealand. Locate this on Figures 3-12 and 4-7. (Based on data of L. R. Sykes, 1966, *Jour. Geophys. Res., 71,* 2981-3006.)

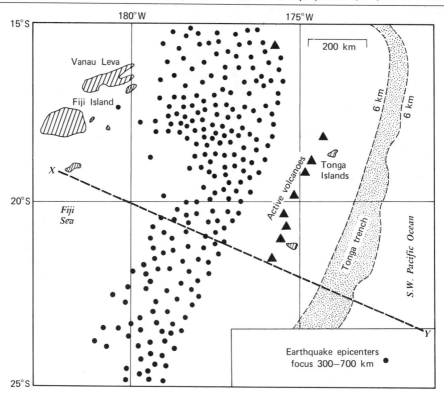

Fig. 4-10 Vertical cross section through line *XY* in Figure 4-9, showing schematically the distribution of earthquake foci down to depths of 700 km. The foci lie close to line *WZ* extending downward from the ocean trench. (Based on data of L. R. Sykes, 1966, see Figure 4-9.)

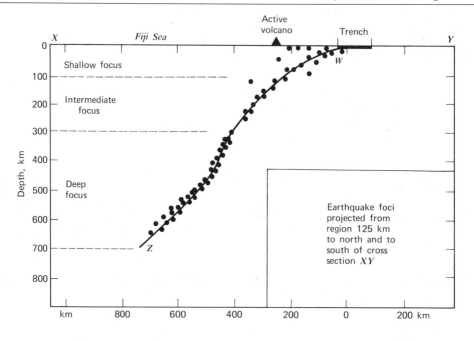

Given a surface dipping into the earth's interior, we can plot contours for specific depths, as shown in Figure 4-11, in precisely the same way as we plot contours for the surface of the solid earth in Figure 3-9. Figure 4-11 is thus a contour map of the nearly planar surface within the earth, which is the locus of earthquake foci. We can construct cross sections through it just as we did through the contoured map of the solid earth surface in Figure 3-9.

The surface representing the earthquake zone mapped in Figure 4-11 displays the standard pattern for the distribution of shallow-focus, intermediate-focus, and deep-focus earthquakes beneath volcanic island arcs and young mountain ranges associated with deep ocean trenches (see Table 4-3). Not all localities exhibit such a narrow zone as the Tonga trench does, but

everywhere the distribution of shallow-focus, intermediate-focus, and deep-focus earthquakes indicates the presence of an earthquake zone dipping down from the ocean trench beneath the island arc or mountain range. The angle of dip may vary between 40 and 60°.

SUMMARY

The earth quakes in response to waves that transmit energy in all directions from an underground focus where the lithosphere fractures. The movement of lithosphere plates causes earthquakes. Adjacent plates are slowly bent until they fracture, and relative movement is accomplished by a jerk along a fault surface along the plate boundary.

Fig. 4-11 The distribution of earthquake foci in the Tonga region represented as a surface dipping down beneath the Tonga trench, with depth contours to the surface drawn at intervals of 100 km. Line WZ in Figure 4-10 is on this surface. (Somewhat simplified representation of data from L. R. Sykes, 1966, see Figure 4-9.)

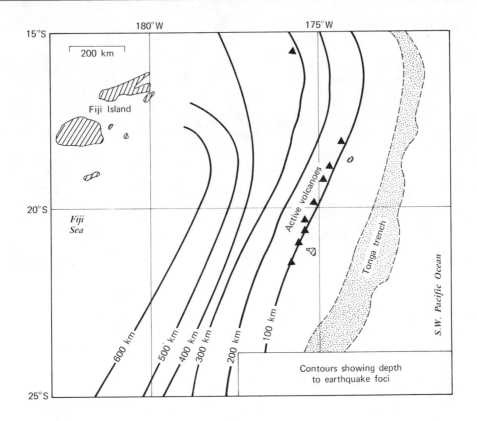

The intensity of an earthquake is defined on an arbitrary scale based largely on the damage to buildings. The Richter scale of earthquake magnitude is based on the energy released at the earthquake focus. Each year, about 30,000 earthquakes are felt by people, 18 of these are strong enough to cause widespread damage, and only 1 or 2 cause devastation.

The effects of earthquakes associated with plate boundaries are illustrated by examples from Peru and California. Destruction of buildings and loss of life occurs in two stages. The first is when the vibration topples buildings; secondary effects include avalanches, landslides, "tidal" waves, floods caused by burst dams, and fires caused by burst gas mains. Major earth-quakes also occur within the relatively stable plates, as illustrated by several examples in North America.

Earthquakes are concentrated in belts following closely the crest of submarine mountain ranges, the chains of geologically young mountain ranges, and their extensions along volcanic island arcs. Deep-focus earthquakes with depths between 300 and 700 km are associated only with ocean trenches. The distribution of shallow-focus, intermediate-focus, and deep-focus earthquakes defines the positions of surfaces of earthquake activity dipping down from ocean trenches beneath the adjacent island arc or mountain range.

SUGGESTED READINGS are at the end of the book.

chapter **5**

Earthquakes and the Theory of Plate Tectonics

Earthquakes are signals of activity. They tell us the story of plate tectonics as it is happening today. The general concept of plate tectonics was introduced in Chapters 1 and 2, and data about the distribution of earthquake foci were presented in Chapter 4. I indicated the relative directions of movement of some plates (Figure 2-3) in connection with Hess's "geopoetry" of sea-floor spreading in Chapter 2, but I have presented no evidence in support of this, or for the divergence or convergence of plates associated with the specific geological features reviewed in Chapter 3. In this chapter, I will discuss earthquakes in more detail and interpret the data in terms of plate tectonic theory.

Earthquakes provide us with three very important lines of evidence. First, the distribution of earthquake epicenters, especially for shallow-focus earthquakes, is assumed to delineate the plate boundaries. Second, the distribution of earthquake foci gives evidence for the existence of lithosphere plates extending down through the asthenosphere and of the attitude of the lithosphere plates at depth within the earth. Third, the study of earthquake waves tells us about the direction of movement of each plate relative to its neighbors. The direc-tions of relative motion determined from earthquake studies are consistent with the existence of three types of plate boundaries: divergent boundaries where new lithosphere is generated; convergent boundaries where lithosphere is destroyed by movement into the earth's interior; and transform fault boundaries where lithosphere is neither created nor destroyed.

STABLE PLATES AND ACTIVE BOUNDARIES

The earthquake epicenters plotted in Figures 4-6 and 4-7 are concentrated in belts that mark the active boundaries of stable, rigid, lithosphere plates. Figures 1-2 and 5-1 show the positions of the plate boundaries. There are six major plates named in Figure 5-1, with several smaller plates located between the larger ones. The boundaries of many of the smaller plates have not yet been adequately defined, but the general picture on a worldwide scale seems to be clear. Determination of the directions of movement of some of the plates, depicted by arrows, is based on evidence reviewed in the following section.

Fig. 5-1 Distribution of lithosphere plates and plate boundaries. See Figure 1-2 for polar views. Compare with Figures 3-12 and 4-6. Arrows show the movements of plates relative to adjacent boundaries.

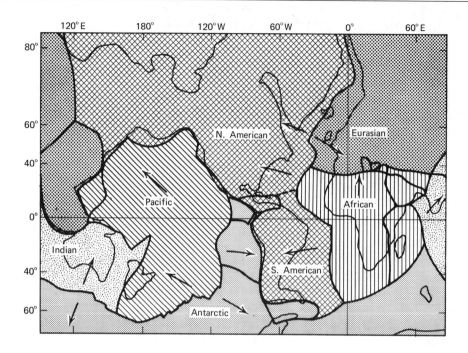

I noted in connection with Figures 3-7 and 3-8 that the boundary between continent and ocean, the shoreline, has no great significance with respect to the surface of the solid earth. The boundary between continental and oceanic platforms is the continental slope, situated some distance out to sea from the shoreline. Now we see in Figure 5-1 that the continental slope, as striking a feature as it is (Figure 3-10), has little significance in plate tectonics unless it coincides with an earthquake belt. Most plates include both continental platform and ocean basin floor. The Pacific plate is the only major plate that includes no continental material.

The plate boundaries are depicted in Figure 5-1 as bold lines. This works very well for the oceanic ridges, both on a large scale (Figure 4-6) and in the detail of small scale (Figure 4-8). But, in other environments, the earthquake belts are less well defined, and it becomes more difficult to draw a single line through the belt represented by epicenters.

In some regions, the band of earthquake epicenters becomes wide because there are successive bands of intermediate-focus and deep-focus earthquakes alongside a band of shallow-focus earthquakes. This is indicated in Figures 4-9 to 4-11. The cross section through the region (Figure 4-10) shows that the surface representation of the boundary between two plates corresponds to the band of shallow-focus earthquakes. This always correlates directly with the deep ocean trenches alongside volcanic island arcs, or young mountain ranges. In some regions, therefore, the plate boundaries are located from a combination of earthquake maps and knowledge of the physical features of the surface of the solid earth (Figure 3-12).

In other regions, where there are no intermediate-focus or deep-focus earthquakes, the zone of shallow-focus earthquake epicenters exhibit considerable broadening. I noted this for the Himalayas, for the western United States, and for the African rift valleys. Comparison of Figures

4-6 and 4-7 shows that this is also the situation for the zone extending from the Himalayas through the Mediterranean region. For these regions, it is quite obvious that the lines drawn as plate boundaries in Figure 5-1 are not based exclusively on the distribution of epicenters. Within a broad belt of epicenters there may be a narrow zone in which major earthquakes are more abundant, and this zone is then identified as the plate boundary. Geological evidence also may be useful.

Figure 4-1 illustrates schematically the movement across a fracture plane, or fault, which we termed a plate boundary. The concept of a simple fracture plane as the boundary between plates appears to be a reasonable approximation for some environments, such as ocean ridges and connecting fracture zones, but there is evidence that across a continental plate, a whole series of subparallel fracture zones produces a complex boundary region instead of a specific boundary. The interaction of two plates of continental lithosphere evidently differs from that of plates involving oceanic lithosphere.

At different times, relative movements between the plates may be accomplished along different faults within the broad fracture zone. The net effect then is that the large plates appear to be separated by a whole series of small plates, which together constitute the boundary.

In some regions, one of the fault planes can be identified as the most active boundary. An example is the San Andreas fault in California. The 1906 San Francisco earthquake was caused by fracture along this fault plane, but the 1971 San Fernando earthquake was caused by fracture along one of the fault planes parallel to it.

Earthquakes are not restricted to the plate boundaries. The supposedly stable plates do experience them. This is obvious from a glance at Figures 4-6 and 5-1, and I have previously outlined the effects of the great New Madrid earthquake and other earthquakes with epicenters well within the North American plate (Figure 4-5).

The whole assembly of plates sliding over the earth's interior is in a state of strain. Most of the strain is released in earthquakes near the margins, where adjacent plates are grinding together, but strain accumulated within the plate may be more effectively released by a minor readjustment away from the plate boundary. This is accomplished by fracture along a plane of weakness within the plate. Other earthquakes associated with stable plates are caused by volcanic activity.

DIRECTIONS OF MOVEMENT AT PLATE BOUNDARIES

The earthquake waves emanating from the focus where the first motion occurred along the fracture plane carry with them vital information about the direction of the initial movement of one block relative to the other. The waves from each earthquake are recorded by sensitive instruments in a worldwide network of receiving stations. Detailed analysis of the records makes it possible to work out the displacement direction on the fault surface at the earthquake focus.

Movement on fault planes can be classified into the three types illustrated in Figure 5-2. These correspond to three types of plate boundaries (Table 5-1). Figure 4-1 shows development of a fracture where one block is displaced sideways relative to the adjacent block. This is also represented in Figure 5-2a, and shown in the map in Figure 5-2b.

In regions of tension, if the lithosphere is being stretched, a fracture could develop, as illustrated in Figures 5-2c and 5-2d. One block moves downward relative to the other along an inclined fault plane that is usually not far from vertical. The net effect is that one block moves away from the adjacent block. The blocks diverge from each other along the divergent boundary.

In regions of compression, one block can be forced to move closer to the adjacent block if fractures develop, as shown in Figures 5-2e and 5-2f. One block overrides the other by moving upward along an inclined

Fig. 5-2 Three types of faulting. The top diagrams represent blocks of the lithosphere that have broken along a fracture plane. The arrows show the directions of relative movement for each example. The lower diagrams are schematic maps looking down on the blocks, showing the relative movements of the lithosphere on either side of the fault line. Compare Table 5-1 for correlation with plate boundaries.

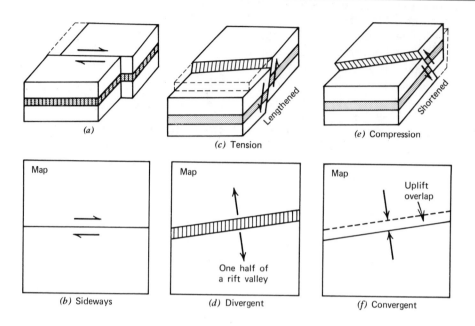

Table 5-1 Classification of Plate Boundaries

Boundary	Ocean-ocean	Ocean-continent	Continent-Continent
Divergent (tension)	Ridge crest. S earthquakes[a], narrow belt. Submarine lavas		Rift valley. S earthquakes, wide zone. Volcanoes
Convergent (compression)	Ocean trench and volcanic island arc	Ocean trench and young mountain range	Young mountain range
	SID earthquakes, wide belt	SI±D earthquakes, wide belt	S±I earthquakes, wide zone
	Volcanoes	Volcanoes	No volcanoes
Transform (neither tension nor compression)	Fracture zone of ridges and valleys. S earthquakes, narrow belt only between offset ridges		Fault zone S earthquakes, broad zone
	No volcanoes		No volcanoes

[a]S, shallow-focus; I, intermediate-focus; D, deep-focus earthquakes.

fault plane. The angle of slope of the plane may vary considerably. The blocks converge toward each other along a convergent boundary.

The limiting examples given in Figure 5-2 show relative movement on fault planes restricted to the vertical or horizontal planes. Usually, the relative motion involves some combination of the motion in Figure 5-2a with that in Figure 5-2c or 5-2e. But, on many faults, the motion tends to be dominantly horizontal, (Figure 5-2a) or dominantly vertical (Figures 5-2c and 5-2e).

Figure 5-2 shows how determination of the displacement direction on a fault plane by the study of earthquake waves gives a picture of the relative movement of the blocks of crust on either side. If the displacement direction is determined, the effective movement across the fault boundary at the surface is in the sense shown in the maps. The three maps show the relative directions of movement resulting from convergence or compression (Figure 5-2f), from divergence or tension (Figure 5-2d), and from lateral movement with neither convergence nor divergence (Figure 5-2b).

Examination of many earthquake records since the middle of the 1960s has confirmed that earthquakes associated with plate boundaries tend to correspond to one of these three types, and the distribution of these types gives strong support to Hess's geopoetry of sea-floor spreading (Figures 2-1 and 2-3).

The earthquakes along ocean ridges indicate displacements, as in Figure 5-2c, and the crustal blocks therefore are divergent, as in Figure 5-2d. For earthquakes along the fracture zones displacing segments of the ocean ridges, as in Figure 4-8, the displacement is as in Figure 5-2a, and the crustal blocks are therefore moving sideways relative to each other. The shallow-focus earthquakes on the landward side of ocean trenches indicate displacement, as in Figure 5-2e, and the crustal blocks are therefore convergent, as in Figure 5-2f.

Figure 5-3 illustrates the directions of relative movements of plates on either side of plate boundaries corresponding to ocean ridge and offsetting fracture zone. Figure 5-3a is a three-dimensional version of a part of Figure 4-8. It shows the lithosphere, about 100 km thick, overlying the asthenosphere. It shows one fracture zone, separating the ocean ridge into two portions. Figure 4-8 shows the distribution of earthquake epicenters. Figure 5-3a shows some earthquake foci within the lithosphere beneath the ocean ridge. Note that there are no earthquake foci associated with the fracture zone below d; these are restricted to the portion bc of the fracture zone (see Figure 4-8).

The earthquakes along the ocean ridge indicate fracture motions corresponding to Figure 5-2c, which shows that the ridges are divergent plate boundaries. This is consistent with the concept of sea-floor spread-

Fig. 5-3 The relative movement of plates associated with ocean ridge and fracture zones. (a) Block diagram. (b) Map of surface. Compare Figures 4-8, 5-2a, and 5-2b.

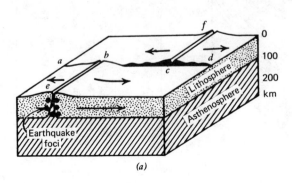
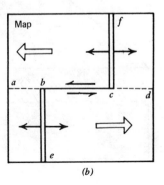

(a)   (b)

ing (Figures 2-1 and 2-3). The divergence of the plates is shown by the arrows in Figure 5-3. The earthquakes along the portion *bc* of the fracture zone indicate motions corresponding to Figures 4-1 and 5-2*a*. The map in Figure 5-3*b* shows how this fits in with the divergent plate boundaries. According to the earthquake epicenters in Figure 4-8, the plate boundary in Figure 5-3*b* follows line *ebcf*. Portions *eb* and *cf* are divergent boundaries. The sense of movement of the plates on either side of boundary *bc*, imposed by divergence from the ridges, produces the sideways displacement indicated by the earthquakes occurring along *bc* (Figure 5-2*a*).

The arrows showing lithosphere plates diverging from the ridge boundaries show that the sense of movement is the same on either side of lines *ab* and *cd*. The absence of earthquakes along portions *ab* and *cd* of the fracture zone also shows that there is no differential movement along these portions, and we conclude that they are not active. The surface relief that identifies them as part of fracture zone *abcd* is inherited from an earlier period of activity, when these parts of the plates were in region *bc* between the offset ocean ridges.

It was only in 1965 that this kind of fracture *bc* was identified by J. T. Wilson and termed a transform fault. Faults with essentially lateral displacement (Figures 4-1 and 5-2*a*) were well known, but transform faults are distinctive because they terminate abruptly at both ends. Despite this, the plates on either side are displaced through great distances by the process of sea-floor spreading. Figure 5-3 shows a transform fault connecting two ocean ridge segments, but one or both ends of the active portion of a transform fault could be connected to a plate boundary of compressive type.

Figure 5-4 illustrates the directions of relative movements of plates on either side of plate boundaries corresponding to an ocean trench. Figure 5-4*a* is a three-dimensional version of Figures 4-10 and 4-11. It shows the lithosphere layer, about 100 km thick, overlying the asthenosphere. Figures 4-9 to 4-11 show that the shallow-focus earthquakes occupy a zone near the ocean trench. These earthquakes indicate motions corresponding to Figure 5-2*e*, which shows that the trenches are convergent plate boundaries. This is consistent with the concept of sea-floor spreading (Figure 2-3). The convergence of plates is shown by the arrows in Figure 5-4*b*.

Figure 5-4*a* also shows the deeper earthquake foci concentrated in a narrow zone dipping down through the asthenosphere from the ocean trench (Figures 4-10 and 4-11). In Figure 4-10 just the earthquake

Fig. 5-4 The relative movement of plates associated with ocean trench (and volcanic island arc). *(a)* Block diagram, compare Figures 4-10 and 5-2*e*. *(b)* Map surface, compare Figures 4-9 and 5-2*f*.

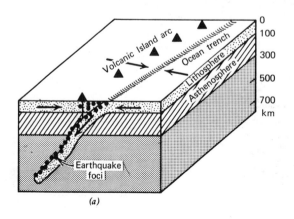

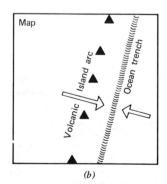

foci are plotted. The distribution of earth-quakes can be used to delineate the plates of cool, rigid lithosphere, and these deep earthquake zones are strong evidence for the existence of lithosphere slabs extending down into the earth's interior through the asthenosphere. Careful study of earthquake waves that have passed through this part of the asthenosphere shows that the material just below the deep zone of earthquake foci has properties different from the rest of the asthenosphere. It behaves in the same way as cold lithosphere instead of as the warmer asthenosphere, and in Figure 5-4a the inferred position of the deep lithosphere layer is shown. The combination of this information with the distribution of earthquake foci gives the general picture of the lithosphere plate extending through the asthenosphere with the earthquake foci occupying a zone along its upper edge, as shown in the sketch. The evidence from the earthquake waves emanating from this deep zone of foci indicates that the directions of relative motion usually occur in a plane parallel to the earthquake zone, corresponding to Figure 5-2e. This indicates that the lithosphere slab is moving downward, as shown by the arrow. This gives a picture of one lithosphere slab bending beneath an ocean trench and moving down into the earth's interior.

The simple, regular picture illustrated in Figure 5-4 does not hold for all regions. In many places, the deep zone of earthquake foci (Figures 4-10 and 4-11) is bent and contorted, especially in the deeper regions, which suggests that the lithosphere slab itself may become bent and contorted after it passes through the asthenosphere.

A continent-continent convergent plate boundary differs from that in Figure 5-4. As we will see in Chapter 11, there is evidence that India once had independent existence as a continent in the southern hemisphere and that it drifted north until it collided with Asia, producing a convergent plate boundary. But continental lithosphere is composed of rock material too light to sink into the earth, as the lithosphere does in Figure 5-4; it has to remain near the surface, like the scum on a pot of boiling soup (Figure 2-1b). Relative movement of the plates and compression along the boundary continued after the continents had collided, and this caused the Asian plate to override the Indian plate (Figures 5-2e and 5-2f). The result was that this plate boundary is marked by a double thickness of continental rocks, which accounts for the enormous elevation of the Himalayas and the Tibetan plateau and for the wide belt of shallow earthquakes. This gives us a fourth type of boundary to add to those illustrated in Figures 5-3 and 5-4 (Table 5-1).

RELATIVE MOVEMENT OF PLATES ON A SPHERE

Figure 5-5 is a sketch of the Pacific plate and its boundaries, with the arrows representing the direction of movement of the plate relative to its neighbors, as determined for many shallow-focus earthquakes. The directions of relative movement indicated are consistent with those expected for the boundaries of divergent, convergent, and transform type, according to the tectonic features of the boundaries shown in Figure 3-12.

In a general way, the directions of movement indicate motion of the Pacific plate away from the divergent boundaries of the southeast toward the convergent boundaries of the northwest but, because the Pacific plate is bordered by several different plates, the relative motions indicated in several regions differ from this general trend. The large arrow shows the direction of the Pacific plate relative to its northern and northeastern neighbors. It has been established that the relative movements indicated by most of the earthquakes along these boundaries are consistent to within ±20°. According to plate tectonics, the rigid lithosphere migrates across the asthenosphere with no significant distortion, so that the movements associated with these earthquakes represent relative movement of the plate as a whole. The di-

Fig. 5-5 The Pacific plate and its boundaries, see Figures 5-1 and 1-2. The arrows show the direction of movement of the plate relative to its neighbors, based on earthquake studies. The large arrow shows the movement of the plate relative to its northern and northeastern neighbors.

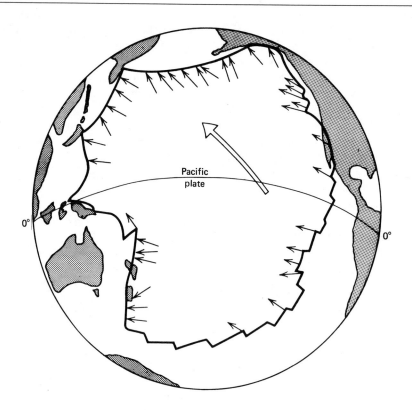

rection of relative movement of the Pacific plate cannot be illustrated by a single arrow and cannot be completely defined unless we consider the movement of plates across a sphere.

Imagine first that the northern hemisphere of the earth consists of a large continuous plate, fitting like a shell over the top half of the earth. This shell can be rotated about the N-S axis, independently of the sphere beneath it, with the results shown in Figure 5-6a. Each point on the plate traces out a circle concentric about the pole, P. These are small circles, corresponding to lines of latitude, except for the edge of the plate, which corresponds to the equator; this is a great circle, with its center at the center of the earth. Note that all points are moving due east.

In order to define the motion completely, we need to know, in addition, the rate of movement. This is given by the angular velocity of rotation about the axis. Angular velocity is defined as the number of degrees of rotation per unit time. For a given angular velocity, the speed of movement of points, in terms of distance covered, increases with distance from the pole, P. The point P rotates, but it does not cover any distance, so its speed is zero. Points on the equator travel the distance of the complete circumference in a single rotation, and they travel the fastest.

Now consider a small portion of the hemisphere, shown in Figure 5-6a as Plate A. Because it is a part of the shell, its motion also is completely defined by the position of the pole of rotation, P, and the angular velocity around P. The pole P is like a pivotal point around which Plate A rotates. This is true even if we abandon the rest of the hemisperical shell and consider the movement of Plate A by itself. This is a basic mathematical theorem for the movement of plates on a sphere. The example I chose for convenience of visualization is a unique situation, with the axis of rotation of the hemisphere coinciding with the axis of rotation of the earth. In the context of plate tectonics, this axis is called the axis of spreading, and the point P is called the pole of spreading. These have nothing to do with the rotation axis of the earth or the north pole of the earth, geographic or magnetic.

In Figure 5-6b, the hemispherical plate has been rotated through 90° so that the axis of spreading is now at right angles to the rotation axis of the earth, N-S, and the pole of spreading, P, is on the equator. The edge of the hemisphere coincides with a line of longitude, passing through the north and south poles. As before, the movement of every point on the hemisphere is completely defined by the position of the spreading pole, P, and the angular velocity of rotation about the spreading axis. In this example, we see that for each point on the concentric small circles of the shell, the direction of motion changes constantly with respect to the earth's geographic coordinates, as defined by the axis N-S and the earth's lines of latitude and longitude. A point at the equator moves initially to the north, then turns gradually toward the east, then down to the south, underneath P to the west, and then up to the north again. At any instant in time, each point on the hemisphere is moving in a different geographic direction (except for the circumference at the edge of the shell, which is a great circle).

Now consider Plate A, a small portion of the hemisphere; its motion is defined completely by the spreading axis and by the angular velocity of rotation about the axis. Its direction of motion has to be defined with respect to the pole of spreading because each point on its surface moves in a different geographic direction. This explains the statement that the motion of the Pacific

Fig. 5-6 Rotation of a hemispherical shell on a sphere. The movement of Plate A is controlled by the movement of the hemisphere, which is defined in terms of the angular velocity about an axis (the spreading axis).

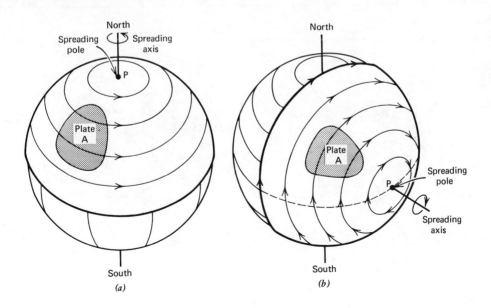

plate cannot be illustrated properly by a single arrow, as in Figure 5-5.

Figure 5-6 shows two special cases, where the spreading axis is coincident with, and at right angles to, the rotation axis of the earth. The conclusions from Figure 5-6*b* are valid for any intermediate situation. For any moving plate on the surface of the earth, it is possible to find a pole P that acts as the pivotal point around which the plate rotates.

Figure 5-7 is a more general example, illustrating the movement of two plates relative to each other. The boundary between Plates 1 and 2 in Figure 5-7*a* is a spreading ridge offset by a series of transform faults. It is convenient to assume that Plate 1 remains stationary, and to consider the movement of Plate 2 relative to Plate 1. Whatever the relative direction of Plate 2 across the surface of the sphere, a pole P around which its motion is centered can be located. This defines the axis of spreading. If relative motion is to continue between Plates 1 and 2 about this spreading axis, then the transform faults where the two plates are in contact must lie on small circles concentric about the pole P.

The shaded areas in Figure 5-7*b* illustrate the generation of new lithosphere along the divergent plate boundary as Plate 2 moves away from Plate 1. The angular velocity around the spreading axis defines the rate of surface spreading along the plate boundary, and the figure shows that the rate of separation of the plates increases with distance from the pole of spreading.

Figure 5-7 illustrates how the pole of spreading can be determined for any given pair of plates from the geometry of the transform faults. There can only be one pole fitting the whole series of concentric circles traced out by the faults and by the direction of relative motion between the plates. In Chapter 10, we will see that the position of the pole can also be determined by independent measurements of spreading rates at various positions along the plate boundary.

Comparison of Figures 5-5 and 5-7 is instructive. The Pacific plate may be considered equivalent to Plate 2 in Figure 5-7, and it is quite obvious that the axis of spreading for the Pacific plate does not correspond to the axis of rotation of the earth, and the pole of spreading does not corre-

Fig. 5-7 (a) Plates 1 and 2 are separating along the divergent boundary (compare Figure 5-3) with separation defined by rotation about the spreading axis. (b) Separation of the plates and the relative movement of Plate 2 is shown by holding Plate 1 fixed. Compare Figure 5-5.

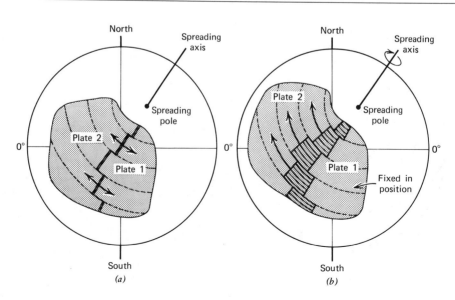

spond to the north geographic pole. The pole for the Pacific plate is a point in the North Atlantic Ocean, between Canada and the southern tip of Greenland. Compare Figures 5-1, 5-5, and 5-7.

Figure 5-5 shows how the direction of relative movement of a lithosphere plate can be determined by the study of waves from earthquakes located at the plate boundaries. It also shows that this direction could be estimated from the types of plate boundaries. The plate moves away from ocean ridges, toward ocean trenches, and parallel with transform faults (Figure 5-7b). The arrows showing the movements of plates in Figure 5-1 were determined from this kind of evidence.

Figure 5-7b shows the generation of new lithosphere at divergent plate boundaries, with obvious increase in surface area. In order to accommodate this material, the surface of the earth must increase in size or lithosphere must be destroyed elsewhere. There is no convincing evidence that the volume of the earth has changed significantly for the last few hundred million years and, according to plate tectonics theory, lithosphere is consumed at convergent plate boundaries. Calculations have shown that for the world as a whole (Figure 5-1), the volume of lithosphere slab moving down into the earth's interior beneath volcanic arcs each year (Figure 5-4) is approximately equal to the volume of new lithosphere generated at ocean ridges (Figure 5-3).

Figure 5-1 shows that all of the plates are moving relative to each other, and the total picture becomes quite complex. The sense of movement of the Pacific plate seems obvious (Figures 5-1 and 5-5), and the South American plate fits the standard model (Figure 2-3) along its boundaries. But what is happening to Africa? It has spreading plate boundaries on the west, east, and possibly the south (Figure 3-12), with no compression boundaries or sinks to accommodate the new lithosphere generated and spreading toward Africa. Some northerly movement of the plate is suggested by the convergent boundary in the Mediterranean region, but this does not account for the east and west spreading boundaries. Apparently, the African plate is being enlarged by the generation of new lithosphere along these latter boundaries and, in the absence of intermediate sites of plate destruction, it seems that the plate boundaries surrounding Africa must be moving themselves, relative to the African plate and to each other.

DURATION OF PLATE MOVEMENTS

The record of earthquakes tells us where the plate boundaries are now, how the plates are moving now, and how they have been moving for the past few tens of years (Figures 4-6, 4-7, and 5-1). The consequences of many earthquakes are recorded in ancient manuscripts of the eastern Mediterranean and the Near East. Recent investigations by a team of geologists, linguists, and classical historians have identified and located more than 2000 major earthquakes that occurred between 10 A.D. and 1700 A.D. From these they have constructed century-by-century earthquake maps that demonstrate that activity has shifted intermittently from one fault zone to another in the complex plate boundary passing through the northeast Mediterranean region (Figure 5-1). The earthquake maps of Figures 4-6 and 4-7 and the historical records extending back through 2000 years represent barely an instant in the immense span of geological time (Figure 1-3).

Recent earthquakes do not tell us if the plate boundaries have changed their positions with time, or how long the plates have been moving, or the precise rates of movement of the plates, or the regularity of the movements—have the plates moved at the same rate in the same directions back through long intervals of time, or have they changed direction and speed steadily or abruptly? Has motion been continuous, or have all plates stopped moving for short or long intervals?

The existence of lithosphere slabs extending to depths of 700 km in the mantle at convergent boundaries (Figure 5-4) indicates that the movements have continued long enough for 700 km of new lithosphere to be generated and transported laterally away from the ocean ridges. Rates of movement are required before this distance can be recalculated as a time, but clearly we are not dealing with short intervals of time.

Rates of sea-floor spreading can be derived from the study of magnetic anomalies in the earth's magnetic field (Chapter 10). These are caused by bodies of rock that became magnetized at their time of formation. In order to understand and evaluate the deductions from the magnetic anomalies, therefore, we must first learn something about the rocks (Chapter 6) and how they become magnetized (Chapter 8), which requires some knowledge of the earth's magnetic field (Chapter 8) and its origin deep within the earth's interior (Chapter 7).

SUMMARY

The earth's surface is covered by six major lithosphere plates and several smaller ones, with plate boundaries delineated by the belts of shallow-focus earthquakes. The relative movement of adjacent plates resulting from fracture can be determined by study of the earthquake waves emanating from the focus where the first motion occurred. The evidence is consistent with the existence of three types of plate boundaries: divergent boundaries where plates move apart at ocean ridge crests; convergent boundaries where plates move toward each other at ocean trenches and geologically young mountain ranges; and transform fault boundaries where plates move past each other with neither divergence or convergence. The deep-focus earthquake zones at convergent boundaries delineate lithosphere plates extending as deep as 700 km into the earth's interior from ocean trenches.

Earthquakes tell us the present positions of plate boundaries and the relative directions of plate movements in recent years. These movements have continued long enough for 700 km of lithosphere to sink into the earth's interior, which is presumably equivalent to the time for the generation of 700 km of new lithosphere at ocean ridges.

The motion of a rigid plate across the surface of a sphere is defined by an axis of rotation and the angular velocity of rotation about the axis. For any pair of adjacent lithosphere plates, the relative motions are defind by an axis of spreading and the angular velocity. For a given angular velocity, the rate of separation or convergence of the plates increases with distance from the pole of the rotation axis. In general, at any instant in time, each point on a lithosphere plate is moving in a different geographic direction with a different speed.

SUGGESTED READINGS are at the end of the book.

chapter **6**
The Geological Cycle

The major features of the surface of the solid earth are produced by sea-floor spreading and plate tectonics, but the detailed shaping and sculpture of the surface is caused by reactions between the lithosphere and the two fluid envelopes that surround it. These are the atmosphere and the hydrosphere. The hydrosphere is a discontinuous layer of water covering more than 70% of the surface (Figures 3-1 and 3-3), and it includes water stored within the lithosphere and dissolved or suspended in the atmosphere.

The biosphere consists of all the living organisms that reside near the surface of the earth. Bacteria have been found in the atmosphere 20 km above the surface and in water from oil wells 2 km below the surface. Living organisms are concentrated in the hydrosphere, and especially near the boundaries between lithosphere and hydrosphere, and between hydrosphere and atmosphere. They are also concentrated near the surface layer between lithosphere and atmosphere. The biosphere includes a great variety of life forms and billions of individual organisms. Although the total mass of the biosphere is only 0.0015% of the mass of the hydrosphere at any instant of time, it controls many important chemical reactions involving oxygen, nitrogen, and carbon. These elements are cycled through the biosphere at a very rapid rate compared with geological processes.

So far, we have considered the lithosphere as a homogeneous, rigid, rock layer with no mention of its composition. In fact, it is made up of many individual rock types with a wide range of compositions. These different rocks are the products of physical and chemical processes occurring within the framework of the geological cycle, which has long been a central theme in geology. The most intensely studied geological processes are those related to the interaction of the fluid and solid layers of the earth where these meet and overlap. Yet this is actually a rather restricted part of the total earth science picture described by plate tectonic theory.

The geological cycle consists of the three subcycles shown in Figure 6-1. The hydrological or water cycle traces the movement of water between hydrosphere, atmosphere, and lithosphere. The atmosphere and hydrosphere together wear down the continents and deposit the material in the oceans, thus destroying rocks, distributing them, and forming new rocks in other places. This is one significant part of the

Fig. 6-1 The relationship of the Geologic Cycle to plate tectonics and energy sources.

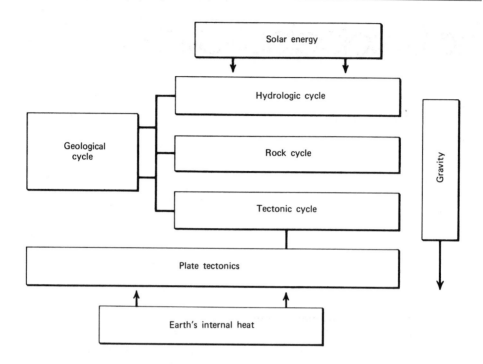

rock cycle. Another part involves the deep burial of rocks, followed by folding and uplift to form mountain ranges. This process is possible because of the tectonic cycle, which causes large areas of the earth's surface to sink slowly through long periods of time and subsequently to rise slowly.

For more than 150 years geologists have been tracing the migration of materials through these cycles and speculating about the cause of the tectonic cycle. Without intermittent uplift of the continents, all exposed rocks would have been worn down and deposited into the ocean long ago. We owe the existence of our civilization to the tectonic cycle. If the land had not been repeatedly uplifted to counterbalance the effects of erosion, the life forms that developed and evolved in the oceans would never have crawled out of the sea; land-based animals would never have gotten onto the ground because there would have been no ground.

The theory of plate tectonics for the first time provides a satisfactory explanation for many aspects of these cycles. In Figure 6-1, therefore, the geological cycle is shown as a superficial consequence of the lateral migration of lithosphere plates. As shown in Figure 6-1, the dominant energy sources for the cycles are the solar energy reaching the earth's surface and the earth's internal heat, which drives the lithosphere plates. The force of gravity makes things move downhill and therefore provides a leveling force important in the various cycles. It is also an important factor in the vertical movements within the earth that are involved in plate tectonics (Chapter 16; Figure 16-2).

THE HYDROLOGIC (WATER) CYCLE

The hydrosphere is a very thin film of water on the surface of the solid earth with extensions into the atmosphere and the lithosphere. The distribution of water in the hydrosphere is illustrated in Figure 6-2. More than 97% of the total volume fills the ocean basins. This reservoir of water covers an area of 362×10^6 km^2 with an average thickness of only 3.7 km (Figures 3-3 and 3-6). Only 0.02% of the water occupies lakes and rivers at any instant in time, but 100 times this amount of water is stored in ice caps and glaciers, mainly in Antarctica and Greenland. The atmosphere contains one thousandth of 1% of the total water either as dissolved vapor or as clouds. This is one twentieth the amount of water in lakes and rivers. The rocks of the lithosphere enclose 0.6% of the total water in small cracks and pore spaces, 30 times as much as that in lakes and streams. This subsurface water normally extends down to depths of about 800 m but, in some regions, it occurs at greater depths.

Figure 6-3a is a schematic representation of the hydrologic cycle, which is concerned with the changes in position and in physical state of water as it circulates between the hydrosphere, the atmosphere, and the lithosphere. The solar energy reaching the surface of the earth causes the liquid water and ice to evaporate and enter the atmosphere. The water vapor is transported over the surface by the general circulation of the atmosphere, and it then follows one of a number of circuits back to the ocean reservoir.

It has been estimated that each year, 336×10^3 km^3 of water evaporates from the ocean surface, and 63×10^3 km^3 evaporates from water and ice on the continents or is given off by plants. This is only 0.03% of the total water in the hydrosphere (Figure 6-2), but notice that it is 30 times the amount of water in the atmosphere and 300 times the amount of water flowing in river channels at any point in time. This indicates that water does not remain long in the atmosphere or in rivers before returning to land or to the ocean reservoir.

Figure 6-3b illustrates the balance of water circulation in the cycle. Of the total water evaporated each year, 84% evaporates from the ocean and 16% from the continents. The same total quantity of water that is evaporated annually condenses to form clouds and is precipitated as rain, hail, or snow but, in the process, the water is redistributed; 75% of the total is reprecipitated over the oceans and 25% over the continents. Thus, 9% of the circulating water is transferred from the oceans to the continents through the atmosphere. This excess of water is returned to the ocean either by flow over the surface in rivers or by slow percolation through the ground.

Figure 6-3c shows what happens to the

Fig. 6-2 Distribution of water in the hydrosphere, expressed as percentages of total volume of water. Note that only 0.03% of the total water is cycled each year.

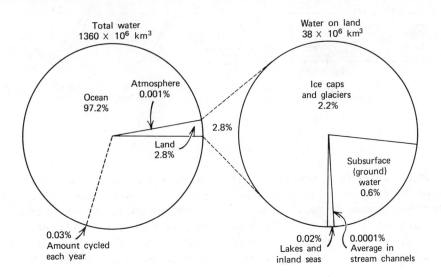

Fig. 6-3 *(a)* Schematic representation of main parts of hydrological cycle. *(b)* Evaporation and precipitation of water in hydrologic cycle, expressed as volume percent of the total cycled per year. Note excess evaporation from ocean and excess precipitation on land. Balance is maintained as shown. *(c)* Shows what happens to the water precipitated on land. Most evaporates, least sinks into ground. Contrast this with total amounts of water in atmosphere and subsurface in Figure 6-2.

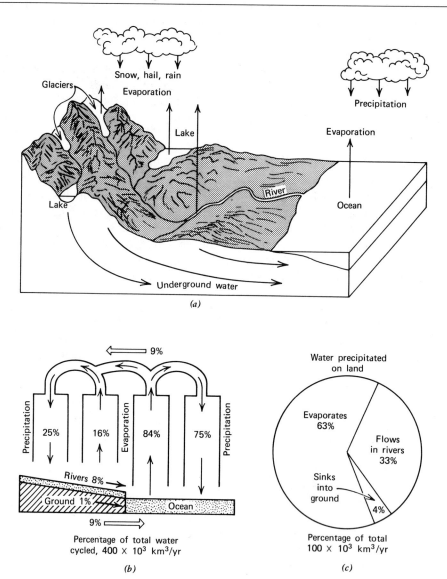

water precipitated on land. Most of it (63%) evaporates right back into the atmosphere, one third flows across the surface and into stream and river channels, and 4% percolates into the ground and makes its way very slowly to the oceans as subsurface wa-

ter. The values in Figure 6-3c become 16%, 8%, and 1%, respectively, when expressed in terms of the percentage of total volume circulated per year, as in Figure 6-3b.

Notice that the annual flow of water through river channels is eight times as great as that transferred through the ground from continent to oceans (Figure 6-3*b*), and the total volume of subsurface water is 6000 times that in rivers at any time (Figure 6-2). The subsurface cycle is a very slow one. Another slow circuit is that involving the formation of ice sheets, where water may be stored for thousands of years before the ice melts and runs back to the ocean reservoir. The return flow via rivers is relatively fast, but the shortest circuit is followed by the large percentage that is cycled directly back and forth between the atmosphere and hydrosphere by evaporation, condensation, precipitation, and reevaporation.

Of particular interest for our study of the lithosphere is the action of the precipitation and running water in eroding the land surface and transporting the material to the ocean. This involves the disintegration of rocks and the formation of new rocks. I will return to this, in connection with Figure 6-5, after reviewing the rock types that constitute the lithosphere.

THE ROCK CYCLE

There are thousands of rocks defined by name, but we can ignore the detailed description of most of these. For our purposes,

Fig. 6-4 Processes involved in the formation of igneous rocks.

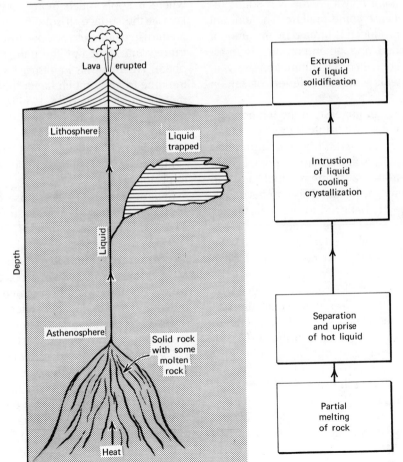

we need to know only the three major classes of rocks: igneous, sedimentary, and metamorphic, and a few specific rock types.

Igneous rocks are those that have solidified or crystallized from a hot, molten condition. The temperature of the liquid material is about 1000°C. The lava erupted from volcanoes is a familiar example. Figure 6-4 illustrates schematically the processes involved in the formation of igneous rocks.

If the temperature rises high enough at depth within the earth, the rock material will begin to melt. The rocks of the earth's interior are composed of aggregates of minerals with different compositions. Therefore the rock does not have a single melting point, like a simple material such as ice, but it melts through a range of temperatures, like a mixture of ice and salt.

When salt crystals are thrown on snow (ice crystals) at temperatures below the freezing point of water, the two solids interact to cause some melting. The amount of liquid produced is limited by the amount of salt added, and the end product is an assemblage of ice crystals (unmelted snow) in contact with a liquid solution of salt in water.

Similarly, an increase in the temperature of rocks in the earth's interior will eventually produce an assemblage of unmelted minerals with some liquid of composition different from that of the original rock and the remaining minerals. This process of partial melting is believed to occur in the asthenosphere layer. The product can be imagined as a tough sponge, as though it were a rock traversed by a network of molten channels.

The molten rock material produced by partial melting is less dense than the surrounding unmelted rock, so it has a tendency to move upward toward the surface. It becomes squeezed out of its sponge at depth, following specific paths of structural weakness toward the surface. If it makes it through the brittle lithosphere, the molten material erupts, building a volcano of successive lava flows that cool and crystallize.

The crystallization process is similar to that of freezing a solution of salt in water.

When the temperature of such a solution is lowered, crystals of ice are formed, and the concentration of salt in the remaining solution is increased. With further cooling more ice crystals are produced, and eventually these are joined by crystals of salt. The end result is that a homogeneous liquid solution of one composition is converted into a solid aggregate of two solids, ice and salt, both with compositions different from that of the original liquid. Similarly, a molten lava flow crystallizes to produce a rock composed of several different minerals.

A lava is called an extrusive igneous rock because it made its way through the lithosphere and was extruded onto the surface. The molten material does not always manage to break through to the surface. Especially in continental environments, the liquid may become trapped in an underground chamber, as indicated in Figure 6-4, where it cools much more slowly than a lava at the surface. It solidifies to produce an intrusive igneous rock because it was intruded into, but not through, the lithosphere. The minerals in intrusive rocks have the time to grow much larger than those in rapidly cooled lavas. A familiar intrusive rock is granite, which forms deep within mountain ranges, but which is probably better known as a polished ornamental stone on the fronts of bank buildings and on tombstones.

Sedimentary rocks are those that have been deposited at or near the surface of the earth through the action of the hydrosphere, the atmosphere, or the biosphere on the materials forming the earth's surface. Most of these rocks were deposited under water, but they include sediments formed on land, beneath the atmosphere. They are formed at normal surface temperatures, in contrast to the hot igneous rocks.

Figure 6-5 illustrates schematically the involvement of the hydrologic cycle in the formation of sedimentary rocks. The land surface, the source region for new sedimentary rocks, is subjected to weathering. The exposed rocks are broken up by physical and chemical processes through the action of air, water, and heat from the sun.

Fig. 6-5 Processes involved in the formation of sedimentary rocks.

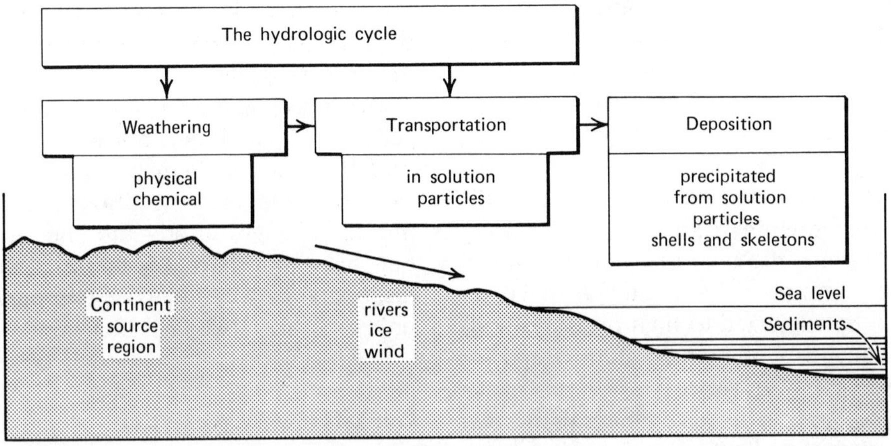

The products of disintegration are then transported to the ocean, mostly by flowing rivers, but also in part by wind action. Winds can carry fine dust particles for hundreds of kilometers across the oceans. Rain and river water dissolve a very small fraction of the weathered rocks, and this soluble portion is added to the dissolved salt of the oceans. The small grains and fragments derived from the source region are transported in rivers either in suspension or by rolling along the river bed. Most of you have seen the muddy waters of a river carrying the land down to the sea. Eventually the particles make their way to the ocean, where they sink because the flow of water is no longer sufficient to keep them moving. They may be deposited at the mouth of the river to form a delta, but smaller particles drift further out to sea before eventually falling to the ocean floor. Successive layers of sediment accumulate on the ocean floor, building up a sequence of sedimentary rocks. Familiar examples of the sedimentary deposits that eventually become consolidated into sedimentary rocks are the muds of river estuaries and the sand beaches along the shore lines.

Figure 6-5 shows schematically the processes of weathering, transportation, and deposition, which lead to the formation of sediments. I have discussed the deposition of particles carried from the land, by river or wind, and the addition of dissolved salts to the ocean. The dissolved materials can be extracted from the seawater solution in two ways. They may be precipitated directly from solution under suitable conditions, like excess sugar precipitating from a too-sweet cup of coffee, or they may be extracted from solution by growing organisms.

Shellfish such as snails and clams extract calcium carbonate in order to grow their shells. There are many other organisms of varied sizes that extract chemicals from the ocean water, including the plankton, hosts of microscopic animals and plants drifting near the ocean surface at the mercy of ocean currents. As these die, they produce a steady rain of shells and skeletons that falls to the ocean floor and is incorporated into the sediment. The sequence of microfossils and mineral particles cemented together by minerals precipitated from the ocean water together become a part of the record of the earth's history, containing evidence about the geological history of the sediment sources, biological history, climatic changes, and many other events.

The third major class of rocks, the metamorphic rocks, are those that have changed from their original state by the action of high pressures and high temperatures, consequent on deep burial within the earth. This can only happen where and when the tectonic cycle is operating.

In Figure 6-5, there is a limit to the thickness of sediments that can accumulate in the ocean because the ocean will eventually fill up. If the ocean basin floor were sinking, however, then this would leave room for continued accumulation of sediments. If sinking were to continue long enough, then the oldest sedimentary rocks at the bottom of the sequence of layers would be subjected to high pressures and, simultaneously, the temperature would increase.

If you heat chemicals or minerals in a test tube, sooner or later you reach a temperature where some kind of chemical reaction occurs. The same thing happens with deeply buried rocks. Chemical reactions occur, the mineral particles making up the sediment react and recrystallize, producing new minerals, which are distinctive for metamorphic rocks.

This is the first stage in the tectonic cycle (Figure 6-6). Subsidence is followed by compression and folding of the metamorphic rocks at depth within the earth, and later the whole region rises. As the metamorphosed rocks are uplifted, the surface is weathered and eroded away, as indicated in Figure 6-5, so that eventually the once deeply buried rocks become exposed at the surface. Metamorphic rocks include not only modified sediments, but any kind of preexisting rock that became involved in the tectonic cycle.

The lithosphere is composed of igneous, sedimentary, and metamorphic rocks. The high-temperature intrusive igneous rocks may become metamorphosed at depth through changes in pressure and temperature after they have crystallized, and the distinction between igneous and metamorphic rocks then becomes blurred.

THE GEOLOGICAL CYCLE

Figure 6-7 summarizes the relationships among the subcycles comprising the geological cycle (Figure 6-1). The cycle begins with the eruption of new material derived from the earth's interior by partial melting. As soon as the lava is exposed to the atmosphere and hydrosphere, the fluid envelopes react with the new rock. The hydrologic cycle causes weathering, transportation, dispersal, and deposition of the igneous rocks at the bottom of an ocean basin. There, the particles and chemical precipitates, together with fossil remains, accumulate in a layered sequence of sedimentary rocks.

Fig. 6-6 Processes involved in the formation of metamorphic rocks and mountain building.

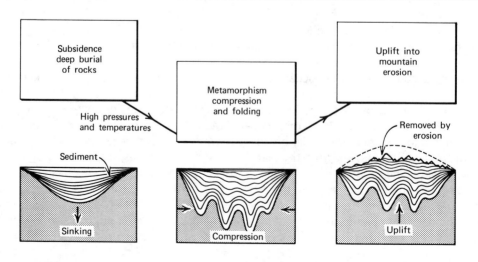

Fig. 6-7 Diagrammatic representation of the Rock Cycle within the Geologic Cycle. (See Figure 6-1.)

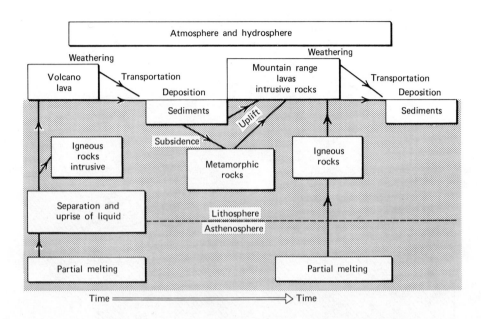

If the tectonic cycle operates, then the ocean floor sinks, the sediments subside, more accumulate above them, and eventually the pile becomes thick enough that they become metamorphic rocks. Compression and folding occurs, and subsequent slow uplift elevates the contorted metamorphic rocks to produce a mighty mountain range. This mountain range is attacked in turn by weathering and dispersed into neighboring ocean basins. The formation of a mountain range in this way is accompanied by the production of igneous rocks, both intrusive and extrusive, which begins the cycle again.

This very schematic version of the geological cycle has been used for more than a century to account for mountain building. Figure 3-12 shows that the geologically young and active mountain chains are narrow belts. Reconstruction of the geological history leading to the formation of these and other, older mountain chains indicates that the operation of the tectonic cycle is apparently restricted to similar elongated belts. These belts are now correlated with convergent plate bound-

aries, and the explanation of the tectonic cycle can be found in the theory of plate tectonics.

THE GEOLOGICAL CYCLE AND PLATE TECTONICS

The whole geological cycle illustrated in Figure 6-7 can now be plugged into the plate tectonic cycle (Figure 2-3). The current theory, illustrated in Figure 6-8a for a convergent boundary of ocean-ocean type, is as follows (Table 5-1).

The cycle begins at the divergent plate boundaries along ocean ridges with the massive extrusion of lavas from the asthenosphere. This produces new oceanic crust and lithosphere. With sea-floor spreading, the lithosphere is transported laterally away from the ridges toward a convergent plate boundary. This is clearly illustrated by the Pacific plate in Figure 5-5. As the plate migrates, it becomes covered with a thin layer of sediments.

At the convergent plate boundary, the oceanic lithosphere plate is bent, and it

Fig. 6-8 Plate tectonics and mountain building. *(a)* Ocean-ocean compressive boundary and development of volcanic island arc. *(b)* Ocean-continent compressive boundary and development of fold mountain range (Figure 6-6) and volcanic chain.

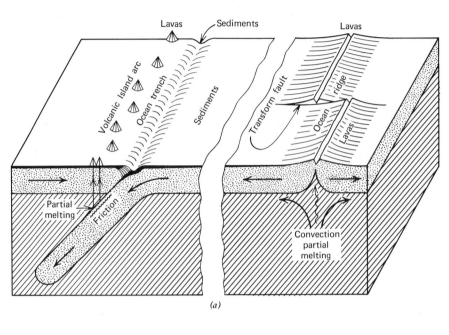

(a)

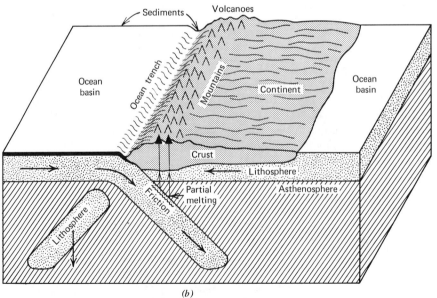

(b)

moves down into the earth's interior, producing an ocean trench at the surface. The surface of the lithosphere plate descending beneath the ocean trench is heated by friction, and this causes partial melting of the rock. Partial melting is followed by the eruption of lavas at the surface, and this produces a volcanic island arc parallel with the ocean trench.

The ocean trench becomes a trap for the oceanic sediments and for additional sediments derived from the adjacent volcanic island arc. The sedimentary rocks are partly dragged down into the interior with the sinking lithosphere and are partly compressed, folded, and plastered against the volcanic island arc. The sedimentary rocks and oceanic crust may be converted into metamorphic rocks.

Eventually, the compression and folding of sedimentary rocks, metamorphic rocks, and volcanic rocks causes unstable conditions and uplift, and the convergent plate boundary becomes the site of a new mountain range. This becomes welded to the margin of the volcanic island arc, which has built up from the products of melting of the oceanic lithosphere plate moving down into the interior. This is the simplest of many sequences of events that are possible. Table 5-1 lists three types of convergent plate boundaries, and each of these can be the site of a mountain range with specific characteristics, in terms of the kinds of rocks present, and the type of folding produced.

A convergent boundary may change from one type to another, which adds complexity to the mountain-building process. For example, continents are carried along by moving lithosphere plates, and sooner or later a continent will be moved into a position against an ocean trench. The plate will try to drag the continent down into the interior, but the continental rocks are much lighter than the rocks composing the oceanic lithosphere. These less dense rocks resist sinking into the more dense rocks. In fact, they are so buoyant that the whole pattern has to change, as shown in Figure 6-8.

Imagine that Figure 6-8a included a continent moving away from the ocean ridge.

When the continent reached the ocean trench, as shown in Figure 6-8b, the lithosphere slab carrying it was no longer able to sink, so it became inactive, and the adjacent lithosphere slab began to sink instead, dipping underneath the continent. The original sunken lithosphere slab produced no more volcanoes because it stopped moving. The volcanic island arc was swept against the continent and incorporated into it by mountain building. A new series of volcanoes was formed along the marginal mountain chain of the continent. Figure 6-8b illustrates an ocean-continent compressive boundary, like that associated with the Andes (Figures 2-3 and 5-1).

The third possibility (Table 5-1) is that a second continent could be brought into collision with the continent in Figure 6-8b. Then the lithosphere would have to stop sinking altogether along that compressive boundary, there would be an enormous pileup between the colliding continents, and the overall pattern of plate movements would have to change. The Himalayan mountain range was produced when the continent of India was carried from the south against the Asian continent, as shown by the reconstructions of continental drift in Figure 12-3 and the cartoon in Figure 6-9.

These mountain-building processes are very slow. A volcanic island arc may be "swept against the continent and incorporated into it," but the rate of sweeping is probably only a centimeter or two per year. It may move 10 m in 1000 years, which amounts to 10 km in 1 million years. But Figure 1-3 shows that there has been plenty of time available for plate tectonics to sweep volcanic island arcs and continents together.

VOLCANOES

In our review of stable plates and active belts, we noted that the signals of activity were earthquakes and volcanoes. We have examined in detail the distribution of earthquakes, and it is apparent from the preceding outline of the geological cycle that volcanic activity is also closely related to plate

Fig. 6-9 Plate tectonics and mountain building. Continent-continent compressive boundary and the development of the Himalayas. (Drawing by J. C. Holden, reproduced by his kind permission. Published in *More about Continental Drift,* by R. S. Dietz, *Sea Frontiers,* 1967, Vol. 13, p. 66-82.)

tectonics. Earthquakes leave no permanent record in the rocks, but lavas and intrusive igneous rocks are preserved. Correlation of volcanic activity and volcanic rocks with present plate boundaries thus provides a basis for interpretation of ancient lavas and other igneous rocks in terms of ancient, formerly active plate margins. This is one important line of evidence employed to trace plate movements back through geological time (Chapter 12).

There are nearly 800 volcanoes classified as active today or known to have erupted in historical times. In addition, there are many thousands of extinct cones and craters sufficiently well preserved that they must have been active in recent times, measured in geological terms. About 500 of the active volcanoes are distributed around the circum-Pacific ring of fire, which correlates closely with the concentration of earthquakes in the same belt (Figure 4-6). In the Alpine-Asian earthquake belt volcanoes are sparse except in the Mediterranean region and Indonesia. Other volcanoes are distributed along the African rift valleys.

There is in addition a great deal of volcanic activity on the ocean basin floors, but no sign of this volcanism can be detected at the surface of the ocean until a volcano rises close to sea level. It has been estimated that 10,000 volcanoes with a height of 1 km or more exist in the Pacific basin; the total volume of lava extruded from the Pacific basin volcanoes during the last 100 million years is far greater than that erupted by continental volcanoes during the same period.

Figure 6-8 shows two types of volcanic activity along plate boundaries that are distinct in terms of location and origin, those associated with ocean ridges, and those parallel to ocean trenches. They can also be distinguished by the character of their volcanic eruptions and by the chemistry of the lavas produced. Two other types of volcanoes are those rising from the ocean basin floors and those erupted on continents away from plate boundaries.

The oceanic lithosphere plate illustrated in Figure 6-8 is punctured by lava rising from the asthenosphere at locations independent of the plate boundaries. Extrusion of the lava on the ocean floor builds a series of volcanoes that forms a linear chain extending in the direction of movement of the plate. The Hawaiian Islands are the peaks of a few of the volcanoes in one of these linear chains. The lavas in ocean-floor volcanoes are very similar in chemistry to the ocean ridge lavas, but there are subtle differences between them. These abundant lavas are called basalts, like the lunar lavas (Chapter 17). The origin of these volcanoes in the middle of stable lithosphere plates remains a matter for speculation, although it is clear that, for some reason, the temperature in a localized region of the asthenosphere is increased. The question of hotspots is reviewed in Chapter 16.

Most active volcanoes on continents are associated with a lithosphere slab sinking beneath an ocean trench at the continental margin. Others, such as the spectacular volcanoes along the African rift valleys, are associated with tension. It appears that millions of years ago the African continent began to split open but did not quite make it. The rift valleys are the scars of that abortive attempt to form a new ocean. Sea-floor spreading began, instead, along the line of the Red Sea. The volcanic activity initiated during the early period of incipient divergence continues to the present time. The chemistry of these continental lavas is very different from that of the dominant lavas of the oceans and from those of island arcs and the mountain ranges of continental margins.

SUMMARY

The geological cycle consists of subcycles: the water cycle, the rock cycle, and the tectonic cycle. Solar energy drives the water cycle, and the earth's internal heat drives the lithosphere plates and controls the tectonic cycle; the force of gravity controls vertical movements. The many rock types comprising the lithosphere are formed as products of these cycles.

There are three major classes of rocks: igneous rocks, like lavas, crystallized from molten material; sedimentary rocks formed through disintegration of continental rocks by air and water, redistribution largely by rivers, and redeposition, usually beneath oceans; metamorphic rocks have been changed in response to increased pressure and temperature caused by deep burial and folding in mountain ranges.

Plate tectonics explains many aspects of the geological cycle. Eruption of lava at divergent plate boundaries generates new lithosphere, which spreads toward a convergent plate boundary and becomes covered by a thin layer of sediments beneath the ocean. The oceanic lithosphere plate sinks into the interior at a convergent plate boundary, producing an ocean trench at the surface, which serves as a trap for additional sediments. Frictional heating of the sinking plate deep in the earth melts the rock, the molten material rises, and it is erupted to form a volcanic island arc at the surface. Compression and folding of the sedimentary and volcanic rocks at the convergent boundary is the first stage in the formation of a mountain range, and this causes metamorphism of the rocks. Complex mountain ranges are produced when the lithosphere plate carries a continent to a convergent plate boundary and when two continents collide. Different types of plate boundaries and plates are characterized by different types of volcanic lavas, an observation useful in unraveling the history of ancient plate movements.

SUGGESTED READINGS are at the end of the book.

chapter 7

Earthquake Waves and the Inside of the Earth

From our vantage position in the biosphere at the earth's surface we can examine the processes and products of the geological cycle and observe in detail the surficial sculpture of the scenery. However, we are less well placed for finding out the composition and structure of the earth's interior.

The earth is a great rotating sphere moving through space. Many of the physical properties of the sphere as a whole can be determined from various lines of evidence. The size, shape, and mass of the earth have been measured with precision, for example. The known volume and mass give a mean density of 5.5 g/cc, which is much higher than the density of the rocks making up the bulk of the outer part of the earth, 2.6 to 3.1 g/cc. We can deduce, therefore, that a significant part of the earth's interior must be composed of material with density considerably greater than 5.5 g/cc. The distribution of mass within the earth can be determined independently from information provided by artificial satellites orbiting the earth. This evidence indicates that the mass is concentrated toward the center of the earth. Apart from this, the main source of information about the physics of the earth's interior is studies of earthquake waves.

We outlined the model for plate tectonics in Chapter 2 (Figures 2-2 and 2-3), and we have embellished this in subsequent chapters. We reviewed evidence from earthquakes delineating the boundaries of lithosphere plates and their extensions into the earth's interior, but we have considered no evidence for the existence of an asthenosphere layer below the lithosphere.

In this chapter we will examine other evidence from earthquakes that confirms the existence of a layer with properties, depth, and thickness appropriate for the asthenosphere. The same earthquake evidence provides a picture of the concentric structure of the earth's interior and reveals complexities in the concept of a simple lithosphere layer. It also indicates the presence of a molten metallic core, and this is the only reasonable explanation for the earth's magnetic field, which has played such a critical role in the development of plate tectonic theory.

CONCENTRIC STRUCTURE OF THE EARTH

Energy from the focus of an earthquake is transmitted in wave form in all directions (Figure 4-2). The waves passing upward to the surface cause vibrations that produce the devastation described in Chapter 4.

Those passing downward follow paths through the earth's interior that are dependent on the properties of the earth materials before they emerge through the surface at points distant from the focus. The velocities of the earthquake waves vary according to the elastic properties of the rocks through which they pass. Analysis of the earthquake waves received at the worldwide network of receiving stations provides information about their velocities and therefore about the properties of the material at various depths in the earth. The earthquake waves provide a kind of X-ray picture of the earth's interior. Let us look at the complete picture first and then outline the evidence that provides the picture.

The study of earthquake waves shows that the earth has the concentric structure shown in Figure 7-1 (see also Figure 17-2).

There are two major discontinuities in the physical properties of the material within the earth that divide the sphere into three portions: the core, the mantle, and the crust. The core-mantle boundary is about halfway through the earth at a depth of 2900 km, and the mantle-crust boundary is very near the surface of the earth at a depth that varies according to the environment. The mantle-crust boundary is called the Mohorovičić discontinuity, usually referred to as the Moho (Figure 7-6). Note that there is *no* major discontinuity at the depth of about 100 km, corresponding to the lithosphere-asthenosphere boundary depicted in Figures 2-2 and 6-8, and in many intervening diagrams.

There is another boundary, less well-defined, at a depth of about 5000 km, separating the inner core from the outer core.

Fig. 7-1 Layers within the earth. See also Figures 7-5 and 17-2.

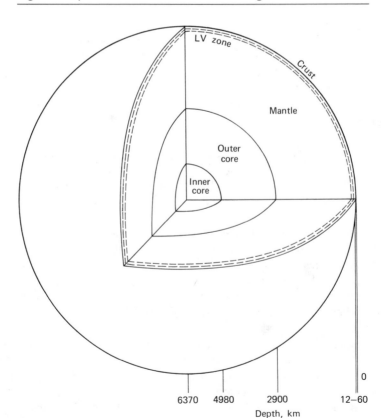

The outer core is not solid, like the rest of the earth, but liquid.

The dashed line in the upper part of the mantle in Figure 7-1 represents a layer called the low-velocity (LV) zone. In this zone, with a variable thickness of 100 to 200 km, the velocities of earthquake waves are lower than in the mantle above and below it (Figure 7-5). The upper and lower boundaries of the low-velocity zone are not sharply defined, like the Moho and the core-mantle boundary, and the depth to the zone varies somewhat from place to place.

EARTHQUAKE WAVES

The release of energy at the focus of an earthquake produces several different kinds of waves. Surface waves are those that travel around the world in the outer layers, and body waves are those that follow paths through the interior. The study of surface waves provides much information about the earth, but it is to the body waves that we turn for determination of the structure illustrated in Figure 7-1.

There are two kinds of body waves, the primary, or P-wave, and the secondary, or S-wave. The P-waves travel faster than the S-waves, so they arrive first at a recording station; this is why they are called primary and the slower S-waves are called secondary. These abbreviations are reminders also that the P-waves are "push-pull" or compressional waves, and that the S-waves are "shake" or shear waves. The difference between these two types of wave is illustrated in Figure 7-2.

Figure 7-2a illustrates schematically the motion of particles back and forth in the direction of travel of a P-wave. A push at one end compresses the particles together and, as the compressed state is transmitted to the adjacent particles down the line in the direction of the original push, it is followed by an interval of expansion where the particles are more spaced out than in the normal state. Sound waves are transmitted in the same way. The same kind of motion occurs in a spring, or "slinky," if it is pushed or pulled at one end and then released, as shown in Figure 7-2b. A line of closely packed passengers standing in the aisle of a moving bus exhibits a similar sequence of compression and expansion as the people sway back and forth in response to acceleration or deceleration of the bus or to a sharp push by an impatient passenger at one end.

Figure 7-2c illustrates schematically the motion of particles in a direction at right angles to the direction of travel of an S-wave. As each particle moves, it drags an adjacent particle with it, and the net effect of wave travel in the direction perpendicular to the particle motion can be visualized by recalling the rope trick illustrated in Figure 7-2d. If you fasten a rope at one end and shake it up and down at the other end, you see a series of waves moving down the rope away from you. The waves keep moving for as long as you supply energy by shaking the rope, but the rope itself does not move at all toward the wall; each particle of the rope just moves up and down.

The velocities of P-waves and S-waves for specific rocks of the earth's crust have been measured directly in the laboratory by generating suitable waves and passing them through selected rock specimens in special apparatus. The wave velocities vary with the properties of the rocks used including, for example, the densities of the rocks. The laboratory measurements on known rocks and other materials thus provide a calibration of material properties against wave velocities. If earthquake wave velocities can be measured at various depths in the earth, then this gives an excellent guide to the properties of the material present.

P-waves can be transmitted through any material, solid, liquid, or gas. As you might anticipate, the P-wave velocity through a solid is different from its velocity through a liquid of the same composition; it is less through the liquid. S-waves, in contrast, can only be transmitted through solids. Figure 7-2c may remind you of the particle motion of water as waves travel across the surface of a pond, but this does not apply to S-waves. S-waves pass *through* a material, not along a surface boundary between two different materials. And it is known from the

Fig. 7-2 The passage of compressional waves (a and b) and shear waves (c and d) through a block of solid material such as rock (a and c). For the compressional or P-waves, the compressed region marked by the arrow (a) passes from left to right, and the material like the marked square moves back and forth in the line of the wave path, suffering alternate compressions and expansions. For the shear or S-waves, each small volume of material like the marked square (c) shakes up and down perpendicular to the direction of the movement of the wave, which is shown by the crest (marked by the arrow) moving from left to right. (Diagrams a and c from O. M. Phillips, *The Heart of the Earth,* 1968, Freeman, Cooper and Co., San Francisco.)

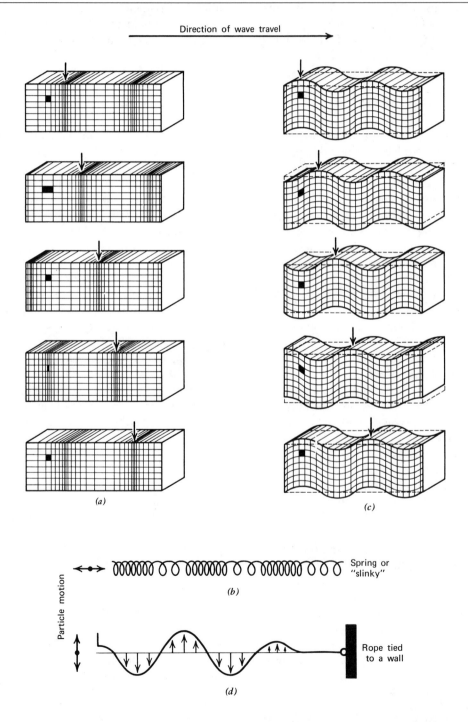

theory of waves and confirmed experimentally that S-waves do not pass through liquids. The shear waves illustrated by the rope in Figure 7-2d are transmitted down the rope because it has properties that permit it to change shape and to maintain shapes. The sideways movement involves shear. Liquids have no shear strength, and they cannot maintain shapes produced by the sideways movement of particles.

Earthquake waves traveling through rock with uniform properties travel in straight lines with constant velocity. In another rock type the velocity is different. In general, it has been established that the velocity of earthquake waves increases in rocks at greater depths, so let us adopt this kind of model to illustrate the behavior of earthquake waves within the earth. Figure 7-3 represents the outer layers of the earth with arbitrary layers composed of rock types A, B, and C in which the P-wave velocities are V_A, V_B, V_C, with $V_A < V_B < V_C$.

An explosion at the surface of the earth creates an artificial earthquake, releasing energy that is then transmitted through the earth just like the waves from an earthquake. This has the advantage that the place and time of the artificial earthquake can be selected for the convenience of the observer.

Consider the P-waves emanating from the explosion E depicted in Figure 7-3 and, for simplicity, neglect the S-waves. The waves follow stright-line paths from E in all directions through layer A. As each ray path crosses the AB boundary, it is refracted, or

bent, as shown for ray f. It follows the new direction through layer B and is then refracted again into layer C. This refraction is somewhat analogous with the bending of light when it passes from air into water. The angle of refraction is a function of the velocities of the wave in the two adjacent layers. P-waves, therefore, are refracted through angles somewhat different from those of S-waves, which have different velocity.

Not all of the energy is refracted through the boundaries between layers. At each boundary, some of the energy is reflected back, as shown for ray g. This is somewhat analogous to the reflection of light from a plate-glass window. For ray h a critical stage has been exceeded because there is only a reflected wave; the angle θ is such that a refracted wave could not enter layer B.

The instruments in the earthquake recording station at R in Figure 7-3 receive vibrations from the explosion at E. The different waves traveling with different velocities will arrive at different times; in addition, similar waves following different paths will arrive at different times. The time that each wave takes to reach R is precisely measured, and skilled investigators can distinguish in the records the arrival of one kind of wave from another. You can see from Figure 7-3 that if the wave velocity in layer A is known, and if the arrival time of wave ray h can be distinguished from the other vibrations recorded by the instrument at R, then it is a simple matter to calculate

Fig. 7-3 Paths followed by waves emanating from an explosion. Note reflection and refraction of the waves at boundaries between layers of rock, A, B, and C.

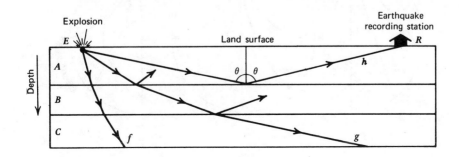

the depth to the boundary between layers *A* and *B*. The time of travel and the wave velocity give the distance traveled, and distance *ER* is already known.

For the real earth, where the rocks are not so regularly arranged and where the wave velocities of the unknown rocks at depth are not known, more elaborate procedures are used to calculate both the velocities of earthquake waves at various depths and the depths to specific boundaries between layers with distinctly different physical properties. A whole series of recording stations at different distances provides additional equations for the solution of the unknown quantities.

EARTHQUAKE WAVE VELOCITY PROFILES THROUGH THE EARTH

Figure 7-4 illustrates schematically the passage of earthquake waves through the earth's interior. Waves of energy emanate from the focus of an earthquake near the surface at point *F*. If the earth were composed of material with uniform properties throughout, then the wave paths would follow straight lines from the focus, as shown in Figure 7-4*a*, with constant velocity throughout. The times taken for a P-wave

and an S-wave to reach a particular recording station at a known distance from *F* permit direct calculation of the wave velocities in the earth.

In fact, it turns out that the body waves travel through the earth from a focus *F* to a recording station *R* in a time shorter than that predicted from the known velocities of waves in the surface rocks. The time difference is greater when the station *R* is further away from the focus *F* (i.e., when the wave path penetrates deeper into the interior). These results show that the earthquake wave velocities are greater at depth than near the surface and that they increase progressively with depth in the earth.

Figure 7-3 shows that as an earthquake wave passes down through layers of rock material with increasing wave velocities, the wave path becomes successively refracted toward the surface. Figure 7-4*b* shows the same effect of refraction for waves passing through the earth with the wave velocities increasing steadily with depth. The wave paths curve back toward the surface. The curved wave path *FR* in Figure 7-4*b* is longer than the straight path *FR* in Figure 7-4*a*; the wave velocity is greater the deeper the wave penetrates.

Figure 7-4*c* represents a cross section through the real earth, showing the inner

Fig. 7-4 Paths followed by waves emanating from an earthquake focus at *F*. (*a*) Linear paths in a sphere with uniform properties. (*b*) Curved paths followed by waves through sphere with wave velocities increasing steadily with depth. (*c*) Wave paths in the earth, with reflections and refractions occurring at depths where the physical properties (and therefore earthquake wave velocities) change abruptly.

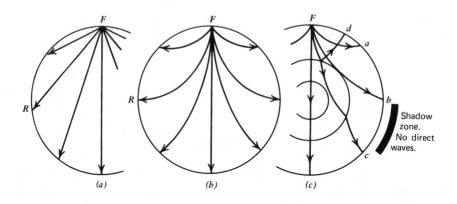

(*a*)   (*b*)   (*c*)

Fig. 7-5 Cross section through the earth showing the concentric layers (Figure 7-1). Graphs showing the velocities of P-waves and S-waves determined for each depth within the earth, based on analysis of the times taken for earthquake waves to travel along paths such as those shown in Figure 7-4c. Note abrupt changes in velocities at the depths of the crust-mantle and mantle-core boundaries, and the fact that S-waves do not exist in the core.

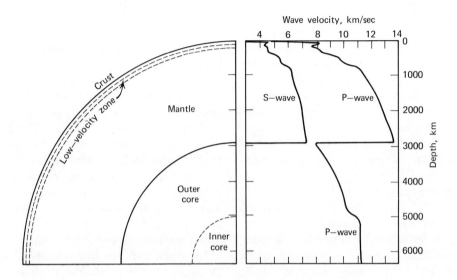

and outer cores and the mantle (compare Figure 7-1). Each earthquake wave that emanates from focus F along paths such as Fa and Fb reaches a recording station somewhere between F and b and penetrates to a specific maximum depth in the mantle on the way. From the measured travel times of many waves to many stations, the velocities of P-waves and S-waves characteristic of the rocks at any given depth in the mantle have been calculated. The results are shown in Figure 7-5.

Earthquake waves penetrating deeper than the limiting wave Fb behave differently. P-wave Fc, for example, is partly reflected back to the surface at a point d. This is evidence for a major discontinuity in earthquake wave velocity along what we have stated is the core-mantle boundary. In addition, wave Fc passing into the core is refracted downward instead of upward indicating a decrease of velocity in the core. Figure 7-5 shows that the P-wave velocity decreases by almost 50% between the deepest mantle and the outer core.

The existence of a shadow zone as indicated between the wave paths Fb and Fc is further evidence that the depth of penetration of path Fb corresponds to a specific boundary within the earth. No P-waves or S-waves are received directly from the focus F in this shadow zone. This is because of the refraction of wave paths such as Fc further away from Fb as they pass through the core.

Another vital item of evidence is that S-waves can penetrate no deeper than a path corresponding to Fb; S-waves cannot pass through the core. This indicates that core material within the surrounding solid mantle is liquid. Figure 7-5 shows the velocities of S-waves increasing with depth and dropping to zero at the core-mantle boundary.

At a depth of about 5000 km, further refraction of the P-wave and an increase in velocity (Figure 7-5) indicate the existence of a solid core within the liquid outer core. The P-wave velocity of the solid inner core is so much lower than that of the deep mantle that the core composition must differ significantly from the deep mantle composition.

There is another major discontinuity in earthquake wave velocities at the Moho between the crust and the mantle. The crust is so thin that this boundary is not distinguishable in Figures 7-4c and 7-5. Its position is indicated in Figure 7-1 and illustrated in detail in Figure 7-6. The depth to the Moho has been determined from refraction and reflection studies of earthquake waves through the crust (as in Figures 7-3 and 7-4c). At the base of the crust, the average velocity of P-waves is 7.0 km/sec, and this jumps to 8.1 km/sec in the uppermost mantle (see Figure 7-5). There is a similar change in S-wave velocities.

Figure 7-5 shows that the variation in velocity of earthquake waves within the mantle is not a regular function of depth. The overall increase in velocity of both P-waves and S-waves with increasing depth occurs in a series of steps down to about 1000 km. This indicates that the upper mantle has a layered structure, but the change in physical properties from one layer to another is much less pronounced than that across the core-mantle boundary and across the mantle-crust boundary.

A decrease in velocity of both P-waves and S-waves at a depth between about 100 and 200 km is shown in Figure 7-5. This region is defined as the low-velocity zone of the mantle, and we will review this in connection with Figure 7-6.

THE CRUST, LITHOSPHERE, AND ASTHENOSPHERE

Figure 7-6 shows cross sections through the outer layers of the earth along approximately the same section illustrated in Figure 3-10. Figure 3-10 is a cross section extending almost halfway around the world near latitude 20°S. It shows the relief of the major surface features of the earth as they are encountered in the Pacific and Atlantic Oceans and South America. These include continental and oceanic platforms, conti-

nental slope, oceanic ridge, young mountain range, ocean trench, and volcanic island arc (see Table 3-1). The figure is drawn with a vertical exaggeration of x100.

Figures 7-6a and 7-6b are drawn with vertical exaggerations of x10 and x2, respectively. In neither section can the surface features be distinguished, so Figure 7-6 should be studied in conjunction with Figure 3-10.

Figure 7-6 shows the position of the Moho. The average depth to the Moho beneath the oceans is 10 km, with the crust beneath the ocean water being 5 to 6 km thick. Beneath the continents the average depth to the Moho is 40 km, but this increases beneath young mountain ranges. Beneath the Andes, the crust reaches 70 km in thickness. The structure of the crust and the depth to the Moho beneath the ocean ridges remains somewhat uncertain.

The position of the low-velocity zone in the upper mantle is indicated in Figures 7-1 and 7-5 and is illustrated in more detail in Figure 7-6. There is no abrupt change in the velocity of earthquake waves that can be measured to define the positions of the boundaries of the low velocity zone and, indeed, there may be no sharp boundaries at all. There remains considerable uncertainty about the positions of the upper and lower borders, but it has been clearly established that the depth and thickness of the low-velocity zone do vary according to the geological environment.

Beneath the ocean basins the upper border of the zone rises to about 60 to 70 km below sea level. This is still well below the Moho and definitely within the mantle. The bottom of the zone appears to be less than 200 km below sea level. A narrow channel of the low-velocity zone is shown rising to the ocean ridge. This fits with the general model for plate tectonics and sea-floor spreading, as depicted in Figures 2-1 and 2-3, and it could account in part for the unusual results obtained from the study of earthquake waves in this region.

The upper border of the low-velocity zone is somewhat deeper beneath the continents, 100 km or more, and its lower border may be as deep as 300 km. Beneath the Andes, there is evidence that the low-velocity zone is much deeper, the upper border being as deep as 200 or 250 km.

The low-velocity zone beneath the Andes is interrupted by a layer of the upper mantle dipping through it. This structure has not been worked out in detail from earthquake waves. However, in Chapters 4 and 5, we reviewed the evidence in support of the existence of such a layer of rock extending down into the mantle. The evidence included the distribution of earthquake foci (Figures 4-10 and 4-11), and the observation that earthquake waves passing through this region show that the material below the deep zone of earthquake foci does have properties different from other material at the same depth.

Comparison of Figure 7-6a, determined from the study of earthquake waves, with Figure 2-3, which was invented to explain global tectonic processes, shows that in the plate tectonics model, the asthenosphere is readily correlated with the low-yelocity zone. The lithosphere layer above the asthenosphere includes the crust and the uppermost portion of the mantle. The discontinuity in physical properties represented by the Moho is a discontinuity revealed by earthquake waves, but apparently it has no significance for the dynamic process of plate tectonics.

The rock types reviewed in Chapter 6 constitute the crust, defined as the rock layers above the Moho. These rocks, for the most part, have densities between 2.6 and 3.1. From the measured earthquake wave velocities in the upper mantle (Figure 7-5), we know that the only rock types with appropriate properties have higher densities, between 3.3 and 3.6. The Moho, therefore, is almost certainly a boundary between different rock types, the light rocks of the crust sitting on the denser rocks of the upper mantle. But the rocks of mantle and crust above the low-velocity zone behave as a single, rigid rock layer.

In contrast, the low-velocity layer is not related to any change in composition. It exists because within this particular zone or

layer, conditions are such that the mantle rock properties have changed. The rock is weaker than the more rigid rocks above and below it, and it is capable of slow flow in response to quite small forces if they are applied for long periods.

There are several explanations possible for the change of properties in the low-velocity zone, and they all relate to the changes in temperature as a function of depth within the earth. The favored interpretation at present is that rocks in this zone are very near to the temperatures where they begin to melt, so that they become more pliable. Possibly, incipient melting may have begun, so that the rocks are lubricated by very thin films of molten rock between mineral boundaries. Is this possible? We stated that S-waves cannot pass through liquids, but Figure 7-5 gives velocities for S-waves within the low-velocity zone. This is not a contradiction as long as we are dealing only with incipient melting. The rock remains essentially solid and perfectly capable of transmitting "shake" waves. The traces of liquid between the mineral grains would certainly make the rock more susceptible to flow.

THE CORE

The shape and properties of the core were reviewed and illustrated in Figures 7-1, 7-4c, and 7-5. The solid inner core, surrounded by a liquid outer core, has a composition different from that of the lower mantle and a melting temperature lower than that of the mantle material. The density of the core material is greater than the average density of the earth, greater than 5.5 g/cc.

We have no direct information about the composition of materials deep within the earth. Physical studies of the whole planet and the study of earthquake waves place limits on the properties of earth materials at various depths. This is not definitive, however, because several different materials may have similar properties.

One approach to the problem of the

Fig. 7-6 Cross section through the outer layers of the earth along a line near latitude 20°S, corresponding to part of the cross section shown in Figure 3-10. Compare Figure 2-3. Distinguish between the crust and mantle (Figures 7-1 and 7-5), continental and oceanic crust,

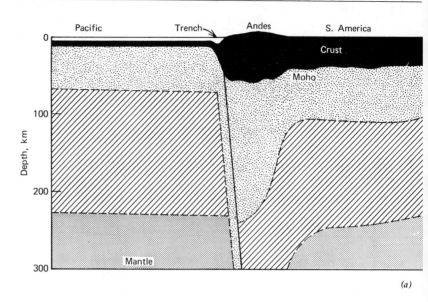

(a)

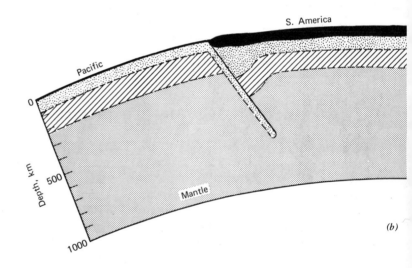

(b)

composition of the earth and its concentric layers is to formulate conceptual models for the origin of the solar system and the earth based on physics and chemistry and to test the deduced compositions for the interior against the measured physical properties. The data employed include the known compositions of the sun and the stars and the detailed study of meteorites.

Meteorites are chunks of material from outer space that made their way through the earth's protective atmosphere without burning up. There are many elaborate theories for their origin, but they have certain features in common. It is generally assumed that meteorites either represent fragments from single planets that broke up in space or fragments from smaller bodies

and the lithosphere and asthenosphere (Figures 2-2, 2-3, 6-8, and 7-5). The distinctions are easily made in part *a*, with vertical exaggeration ×10, but the oceanic crust is barely distinguishable in part *b*. The crust is the upper layer of the lithosphere.

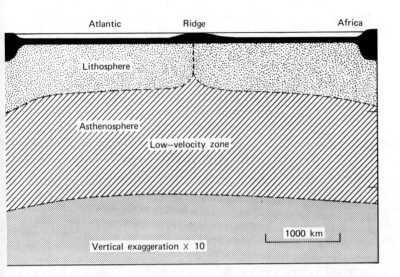

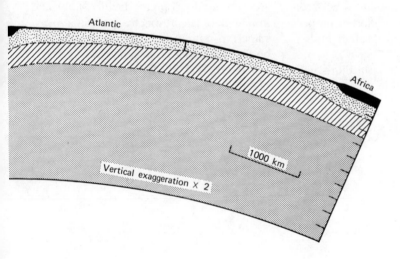

about the size of the moon that broke up by mutual collisions in space.

By analogy with the planetary models for the origin of meteorites, the various types of meteorites that have been discovered are correlated with various portions of the earth's interior on the basis of their properties. Let us consider only the two main classes of meteorites.

The stony meteorites are rocks, not too dissimilar from some rocks known on earth. The iron meteorites consist essentially of an iron-nickel alloy. The properties of the stony meteorites are a reasonable match for the properties of the earth's mantle, and the iron-nickel alloy has properties suitable for the earth's core. No one has suggested any other reasonable material that satisfies the properties of the earth's core so well. Most scientists agree, therefore, that the core consists of solid surrounded by molten iron-nickel alloy. The rotation of this molten metal produces the earth's magnetic field, which we will examine in the next chapter.

SUMMARY

Earthquake waves transmitting energy from the focus where fracture occurs pass through the earth and are recorded by sensitive instruments at a worldwide network of receiving stations. The velocities of the waves at various depths are measured, and this provides the equivalent of an X-ray picture of the earth's interior. P-waves and S-waves have different velocities, and these vary according to the physical properties of the rocks through which they pass. The waves are reflected from surfaces between two layers with markedly different properties. S-waves cannot pass through molten material.

The earth has a concentric structure, with two major discontinuities in physical properties dividing the sphere into core, mantle, and crust. The outer core is molten. The lithosphere layer, about 100 km thick, includes the crust and uppermost mantle, and beneath it is a layer of the mantle about 200 km thick where earthquake wave velocities are relatively low. This is the asthenosphere layer, which is perhaps partially melted and therefore mobile compared with the overlying rigid lithosphere.

The density of most crustal rocks is less than 3.0 g/cc, and that of the whole earth is 5.5 g/cc. Therefore materials within the earth must have densities considerably greater than 5.5 g/cc. Earthquake wave velocities vary as a function of density and, using the measured velocities at various depths, the density of earth materials in the concentric layers has been calculated. The results are consistent with a mantle composed of rocks (at high pressures and temperatures) and a core composed of high-density iron-nickel alloy. These materials are commonly compared with stony and iron meteorites, respectively, that are assumed to represent fragments of planetary bodies that broke up in space.

SUGGESTED READINGS are at the end of the book.

chapter 8

The Earth's Magnetic Field and Magnetized Rocks

Associated with the earth is a magnetic field, which causes certain rocks to become magnetized at their time of formation. The revival of interest in the hypothesis of continental drift in the 1950s and the quantification of Hess's geopoetry of sea-floor spreading in the 1960s were both due largely to the measurement and interpretation of the magnetic properties of rocks of various ages, a study called paleomagnetism. This is concerned with the fossil record of the earth's magnetic field. Before 1950, paleomagnetism was generally regarded as a rather obscure, specialized topic that held little interest for most geologists and geophysicists. Now it is a topic that all earth scientists have to follow with respect and a certain degree of obeisance. Many geologists are not familiar with the physics of magnetism, but they have learned that the earth's magnetism provides as many clues about the earth's history as the conventional approaches of historical geology, and it provides specific dates and rates for various processes.

MAGNETISM

We all know about magnetism. It involves that great favorite of science fiction, a force

field. A magnet has power. It has the power to push away another magnet or to attract a piece of iron. Everyone who has played with magnets or magnetic toys has felt the force of attraction or repulsion between them.

A force is a vector quantity. It is defined not only by a size or an amount, as for a scalar quantity such as volume, but also by a direction. A speed of 45 mph is a scalar quantity, but a velocity of 45 mph in a direction due east is a vector quantity.

If we bring a magnet toward a needle sitting on a carpet, as shown in Figure 8-1, nothing happens at first. But, as the magnet is brought closer to the needle, a position is reached where the needle jumps up and becomes firmly attached to the magnet. This happens because when the magnet is close enough, the needle becomes magnetized by the magnetic force field. The magnetized needle generates its own field, modifying the magnet's original field. The needle is attracted toward the magnet by the combined influence of both magnetic fields and, when the magnet comes close enough, the attractive force becomes greater than the weight of the needle. Then the needle flies up off of the carpet to the magnet. It stays there, firmly attached, until it is forcibly pried loose. If the magnet is

Fig. 8-1 A magnet is surrounded by a magnetic field, which is capable of magnetizing a nail on a table. The magnet can then attract the magnetized nail to it.

weak and the needle is very heavy, the attractive force produced by the combined magnetic fields may not be sufficient to lift the needle from the carpet. Similarly, if the needle is tied down, it will modify the magnetic field, but it will not be able to move.

Figure 8-1 illustrates what we need to know before we can examine the magnetic evidence for sea-floor spreading and continental drift. We need to know something about the earth's magnetic field, about how some metals and rock types become magnetized, and about the influence of magnetized rocks on the earth's magnetic field.

MAGNETIC FIELDS

The field around a magnet is simply the region in which magnetic forces act on other magnetic bodies. The existence of the earth's magnetic field is demonstrated in familiar fashion by a compass. A compass is a small magnet, pivoted in such a way that it is free to rotate in a horizontal plane. Under the influence of the earth's magnetic field, the compass needle always lines itself up in a north-south direction. The end of

the compass needle pointing north is defined as the north pole of the magnet, or the north-seeking pole, and the other end is defined as the south pole of the magnet.

One of the laws of magnetism is that like magnetic poles repel each other and unlike poles attract each other. This is easily illustrated by placing two compasses on a table. If they are far enough apart they are dominated by the earth's magnetic field and they both point to the north. If they are moved toward each other, however, there comes a stage when each is influenced by the magnetic field of the other, as illustrated in Figure 8-2. The magnetic force field clearly increases in strength as the distance between the compasses decreases.

If they approach as in Figure 8-2a, both needles rotate toward the west. If they approach as in Figure 8-2b both needles rotate toward the east. This is readily explained by the attraction between the north pole of one compass and the south pole of the other. The forces between the other poles of the compass needles are less effective because the distances between them are larger. If the compasses approach as in Figure 8-2c, the needles quiver as they are brought within each other's magnetic sphere, but they remain pointing to the north. This is because the north poles repel each other and the south poles repel each other, and as long as the distances between the two pairs of like poles remain the same, the two forces of repulsion balance each other out. A slight displacement of either compass from the balanced position in Figure 8-2c, however, will cause them to rotate into one of the positions illustrated in Figures 8-2a and b.

A magnetic field is continuous in three dimensions around a magnet, but it is conveniently represented by imaginary lines of force that show the direction of the force field at any point. A compass needle becomes lined up parallel to one of the earth's magnetic lines of force. Small compass needles may be used similarly to map out the lines of force surrounding a magnet, as shown in Figure 8-3a.

The magnet is strong enough that for a considerable distance around it, the influence of the earth's magnetic field is negligi-

ble in comparison. The needle of each compass placed in the magnetic field is subjected to forces of attraction and repulsion. The compass's north pole is attracted toward the magnet's south pole and repelled from the magnet's north pole; the forces are in the opposite sense for the compass's south pole. The effects of these combined forces are illustrated by the positions of the compass needles in Figure 8-3a. Needle a points directly away from the magnet's north pole, and needle b points directly toward the magnet's south pole. Needle c, equidistant from the magnet's north and south poles, lines up parallel to the magnet but with opposite polarity, because it is equally attracted toward the magnet's south and north poles. The north pole of compass d is strongly repelled from the magnet's north pole and less strongly attracted by the more distant south pole of the magnet. The net result is that it adopts a position pointing away from the magnet's north pole, but displaced slightly toward

Fig. 8-2 Compass needles can be used to demonstrate that north magnetic poles are attracted to south magnetic poles (a and b), and that north magnetic poles or south magnetic poles repel each other (c).

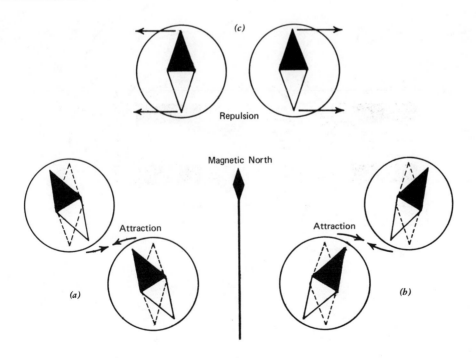

Fig. 8-3 *(a)* Distribution and directions of lines of magnetic force in the field around a magnet, indicated by the orientations adopted by compass needles. The magnetic field is distributed in three dimensions all around the magnet. *(b)* When two magnets are lined up with magnetic poles in the same direction, the concentration of lines of force is increased between them (giving increased magnetic field strength), and the magnets repel each other. *(c)* When two magnets are lined up with magnetic poles in opposite directions, the concentration of lines of force is decreased between them (giving decreased magnetic field strength), and the magnets attract each other.

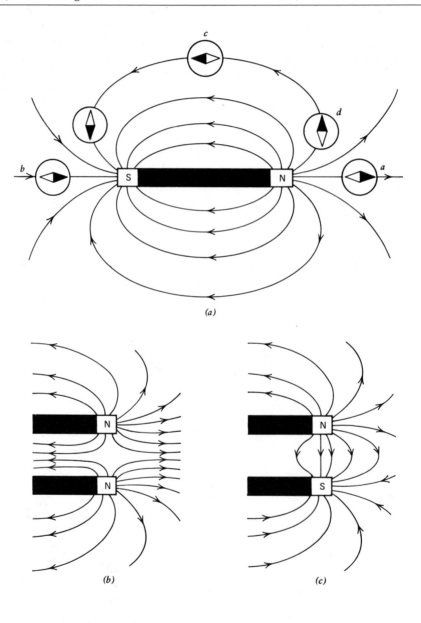

(a)

(b)   *(c)*

the magnet's south pole. Each compass needle points in the direction of the magnetic force field. The compass needles show that the lines of force are not straight, except for the line passing through the axis of the magnet, between a and b. Each direction indicated by a compass needle is a tangent to an imaginary line of force.

Figure 8-3a shows not only the distribution of lines of force around a magnet but also the directions of the forces. At any point, the direction of the field is defined by the force on a north magnetic pole, and this is given by the direction indicated by the north pole of the compass needle. Note that each imaginary line of force extends continuously from a point near the north pole of the magnet to a point near the south pole. Lines of force cannot cross each other. The complete magnetic field is represented by lines of force like those in Figure 8-3a, but distributed in three dimensions. Imagine Figure 8-3a being rotated about its north-south axis; this will give you an impression of similar lines of force extending above and below the magnet.

The magnetic field at any point is defined by two quantities: the direction of the field, which is tangential to the line of force passing through the point, and the strength or intensity of the field. We have seen that the strength of the magnetic field decreases as distance increases away from the magnetic poles. The strength in different regions is represented by the relative concentrations of the lines of force. Figure 8-3a shows that the lines are closest together near the poles and that they spread out with distance from the poles.

The magnetic field patterns illustrated in Figure 8-3 can be examined directly by scattering iron filings on a card placed over magnets. The filings become magnetized and, if the card is shaken gently, the many tiny magnets orient themselves in the directions of the magnetic field; the general pattern produced illustrates the configuration of the field. The filings cluster together where the field strength is greatest, near the magnetic poles. Many of them near the poles will stand on end, giving a visual picture of the lines of force emanating from the poles in three dimensions. Magnetic field intensity is defined in terms of the force of attraction or repulsion that a field exerts at a given point. We will use the unit of force defined as a gamma (γ). This is a very small unit, but the earth's magnetic field is weak. The magnetic variations that turn out to be significant for interpretation of plate movements are very small, amounting to only 1 or 2% of the earth's field strength.

MAGNETIZATION

It is impossible to take a bar magnet such as that in Figure 8-3a and cut off the north pole. If the bar is cut in half, the severed north pole develops a new south pole, and the severed south pole develops a new north pole. Any fragment of a magnet, however small, becomes a complete magnet itself with a north and a south pole, as illustrated in Figure 8-4. From this we can conclude that magnetism is a property of the atomic structure of the material composing the magnet. The material is magnetized only when the atomic groups are lined up in a more or less parallel arrangement. The magnetism is destroyed if a magnet is heated, because heating causes the atoms to vibrate more vigorously; the regular organized structure that causes the magnetism is then lost.

Some materials have atomic structures that change under the influence of a magnetic field. Small atomic groups become oriented along the lines of magnetic force as if they were compass needles (Figure 8-3a). If this occurs, the material has become magnetized, because then its modified atomic organization produces a north and a south magnetic pole. This may be a temporary change that disappears when the magnetic field is removed. Under some circumstances, however, the change induced in the orientation of the atomic particles persists, and the material retains its magnetic properties after the magnetizing field has been removed. Some rocks become magnetized in this way by the earth's mag-

Fig. 8-4 Each time a magnet is cut in half, each portion becomes a complete magnet with its own north and south poles. The fact that this continues down to magnet fragments of minute size indicates that magnetism is caused by the atomic properties of the material.

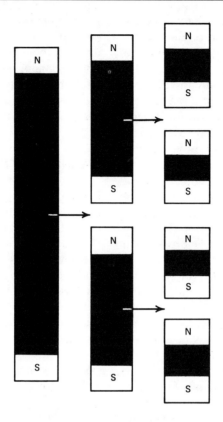

netic field as they are formed. They then retain a fossil record of the earth's magnetic field as it existed at their place and time of formation.

THE EARTH'S MAGNETIC FIELD

Early navigators found their way over the oceans by using a special rock that was influenced by the invisible and then unimagined magnetic field of the earth. Lodestone is a rock now known to be composed largely of an iron oxide mineral that is magnetic. It was discovered that if a piece of lodestone was suspended so that it could rotate freely, it always oriented itself in the same position with respect to the geographic north pole. The north-seeking part of the lodestone was marked, and this

crude compass was then capable of guiding mariners in fog-bound oceans.

It was not until 1600 that William Gilbert examined in detail the behavior of lodestone and iron needles and concluded that the earth has a magnetic field of its own similar to that of a giant magnet. The lines of magnetic force are as depicted in Figure 8-5. Gilbert determined their distribution at the surface of the earth, and their distribution in space has been confirmed by measurements from orbiting satellites and other space craft.

Most of the earth's magnetic field is equivalent to the field that would be produced by a magnet in the center of the earth, slightly displaced from the axis of rotation, as shown in Figure 8-5. In fact, there is no bar magnet in the center of the earth, and this is just a convenient, hypothetical concept. We noted in Chapter 7 that the temperature is high enough to melt the nickel-iron alloy of the outer core and, at such high temperatures, magnets lose their magnetic properties. The hypothetical magnet is oriented with its south pole directed toward the north magnetic pole, because this is required to give lines of magnetic force that rotate the north poles of compass needles to the north (see Figure 8-3a).

The earth's magnetic field is defined at each point on the surface by the field strength and its direction. The directions of the field are shown in Figure 8-5a by the arrows on the lines of force. The attitude of the lines of force in space can be defined by the angle they make with the horizontal plane. In order to visualize a horizontal plane in the southern hemisphere of Figure 8-5a, it is necessary for the viewer either to turn his head or to rotate the diagram. Re-

taining a sense of orientation during these operations can be difficult. Therefore the operation has been performed graphically in Figure 8-5 for three locations; one in the northern hemisphere, n, one in the southern hemisphere, s, and one near the equator, e. For each location, the horizontal surface has been rotated into its normal position with great enlargement in Figures 8-5b, 8-5c, and 8-5d. The lines of force from these locations in Figure 8-5a have been similarly rotated so that the angles between lines of force and the horizontal have been preserved.

In each location the lines of force are directed toward the north. In the northern hemisphere the field is directed downward below the horizontal plane; near the equator the field is nearly horizontal and, in the southern hemisphere, the field is directed upward above the horizontal plane.

We can map the directions of the lines of

Fig. 8-5 *(a)* Diagrammatic representation of the earth's magnetic field. The directions of the lines of magnetic force at the earth's surface and in space around the earth (measured from satellites) are consistent with the presence of a magnet within the earth in the orientation shown (Figure 8-3). In fact, there is not a magnet within the earth. *(b, c, d)* illustrate in larger scale the lines of magnetic force as they would be measured by a person standing on the earth's surface at points *n, e,* and *s,* respectively.

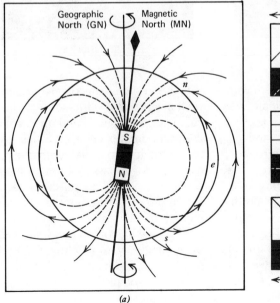

(a)

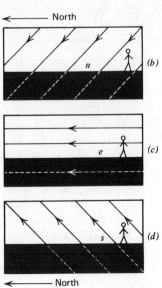

Fig. 8-6 Maps showing the direction toward magnetic north as recorded at the earth's surface by compass needles, and vertical cross sections at the same locations showing the angles of dip of the lines of magnetic force, with respect to the horizontal surface. Compare Figures 8-5*b* and 8-5*d*.

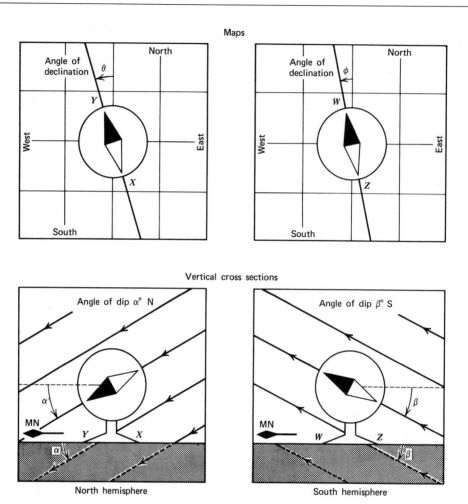

Maps

Vertical cross sections

North hemisphere   South hemisphere

force with a compass and define them with respect to the earth's fixed coordinate system of lines of latitude and longitude. Figure 8-6 illustrates the procedures and results for locations on the northern and southern hemispheres. As shown on the maps, the compass needle rotating in a horizontal plane points toward the north, but not exactly toward the geographic north. This direction is defined by the angle between the compass needle and a line of longitude, measured to the east or west of geographic north. This angle is called the angle of declination.

If the compass needle is suspended in such a way that it is free to rotate in a vertical plane, and if it is lined up in a direction corresponding to *XY* or *ZW*, then it will rotate until it is parallel with the lines of magnetic force passing through the region, as shown in the vertical cross sections in Figure 8-6. An instrument designed for this measurement is called a dip needle. The attitude of the line of force in space is defined by its angle of dip below the horizontal plane. Note in Figure 8-6 that this is independent of the position of the north pole of the compass needle. The angle of dip is simply the angle below the horizontal, measured in a northerly or southerly direction.

The relationships among the three measurements required to define the earth's magnetic field at any point on the surface are illustrated in Figure 8-7, which should be compared with Figure 8-6 for the northern hemisphere. The arrow represents the vector, with its length proportional to the strength of the field. The position of the vector in space is defined by two angles, the angle of declination, $\theta°W$, which locates the line with respect to the earth's coordinate system, and the angle of dip, $\alpha°N$, which fixes the vector in space with respect to the horizontal plane.

The symmetry of the earth's magnetic field, as portrayed in Figure 8-5a, corresponds to the arrangement for the bar magnet in Figure 8-3a. According to this arrangement, each point on the earth's surface equidistant from the north magnetic pole would have the same magnetic field intensity and the same angle of dip. Lines connecting points equidistant from the magnetic poles are concentric small circles, as shown in Figure 8-8a. These same lines, therefore, show the general distribution of lines for equal values of the earth's magnetic field intensity or for equal values of the angle of dip. These can be considered in precisely the same way as contour lines that connect points on the earth's solid surface with the same height (Figure 3-9). In Figure 8-8b, the general distribution of the magnetic contour lines of Figure 8-8a is compared with lines of latitude and lon-

Fig. 8-7 Summary of the information required to define the position of a line of magnetic force in space, with respect to the geographic coordinates and the horizontal surface. Compare Figure 8-6.

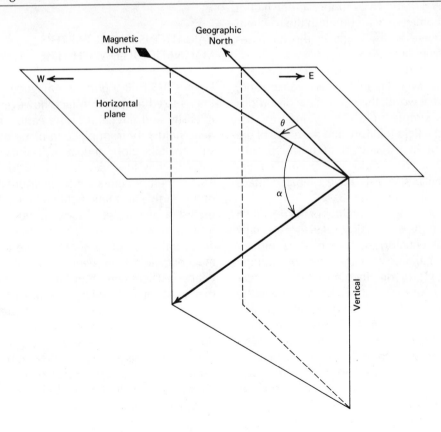

gitude on a map. If the magnetic north pole were coincident with the geographic north pole, the contour lines for the magnetic properties would coincide precisely with lines of latitude. In Figure 8-8*b*, however, the magnetic contours are slightly displaced from the lines of latitude through a small angle corresponding to the angle of declination.

Assume that the lines on the map in Figure 8-8*b* are lines of equal angle of dip. Then, for any point *P*, the map provides the following information: (1) the direction of magnetic north from *P*, and (2) the magnetic latitude and, hence, the distance to the magnetic north. The direction, shown by the compass needle, is perpendicular to the line of constant dip angle (compare Figure 8-8*a*). The relationship among the angle of dip, the magnetic latitude, and the distance to the magnetic pole is shown in Figure 8-8*c*. The distance is zero for an angle of dip of 90°N, increasing to one quarter of the earth's circumference as the angle of dip decreases to 0° (compare Figures 8-8*a* and 8-5*a*). It increases further to one half of the circumference for points in the southern hemisphere as the angle of dip increases again from 0° to 90°S.

The distribution of lines of magnetic force shown in Figure 8-5*a* is idealized and, in the real world, there are many distortions of this simple field. The values for the magnetic field intensity and the angle of dip have been measured at many places throughout the world and, from the measurements, scientists have prepared maps such as those in Figure 8-9.

The lines for equal values of the dip angle in Figure 8-9*a* are fairly regular, especially near the equator; the lines of equal dip are almost parallel with the lines of latitude. The angle of dip increases from 0° near the equator to 90° at two points, one at the north magnetic pole and the other at the south magnetic pole (compare Figure 8-5*a*). For larger angles of dip there are considerable deviations from the simple, regular arrangement depicted in Figure 8-8 but, nevertheless, for any point *P* on the surface, approximate values for the direction to and

distance from the magnetic north pole are given by the line of constant dip angle passing through the point, just as for point *P* in Figure 8-8*b*.

There are two places on Figure 8-9*a* where the angle of dip becomes 90°. These magnetic dip poles, usually described simply as the magnetic poles, are not diametrically opposite each other on the globe. Thus they do not correspond precisely to the position of the magnetic axis shown in Figure 8-5*a*, as they would if the magnetic field could be explained completely by the simple bar magnet model.

The lines for equal values of the magnetic field intensity in Figure 8-9*b* deviate quite significantly from the lines of latitude. These lines for equal values of the field intensity can be considered as contour lines mapping a surface that shows the variation of magnetic field intensity across the globe. Note that the strength is least in the equatorial regions and highest toward the north and south. Each unit of field intensity in Figure 8-9*a* is 10,000 γ.

VARIATIONS IN THE EARTH'S MAGNETIC FIELD WITH TIME

Figures 8-5 to 8-9 portray a magnetic field that is fixed in space. In fact, however, the intensity and direction of the earth's magnetic field vary from place to place on the surface of the globe and, at any point on the surface, they also vary from time to time. This is well documented by measurements of the earth's magnetic field that have been recorded at magnetic observatories for as long as four centuries.

Several kinds of variations in the earth's magnetic field have been recognized and charted. There are short-term changes of magnetic intensity with durations of minutes, hours, days, or years that are explained in part by weak electric currents flowing in the outer layers of the atmosphere. These are influenced by radiation from the sun. There are also long-term changes that include changes in the direction of the field as well as its intensity, with

Fig. 8-8 *(a)* Compare Figures 8-5*a* and 8-6. For an idealized model of the earth's magnetic field, there are lines of magnetic latitude in concentric circles about the magnetic poles. Angles of dip and magnetic field intensities are constant along each of these lines. *(b)* Lines of magnetic latitude compared with geographic latitude and longitude lines on a map. *(c)* The angle of dip is 90° at the magnetic poles, and 0° at the magnetic equator (Figure 8-5). At any magnetic latitude between these limits the angle of dip is given by the graph. For a point with measured angle of dip, the graph gives the magnetic latitude and also the distance to the magnetic pole.

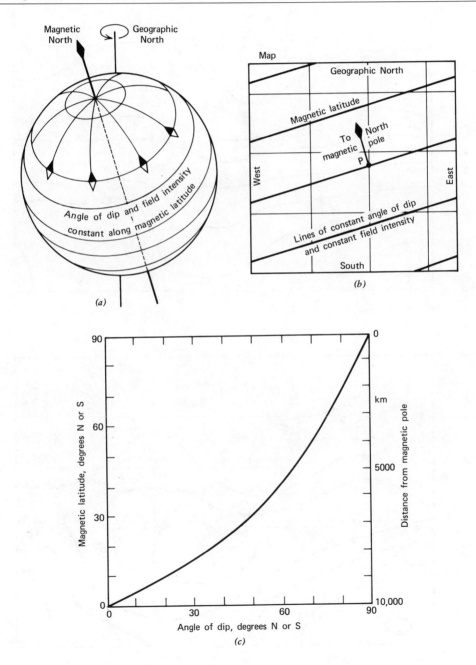

Fig. 8-9 The real magnetic field of the earth is far less regular than the simple model represented in Figures 8-5 and 8-8. (a)Map of lines for constant angle of dip based on measurements (simplified from actual data). Each of these lines should ideally represent a regular magnetic latitude, as in Figure 8-8. (b) Map of lines for constant values of magnetic field intensity (10,000γ), based on measurements (simplified from actual data). Each of these lines should ideally represent a regular magnetic latitude as in Figure 8-8. There are considerable divergencies from concentricity about magnetic poles in both maps.

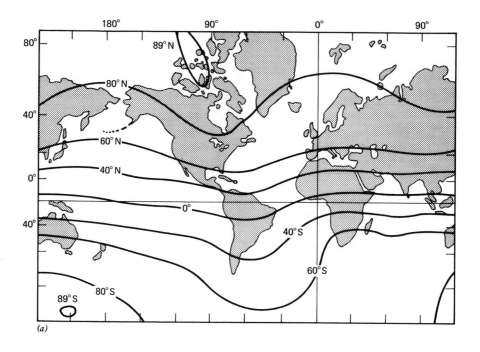

(a)

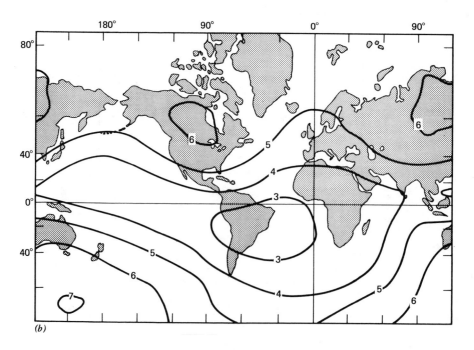

(b)

durations of hundreds of years. It can be shown that these have their origin within the earth.

One of the long-term changes involves a slow migration of the magnetic poles around the geographic poles, although detailed analysis of the measurements from different continental regions indicates that this interpretation is too simple. It is established, however, that through the hundreds of years of scientific records, the earth's magnetic field has remained more or less as indicated in Figure 8-5a. The relative positions of the earth's rotation axis and the magnetic axis have changed, but the angle between them has not exceeded a few degrees. The north magnetic pole has coincided approximately with the north geographic pole for the duration of man's historical records.

CAUSE OF THE EARTH'S MAGNETIC FIELD

There is no completely satisfactory explanation of what causes the earth's magnetic field, although it is certain that most of the field has its origin deep within the earth. The field is definitely not caused by magnetized rocks in the core or mantle because the temperatures are too high. We have noted that a bar magnet loses its magnetism when it is heated.

The only plausible explanation involves electric currents flowing in the earth's metallic core. Figures 7-1 and 7-5 show that the outer core is molten. According to the dynamo theory, the rotation of the earth causes some kind of convective motion of the fluid in the outer core, that is associated with weak electric currents. The interaction of the electric currents and the mechanical rotation of the fluid generates a self-sustaining magnetic field. Continued motion of the electrically conducting fluid through the magnetic field keeps the electric currents flowing, and so on. The dynamo keeps operating.

The long-term variations of the earth's magnetic field appear to be reasonably consistent with convective motions in a fluid outer core, but inconsistent with electric currents flowing in a solid portion of the earth's interior. We will see in Chapter 9 that the earth's magnetic field has reversed completely at intervals through geologic time. The relatively short time required for a reversal (Figure 9-9) is more readily comprehended in terms of rotation of a fluid than in terms of solid materials. The problem is extremely complex, and it remains one of the more challenging problems awaiting solution by a bright young geophysicist; many of the older geophysicists have already tackled the problem without complete success.

MAGNETIZATION OF ROCKS

Figure 8-10 shows a schematic cross section through a volcano in a north-south orientation, and the dashed lines represent the lines of force of the earth's magnetic field. The cross section shows a recent lava flow, which emerged from a crack near the crater and flowed down the volcano in a southerly direction. As it flowed, it cooled and began to crystallize. Solidification was completed at about 900°C, by which time the rock consisted of a mesh of small interlocking crystals. This is indicated by the enlarged cross section through a lava sample, as depicted in inset a.

These crystals are of various minerals with different compositions. Among them is the mineral magnetite, an iron oxide that is the same as the lodestone of ancient mariners. This mineral has the capability of becoming magnetized. At high temperatures, however, the atoms composing the magnetite are oriented in random directions because they vibrate around their average positions in the mineral structure. Under these conditions, the minerals cannot be magnetized by the earth's magnetic field. This is true down to a temperature near 500°C.

With further cooling, the amount of atomic vibration decreases because the available thermal energy is less, and the influence of the magnetic field begins to exert itself. Small atomic groups within the mineral begin to line up parallel to one another in the direction of the magnetic lines of force. Once they are lined up, with the temperature continuing to fall, it becomes difficult for the atoms to vibrate out of this orientation. By the time the temperature reaches about 450°C, the arrangement of atoms in the magnetic mineral is locked in, and each mineral has become a small magnet, with polarity parallel to the earth's magnetic field. This is indicated in inset *b*. The net result is that after cooling, the lava as a whole retains a record of the earth's magnetic field at the time it cooled through the temperature interval of 500 to 450°C, as shown in inset *c*.

The atomic arrangement is fixed in the minerals, so that their magnetization is permanent. It is also stable, since it requires a high temperature to provide enough thermal energy to cause the atomic groups to vibrate away from their orientated positions. The minerals themselves are locked into position within the rock as a whole, so the magnetized direction of the rock is also permanent and stable.

Figure 8-11 illustrates schematically the deposition of sediments on the ocean floor. The dashed lines represent the lines of force of the earth's magnetic field. Mineral grains and the shells and skeletons of marine organisms settle slowly through the water to the bottom. Included among the mineral fragments transported from land by river or wind is a small proportion of magnetized grains. These include grains like those magnetized in the lava flow of Figure 8-10, which are subsequently broken loose when the lava was subjected to weathering. Each magnetized grain behaves like a tiny magnet.

A magnetized mineral grain, if suspended freely in the water, would orient itself in the direction of the lines of magnetic force. While it is sinking, however, its orientation at any point in time is more likely to be controlled by the action of ocean currents on its surface than by the earth's weak magnetic field. This is indi-

cated schematically in the right side of Figure 8-11, where the magnetized grains in water have random orientations. When they settle on the bottom, the magnetized grains then have an opportunity to align themselves at least approximately in the direction of the magnetic field because they are, in effect, suspended and able to rotate in the very loose and unconsolidated sediment. This preferred orientation is shown in the right side of Figure 8-11. Controlled experiments in the laboratory have confirmed that this alignment of magnetized grains does occur during deposition.

When the sediments are later buried by younger sedimentary layers, consolidated, and hardened, the orientated magnetized grains are locked into these positions. The combined effect of the small proportion of magnetized grains is to impart to the rock as a whole a direction of magnetization corresponding approximately to the earth's magnetic field at the time of sedimentation. Most sediments contain a very small proportion of magnetized grains, and their orientation in the direction of the earth's magnetic field is not perfect. For these reasons, the intensity of magnetization of most sedimentary rocks is about 100 times weaker than that of lavas.

The mineral grains and fragments of marine organisms settling on the ocean floor are joined by minerals precipitated from solution in the ocean water, and some of these may become magnetized. Chemical and physical changes may occur in sedimentary rocks during the long process of burial and consolidation, and the resultant effect may be to change the direction of magnetization of the rock as a whole compared with that existing at its time of deposition. The magnetization of sedimentary rocks is thus less stable than that of lavas, as well as being weaker.

The main features of the magnetization of lavas and of sediments are compared in Table 8-1. The direction of magnetization in both kinds of rocks preserves a fossil record of the direction of the earth's lines of magnetic force at the time and place of formation of the rock, provided that the fos-

Fig. 8-10 Vertical cross section through a volcano showing directions of the earth's lines of magnetic force. Lava erupts, flows, and crystallizes. Inset *a* represents the interlocking crystals in the lava after it has solidified but while it is still hot; the minerals are not magnetized; (*b*) shows the same rock when it has cooled to 450°C; some minerals have become magnetized in the direction of the earth's magnetic field; (*c*) represents a larger portion of the lava flow after it has cooled; it contains many minerals magnetized as shown in inset *b* and, consequently, the rock as a whole is magnetized in the direction shown.

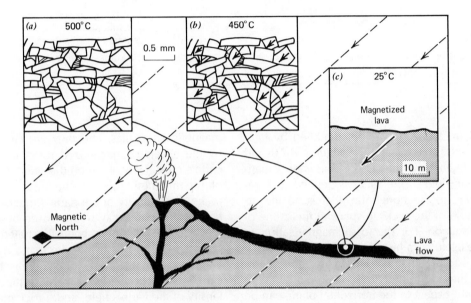

Fig. 8-11 Vertical cross section through ocean and submarine sediments, showing mineral grains settling slowly to the ocean floor, and the direction of the earth's lines of magnetic force. The inset shows that some of these falling particles have been magnetized at some earlier stage in their history (as in Figure 8-10) and, as these settle on the sediment surface, they become oriented in the direction of the earth's magnetic field. This magnetic direction is recorded by the sediment layer as a whole when it becomes compacted and consolidated into a sedimentary rock.

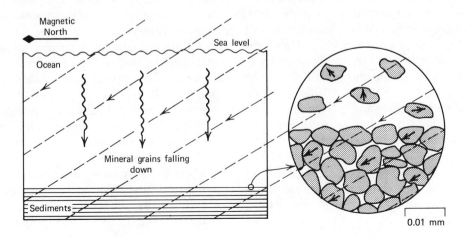

Table 8-1 The Magnetization of Lavas and Sediments Compared

	Lavas	Sediments
Source of minerals	Minerals locked in as lava solidifies	Magnetized mineral grains settle through water
Cause of magnetization	Minerals magnetized as they cool through 500 to 450°C	Magnetized grains aligned as they settle at bottom
Later changes	Lava cools with no further changes	Magnetized minerals locked in position by sediment consolidation. Possible remagnetization
Intensity of magnetization	Weak	100 times weaker
Stability of magnetization	Stable	Variable

sil magnetization is not modified by later events. It is clear from Figures 8-10 and 8-11 that the direction of the earth's magnetic field in the rocks is defined with respect to the horizontal, that is, to the attitude of the rocks in space at their time of formation. For the fossil magnetization to be useful as a historical record, therefore, it is essential that if the rocks are subsequently tilted or folded, their original orientation with respect to the horizontal plane can be deciphered from geological studies.

ANOMALIES IN THE EARTH'S MAGNETIC FIELD

According to Figure 8-5a, the earth's magnetic field corresponds precisely to the field that would be produced by a simple bar magnet within a sphere. The lines of equal magnetic intensity for such a field would follow smooth concentric circles on the earth's surface, as indicated by Figure 8-8. In fact, Figure 8-9b shows that considerable differences exist between the actual magnetic intensity map and the idealized version. The irregularities are of a large scale, extending over thousands of kilometers, and their origins probably lie deep within the earth where the magnetic field is generated.

In addition to these large deviations from the idealized magnetic field, there are many local irregularities that cannot be shown on a map drawn to the scale of Figure 8-9b. These irregularities are called

magnetic anomalies, and they are caused by the effects of magnetized rocks in the earth's crust. Depending on the orientation of the magnetization of the underlying rocks, the intensity of the earth's magnetic field at any point may be increased or decreased. The pairs of magnets illustrated in Figures 8-3b and 8-3c indicate that a rock magnetized in the same direction as the earth's magnetic field will increase the intensity of the earth's field, and a rock magnetized in a direction opposite to the earth's magnetic field will decrease the field intensity. Some rocks, such as deposits of iron ore, are so strongly magnetized that they produce very large anomalies. Locally, the earth's magnetic field may become insignificant compared with the field associated with an iron-ore body.

For any selected region under consideration, the average regional magnetic field can be computed from values measured at intervals throughout the area. The value of the magnetic intensity measured at a specific point may be higher or lower than the value of the average regional field computed for that point. The difference between the measured field and the regional field is defined as the magnetic anomaly at the point. The magnetic anomaly may be positive or negative.

Magnetic anomaly = measured field intensity − intensity from regional average.

Figure 8-9a shows that the regional magnetic field intensity may range from

25,000 to 60,000 γ or more. In continental regions magnetic anomalies may vary from hundreds to thousands of gammas. In oceanic regions, in contrast, the measured magnetic field is remarkably smooth, and the anomalies rarely exceed a few hundred gammas. Typically, the deviations amount to no more than 1 to 2% of the average regional field. The reason for this is that the intensity of the field caused by a magnetized rock decreases with distance from the rock. The 4 to 6 km of ocean water between the rocks of the ocean floor and the recording instrument at the ocean surface reduces their effect on the earth's magnetic field very considerably. Magnetic anomalies measured near the ocean floor with special submarine instruments are much larger.

The general relationship among submarine rocks, magnetic intensity profiles, and magnetic anomaly profiles is illustrated schematically in Figure 8-12. The top row shows three maps of a portion of an ocean floor, with an area of 50 x 40 km. In 1973, the ocean floor in this area was covered by a uniform layer of sediment with no distinctive features. In 1974 lava was erupted through a fissure that opened in the ocean floor, and the lava filled a shallow depression about 10 km wide and more than 50 km long. As the lava cooled, it was magnetized in the northerly direction indicated by the arrow. In 1975, a similar submarine lava flow was erupted, and this filled another shallow valley further to the east. This, too, was magnetized in a northerly direction.

The middle row of diagrams shows three magnetic field intensity profiles measured by an oceanographic research vessel as it sailed along path AB in 1973, 1974, and 1975. Figure 8-12d shows that in 1973 there was a steady increase in the intensity recorded from A to B, with little deviation from the dashed line showing the computed average regional intensity variation. Figure 8-12e shows the general increase from A to B persisting in 1974, but superimposed on this is an abrupt increase in intensity directly above the magnetized

submarine lava flow, XY. The dashed line shows the computed average regional intensity variation: this has been shifted to slightly higher values than the 1973 line because of the increased intensity above XY. Figure 8-12f for the 1975 profile shows, in addition to the magnetic intensity peak above XY, a second peak above the lava flow ZW. The average regional gradient has been shifted to slightly higher values, again by the increased magnetic intensity in parts of the area.

The magnetic anomaly profiles in the bottom row of diagrams in Figure 8-12 were obtained from the middle row of diagrams. For selected points on the profile AB, values for the regional magnetic field intensity given by the dashed lines were subtracted from the recorded values given by the solid lines. In 1973 the anomalies were very small. In 1974 we see a positive anomaly of 120 γ directly above the lava flow XY, with small negative anomalies of a few gammas on either side. The 1975 profile shows two positive anomalies of about the same value above the two lava flows, with small negative anomalies between them and on either side.

The positive anomalies in Figure 8-12i are caused by the magnetized lava on the ocean floor, and the small negative anomalies have no special significance. If the oceanographic ship were to complete a series of traverses parallel to AB (Figure 8-12c), giving a series of anomaly profiles, it would be possible to plot an anomaly map with contour lines at selected intervals for specific anomaly values. Figure 8-13a illustrates such a map. With a contour interval of 50 γ, no negative anomaly contours are included. But, if the contour interval were 10 γ, then there would be negative value contours adjacent to the zero lines.

The map of anomaly contours is directly analogous to the contours for the earth's solid surface in Figure 3-9b. The zero anomaly contour corresponds to sea level, the positive anomalies correspond to the hills rising above sea level, and the negative anomalies correspond to the sea floor below sea level. Just as the cross section in

Fig. 8-12 *(a, b, c)* Maps of a portion of the Pacific Ocean floor showing two successive submarine lava flows (a hypothetical example of a not infrequent process). *(d, e, f)* Magnetic field intensity measured at sea level during traverses along line *AB*. *(g, h, i)* Magnetic anomalies determined from the magnetic intensity profiles and the average regional magnetic values shown above.

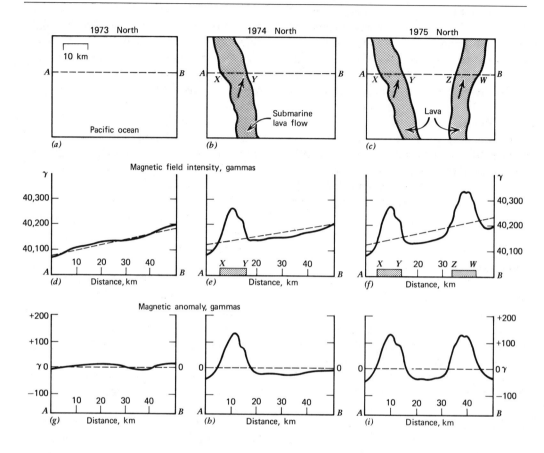

Fig. 8-13 Maps covering the same area as Figure 8-12*c*. *(a)* Contour lines for constant magnetic anomalies, compare Figure 8-12*i*. *(b)* The same data with areas with positive magnetic anomaly marked in black and those with negative anomaly in white.

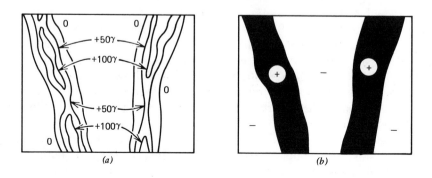

Figure 3-9*d* shows the shape of the solid surface above and below sea level, so the anomaly profile in Figure 8-12*i* shows the shape of the magnetic anomaly surface.

For many purposes, it is convenient to map only the areas of positive and negative magnetic anomalies without showing the additional detail provided by anomaly contours. The normal policy is to make the areas with positive anomalies black in contrast with white areas for negative anomalies, as shown in Figure 8-13*b*. Comparison with Figure 8-12*c* demonstrates that in this example, the positive anomalies map out the extent of the submarine lava flows.

SUMMARY

The earth is surrounded by a magnetic field that is almost equivalent to the field that would be produced by a hypothetical bar magnet in the center of the earth, with its axis slightly displaced from the axis of rotation. The magnetic field varies with time in several ways, but the magnetic axis remains within a few degrees of the rotation axis. The magnetic field is caused by some kind of convective movement in the molten alloy of the outer core, which is associated with weak electric currents.

The earth's magnetic field is defined at each point on the surface by the field strength and its direction. The direction is defined with respect to the geographic coordinates and the angle of dip below the horizontal plane. The strength and angle of dip vary with distance from the magnetic poles. Lavas and some sediments become magnetized at their time of formation, and they retain a fossil record of the direction of the magnetic field at the time and place of origin. The direction of fossil magnetization of a rock provides the following information: the direction of magnetic north when the rock was magnetized and the distance of the rock from the magnetic pole.

Magnetized rocks disturb the earth's magnetic field, producing irregularities, or anomalies, where the intensity and angle of dip depart from the average value computed for the general area. The intensity of the magnetic field at any point may be increased or decreased by the underlying magnetized rocks, depending on the orientation of their direction of magnetization. In oceanic regions, the magnetic anomalies rarely exceed 1 to 2% of the average regional field but, in continental regions, they are very much larger.

SUGGESTED READINGS are at the end of the book.

chapter 9

The History of the Earth's Magnetic Field: Polarity Reversals

Changes in the earth's magnetic field are recorded by rocks that became magnetized as they were formed. The record of the magnetic field at the time and place of formation of a rock persists even if the continent containing the rock drifts to another location on the earth's surface, where the existing magnetic field differs from the fossil magnetic record in the rock.

From studies of the fossil magnetism of rocks of various ages, it has been established that the earth's magnetic field has reversed completely and repeatedly at intervals through geological history. The north magnetic pole has become the south magnetic pole, and vice versa. First, we will compare the earth's magnetic field under conditions of normal and reversed polarity, and then we will examine the evidence leading to the discovery of reversals and the calibration of the polarity reversal time scale. Then we can review some of the applications of the magnetic polarity reversals to plate tectonics.

NORMAL AND REVERSED POLARITY

Figures 8-5 and 8-6 summarize the main features of the earth's magnetic field at the present time, which is defined as a time of normal polarity. These features are reproduced in the left side of Figure 9-1. The right side compares the arrangement of lines of magnetic force during a period of reversed polarity.

For the idealized conditions represented in the figure, the change from normal to reversed polarity involves the complete reversal of the magnetic field (Figures 9-1a and 9-1b), as indicated by the reversed direction of the arrows on the lines of magnetic force and as represented by the reversal of the hypothetical magnet within the earth. The magnetic axis remains in the same position with respect to the rotation axis of the earth, but the north magnetic pole has changed positions with the south magnetic pole. Notice that the geometrical arrangement of the lines of magnetic force in space remains the same.

Consider point A on the northern hemisphere with normal polarity. Figure 9-1c shows that the compass needle points toward the magnetic north pole, in a direction $\theta°$ from the longitude line that passes through the geographic north pole; the angle of declination is $\theta°E$. The dip needle at A, oriented so that it is free to rotate in a vertical plane toward the magnetic poles, shows that the lines of magnetic force dip

Fig. 9-1 Comparison of magnetic properties during normal and reversed polarity epochs. Compare Figures 8-5 and 8-6 for normal polarity.

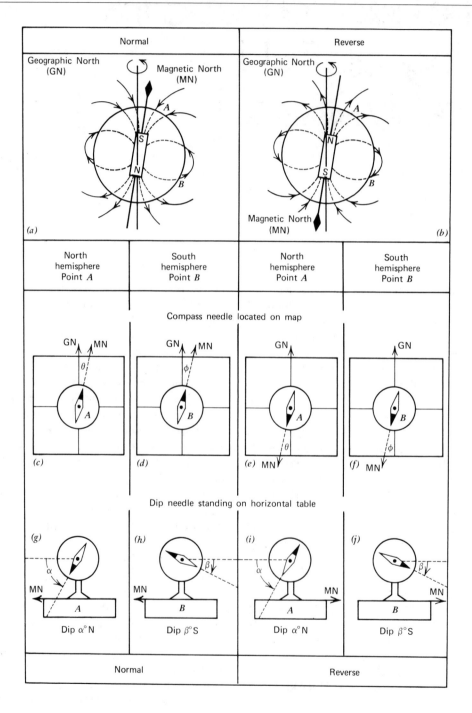

below the horizon through $\alpha°$ in a northerly direction (Figure 9-1g).

Now consider A with reversed polarity. The compass needle here points south, in a direction $\theta°$ from the longitude line that passes through the geographic south pole (Figure 9-1e). The dip needle is reversed compared with its position with normal polarity, but the geometrical positions of the lines of magnetic force in space at this point remain the same. Therefore the angle of dip is still $\alpha°$N, despite the fact that the direction of the magnetic force has changed (Figures 9-1g and i).

Consider now point B on the southern hemisphere. Figure 9-1 shows that the change from normal to reversed polarity involves no change in the position of the lines of magnetic force (Figure 9-1a and 9-1b), which means no change in the angle of dip $\beta°$S (Figure 9-1h and 9-1j). What does change is the direction of the magnetic force along the lines of force, as shown by the direction of the compass needle (Figure 9-1d and 9-1f), and of the dip needle (Figure 9-1h and 9-1j).

A complete reversal of the earth's magnetic field such as that illustrated in Figure 9-1 can probably be explained in terms of complex motions in the rotating, electrically conducting molten metal in the outer core, as we noted in Chapter 8, although no one has yet solved the problem of precisely how it happens.

During the first half of this century, only a few geologists and geophysicists were interested in the fossil magnetization preserved in some rocks. Geophysicists have collaborated with archaeologists and anthropologists to study the magnetic properties of pottery and fireplaces at sites occupied by primitive man. From measurements of the angles of declination and dip Figures 8-7, 9-1c, and 9-1g) they have been able to estimate the ages of the fireplaces and baked earthenware. The estimates depend on determining the variations in angles of declination and dip through several thousand years. We have not reviewed the relatively short term fluctuations that occur in these properties, nor will we consider them further because they are insignificant in the time scale of millions of years that concerns us in the story of sea-floor spreading and continental drift. Polarity reversals have proved to be changes that can be calibrated in terms of this much longer time scale.

CALIBRATION OF THE POLARITY REVERSAL TIME SCALE

The esoteric study of fossil magnetism aroused increasing interest during the 1950s when results were published and interpreted in terms of continental drift. Arising from the flurry of research activity that developed was the rather puzzling discovery that reversely magnetized rocks were quite common. It had been known since 1906 that some rocks were magnetized in precisely the reverse direction of the earth's present magnetic field, but it had generally been assumed that these were rarities. It was established in the 1950s that approximately half of the rocks studied were normally magnetized, and the other half were reversely magnetized.

The suggestion that reversely magnetized rocks might have been formed at a time when the earth's magnetic polarity was reversed compared with the present field was greeted with scepticism. The idea was really too radical for conservative scientists. Alternative explanations were explored, involving the properties of the magnetic iron-bearing minerals in rocks. It was discovered that some rocks would, indeed, become magnetized in a direction opposite to the applied magnetic field under special circumstances. But, as more and more reversely magnetized rocks were found, it became apparent that they could not all be explained by appealing to special circumstances. Several lines of evidence were adduced from magnetized rock masses to support the concept that the earth's magnetic field periodically reversed itself, as illustrated in Figure 9-1.

Many, probably most, geologists remained unimpressed or completely disinterested in the mounting evidence for reversals of the earth's magnetic field until it became clear that the reversals might provide a means for dating events. This prospect arose following the combination of paleomagnetic studies with another research topic that blossomed during the 1950s, that of geochronology, or dating rocks from their content of radioactive elements and their decay products (Chapter 1).

Three members of the U. S. Geological Survey set about the task of collecting lava specimens from all over the world and measuring for each specimen its date, or age of formation, and its direction of fossil magnetization. A. Cox, B. Dalrymple, and R. Doell obtained the kind of results shown in Figures 9-2 and 9-3. These results led to determination of the dates when the polar-

ity of the earth's field changed. Their first, preliminary polarity reversal time scale was published in 1963. More detailed polarity sequences with better defined reversal boundaries were published in the following years and, by 1966, the evidence was sufficient to persuade most scientists that periodic reversals of the earth's magnetic field are an essential feature of the earth's history.

There are many places in the world where volcanic activity has occurred through a period of millions of years. A volcano erupts for a time and then is dormant for a long period before it bursts into eruption again. Most volcanoes are thus built up of successive lava flows of progressively younger ages. Between eruptions the volcanoes are subject to erosion, and some of

Fig. 9-2 Schematic representation of the history and growth of a volcano, and its preservation of the ancient directions of the magnetic field. Stages of growth are shown by eruptions at three times during the past 2 million years *(d, c, b)*. Lavas from each of these eruptions become magnetized, as shown, and measurement of the direction of fossil magnetization and the age of each lava layer permits the construction of a polarity time scale, as shown on the left of the figure. See Figure 8-10.

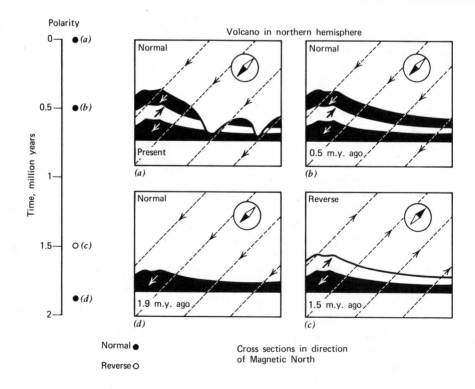

Fig. 9-3 The polarity reversal time scale shown at the right of the figure was determined by the combination of results such as those in Figure 9-2 from volcanic regions in many places. This is a simplified and schematic representation of actual measurements. The results from Figure 9-2 could represent data from Iceland, and they are so listed. The other columns represent the type of results obtained from other places. The range of error in the dating measurement for rocks of various ages is shown on the left. Compare with Figure 1-3.

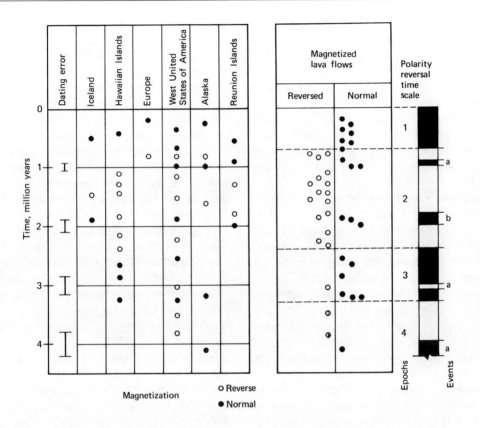

the rock is stripped off and removed through transportation by rivers and wind (Chapter 6). Stream valleys carved into the flanks of the volcano and the lava flows extending around it expose for the geologist samples of the older lava flows that had once been covered by the younger lava flows.

A geologist can therefore survey a volcanic region, map the sequences and extents of lava flows of various ages, and collect rock specimens from the various lava layers for detailed investigation in the laboratory. Figure 9-2 illustrates the type of history that can be unraveled by a combination of geological mapping, paleomagnetic measurements, and age determinations.

Figure 9-2a represents in schematic fashion a north-south cross section through a partially eroded volcano surrounded by lava flows. This could represent one of many similar eruption centers on Iceland, in the northern hemisphere. Geological mapping shows that there were three periods of eruption, represented by the three layers shown. Specimens from each layer were collected by a geologist and returned to the laboratory. There the dates of formation of the layers were found to be 1.9 million years ago, 1.5 million years ago, and 0.5 million years ago.

For each specimen the direction of fossil magnetization also was determined and recorded as a direction dipping below the horizontal or above the horizontal. This can only be done, of course, if the position of the horizontal plane was marked on each rock specimen before it was removed from its position in the lava flow. Careful orientation of rock samples with respect to present geographical coordinates is essential for all paleomagnetic studies. The direction of the fossil magnetization then gives the direction of the earth's magnetic field at the time of formation of the rock specimen, relative to the present position of the rock. For a rock in the northern hemisphere, a magnetization direction pointing down corresponds to a period of normal polarity (Figure 9-1g), and a magnetization direction pointing up corresponds to a period of reversed polarity (Figure 9-1i).

Figure 9-2a shows the ages and directions of magnetization of the three lava flows that were mapped. Figures 9-2d, 9-2c and 9-2b illustrate schematically the successive flows and the growth of the volcano, as deduced from the study. The first eruption, 1.9 million years ago, produced extensive lava flows and a small volcano. The arrow showing the direction of fossil magnetization indicates that the magnetic field 1.9 million years ago was normal, and the lines of magnetic force were therefore as shown in Figure 9-2d. The second period of eruption, 1.5 million years ago, covered the earlier rocks with new lava flows. The direction of magnetization of these lavas shows that the eruptions occurred during a period of reverse polarity. The fossil magnetism of the older lavas underneath the new flows remained unchanged by the polarity reversal that occurred at some time between 1.9 and 1.5 million years ago. The third eruption, 0.5 million years ago, occurred during a period of normal polarity and produced a third layer of lavas above the other two. Each cross section is keyed to the time scale on the left-hand side of Figure 9-2, and the polarity is represented by the circles, closed for normal, and open for reversed.

Figure 9-2 gives us three values for the polarity of the earth's magnetic field at times back to 1.9 million years ago, but the intermittent volcanic eruption leaves large gaps in the time scale. Application of similar methods to many other volcanic provinces, however, in many parts of the world, gradually filled in the gaps with determinations of normal or reversed polarity.

Figure 9-3 illustrates the type of results obtained for six different localities. The results from Figure 9-2 are shown in the first column. The collective results show a pattern of time periods with normal polarity separated by time periods with reversed polarity. On the right side of Figure 9-3 results for lavas of many ages from many places are plotted simply as normal or reversed, and these are used to locate the boundaries between periods of normal and reversed polarity. It was on the basis of this kind of evidence that Cox, Dalrymple, and Doell in 1964 concluded that the dated normal and reversed magnetization directions in lavas from various locations were the same throughout the world and, therefore, consistent with global reversals of the magnetic field, as illustrated in Figure 9-1.

Four main *polarity epochs* are distinguished with short *events* of opposite polarity within them. The polarity epochs and events were originally named after scientists who pioneered in magnetic research but, as the history of polarity reversals was extended back through time (see Figure 9-4) and the number of known events increased, this system of nomenclature became cumbersome. We also tend to run out of scientists' names. A simpler system numbers the epochs as 1, 2, 3, etc., back from the present. All even-numbered epochs are therefore of reversed polarity. Events within each epoch are designated a, b, c, etc., with increasing age.

Note that for the shorter polarity events in Figure 9-3, there appear to be simultaneous normal and reverse polarity represented by different lava specimens. This is explained by the uncertainty in the radioactive dating method. The accuracy of the only suitable dating method is ±5%. The possible range of error for individual measurements on rocks of various ages is shown in Figure 9-3. For a rock 2 million years old, this

Fig. 9-4 An estimate of the sequence of polarity reversals through the past 500 million years (Figure 1-3), incorporating results from Figures 9-3, 10-9*b*, and records of fossil magnetism in rocks of many ages from many continents. The evidence available indicates the occurrence of long polarity intervals during which the polarity underwent reversals, separated by intervals when the polarity remained constant.

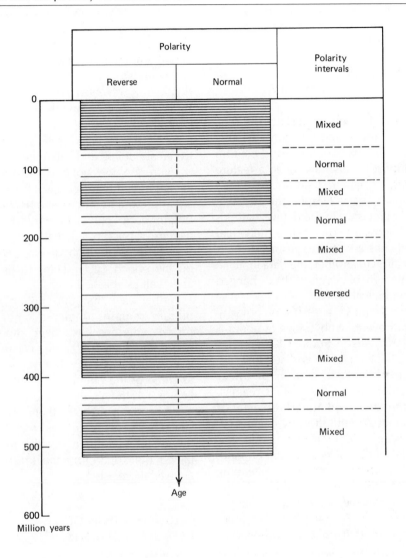

amounts to ±100,000 years. The short events last for only 100,000 years or so, and it is therefore impossible with the only dating technique available to distinguish the exact positions of some of the reversal boundaries. The occurrence of normally magnetized rocks with age near 2 million years shows that there is an event of normal polarity near 2 million years, despite the fact that the same time slot is apparently occupied by rocks from other locations with reversed polarity.

The polarity reversal time scale in Figure 9-3 has been calibrated by paleomagnetic studies of dated lavas from land. It extends back in time only to 4 million years. The possible error in dating a lava 4 million years old is as high as ±200,000 years. For a rock 10 million years old, the accuracy is ±0.5 million years, giving a possible range of 1 million years, which is longer than any of the polarity epochs shown in Figure 9-3. For older rocks there is therefore more and more overlap on the time scale, with apparently contemporaneous rocks having different polarities. It becomes impossible to distinguish the separate polarity epochs.

POLARITY REVERSALS
THROUGH GEOLOGICAL TIME

The recorded magnetic properties of lavas, other igneous rocks, and sediments of many ages from many parts of the world confirm that polarity reversals have been a feature of the earth's magnetic field for more than 1,000 million years. Despite the fact that insufficient precision in the dating methods makes it impossible to correlate individual polarity epochs and events from the rocks of one location to another, the data indicate that the pattern of reversals has changed through geological time.

Figure 9-3 shows a series of polarity epochs and events with cycles of about 1 million (10^6) years or about 100,000 (10^5) years. In Chapter 10, we will see that, according to the interpretation of linear magnetic anomalies in the oceanic regions, this pattern of mixed polarities extends back to about 70 million years (Figure 10-9b). This is consistent with the frequencies of polarities of the rocks collected on land. Figure 9-4 shows a time scale extending back to 600 million years. On this scale, the normal and reversed polarity epochs in the time interval 0 to 70 million years cannot be distinguished from each other very clearly. This mixed polarity interval thus appears as a striped area covering both normal and reversed polarities.

For the next 50 million years, from about 70 to 120 million years ago, the majority of rocks that have been studied have normal polarity, and this is represented in Figure 9-4 by a long, normal polarity interval with two reversed zones. Additional reversed rocks may be discovered but, statistically, there is negligible chance that this interval will turn out to have about 50% each of

normally and reversely magnetized rocks. The next interval back in time, however, from 120 to 150 million years, appears to be one of mixed polarity like the recent interval.

Analysis of worldwide data suggests that the pattern of alternating intervals through the past 600 million years is as shown in Figure 9-4. Intervals of mixed polarity alternate with intervals of dominantly normal or dominantly reversed polarity. The longest interval, which is well documented in United States, Russia, Europe, and Australia, is the reversed polarity interval between about 230 and 350 million years ago.

Figure 9-4 does not have the precision of the calibrated time scale of Figure 9-3. It is an approximate representation of insufficient data. Nevertheless, the general pattern of polarity intervals with cycles of 5×10^7 to 10^8 million years superimposed on the epoch cycle seems to be reasonably well established.

Further back than 600 million years the picture remains unclear. Sequences of mixed polarities have been reported for rocks older than 1000 million years, but the dating error of ± 50 million years is as long as some of the polarity intervals shown in Figure 9-4. In fact, the best estimate of true age is probably no better than ± 100 million years, because for such old rocks there are other factors influencing the results obtained, quite apart from the uncertainties in analytical precision.

MAGNETIZATION OF
DEEP-SEA SEDIMENTS

Volcanic activity is an intermittent process and the calibrated polarity reversal time scale is therefore pieced together by evidence from many places, as shown in Figures 9-2 and 9-3. The chance that a lava will be erupted and magnetized at the time of an actual reversal from one polarity to another is rather small. For most deep-sea sediments, on the other hand, the process of sedimentation is continuous. The sediments contain a small proportion of minute grains of magnetized iron-oxide minerals, probably fallen from dust carried over the oceans by the wind. These cause the sediments to become magnetized, and the sediments thus provide a continuous history of the earth's magnetic field through the time of reversal from one polarity to another. The problem is to obtain these sediments from beneath their cover of 5 km of ocean water in a condition suitable for magnetic studies.

Since the *Glomar Challenger* went to sea, many long cylindrical cores of oceanic sediments have been recovered by drilling. Before that, shorter cores through the sedimentary layers were recovered in cylindrical tubes that were lowered to the ocean floor in such a way that they penetrated the soft sediments. Penetration was accomplished either by the action of gravity alone or by an internal piston that was triggered when the tube reached the ocean floor. In this way, cross sections through the oceanic sediments as deep as 15 m or more were obtained.

Reversed magnetization in deep-sea sediments was first reported by C. G. A. Harrison and B. M. Funnell in 1964, about the same time that the polarity reversal time scale was being calibrated from measurements of lavas on land (Figure 9-3). In 1966, two publications by N. D. Opdyke and T. H. Foster, with four coauthors, described many sediment cores from the north Pacific Ocean and the Antarctic Ocean, each with a whole sequence of alternately normal and reversely magnetized layers. They were able to correlate these sequences from one core to another and with the calibrated polarity reversal time scale. The remarkable correspondence between the polarity reversal time scale based on radioactive dating methods for lavas on land, and the reversal scale as measured in centimeters or meters depth in deep-sea sedimentary cores, was one of the influential correlations that turned H. H. Hess's geopoetry of sea-floor spreading into a revolution in earth sciences.

Figure 9-5 outlines the procedures followed for the magnetic study of deep-sea sediments. A cross section through the

Fig. 9-5 Procedures for measuring polarity reversals in deep-sea sediments. *(a)* A drill core is recovered from the ocean floor. *(b)* The cylinder of sediment is cut into slices at intervals. *(c)* The direction of magnetization of the samples is measured. *(d)* Additional slices are cut between pairs of slices with different polarities, and the depth in the core where the polarity reversal occurs is thus defined.

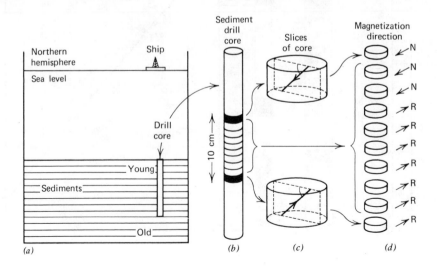

ocean and its floor of layered sediments in the northern hemisphere is represented schematically. A vertical core is collected from the sediments, giving a cross section through the layers. The core is a long cylinder, as shown in Figure 9-5b, with the youngest sediments at the top and the oldest at the bottom. Slices about 1 cm thick are cut from the core at intervals of about 10 cm, and the direction of magnetization for each slice is measured.

There is no way to identify the direction of geographic north on a sedimentary core. The original position of the sediments with respect to the geographic coordinates is lost during the process of cutting and recovery.

What can be measured is the angle of dip and whether the direction of magnetization points downward below the horizontal or upward above the horizontal. In the northern hemisphere, a direction of magnetization below the horizontal indicates normal polarity, and a direction pointing above the horizontal indicates reversed polarity Figures 9-1g and 9-1i).

Figure 9-5c shows a pair of slices with alternate polarities. For precise location of the boundary between the normal and reversely magnetized rocks, additional slices are cut and measured, with results shown in Figure 9-5d. In this way, the whole core is divided into a sequence of normal and re-

Fig. 9-6 Determination of deep-sea sedimentation rates from magnetized sediment cores. (a) Procedures shown in Figure 9-5 give the depths in the sedimentary core where reversals occur, a, b, c, d, etc. (b) The results are represented as shown. (c) Core polarity measured as depth interval is correlated with the polarity reversal time scale, which is calibrated as shown in Figure 9-3. For each core polarity reversal, the depth in the core is plotted against the calibrated reversal time. The line through the points shows that the rate of sedimentation was fairly constant for the past 2.5 million years.

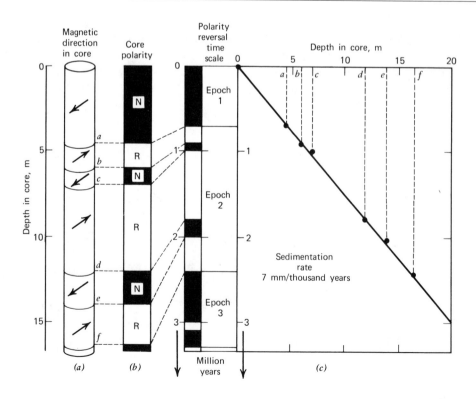

verse magnetization, as shown in Figure 9-6a. The sequence of magnetization in a sedimentary core, which is simply a cylinder cut through layers of sediments, is conventionally represented as in Figure 9-6b. Graphically, this is identical with the conventional method of representing the polarity time scale, as in Figures 9-3 and 9-6c, but the two should be clearly distinguished. One is a calibrated time scale and the other is a sequence of magnetized sediments measured in centimeters. The length scale can be correlated with the time scale, as we will see, despite the fact that there is no satisfactory way to date the sediments by methods involving radioactive elements.

The position of each reversal of polarity in a core, the points a, b, c, etc. in Figure 9-6, are assumed to correspond with the polarity reversals calibrated in Figure 9-3. Therefore, in Figure 9-6c, the depth of each polarity boundary in the core is plotted against the corresponding reversal time. The points lie close to a straight line. This line shows that 21 m of sediment were deposited in 3 million years, which gives an average sedimentation rate of 7 mm per thousand years.

Figure 9-7 compares similar results for four sediment cores collected from various oceans. Each core was analyzed as outlined in Figures 9-5 and 9-6, and the results are represented in Figure 9-7a in terms of the depth intervals of the cores, which are normally and reversely magnetized. The dashed lines connect depths in each core

Fig. 9-7 Comparison of deep-sea sedimentation rates from magnetized sediment cores. (a) Core polarities are determined for several cores as shown in Figures 9-5 and 9-6. (b) The depths of reversals are correlated from core to core and with the calibrated polarity reversal time scale, and points are plotted for each core as in Figure 9-6. A straight line for each core gives a different rate of sedimentation for each.

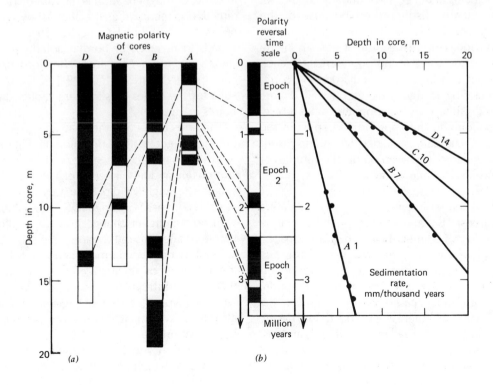

that correspond to the boundaries between successive polarity epochs. These depths are plotted against the calibrated polarity reversal time scale in Figure 9-7b, and the points lie close to four straight lines, each one giving the average sedimentation rate for the location where the core was obtained. Studies of this kind for deep-sea sediments from many oceans give sedimentation rates varying from millimeters to centimeters per thousand years.

Given results of the type shown in Figures 9-6 and 9-7, it appears that in many deep ocean environments the average sedimentation rates remain constant through a few million of years. The graphs can then be used in another way. For any particular level in a sediment core, the age of formation, or deposition, can be read directly from the straight line.

Notice in Figure 9-7a that, in addition to the longer sections of the sediment cores that correspond to the polarity epochs, there are shorter magnetized lengths that correspond to the polarity events. Figure 9-3 shows that in calibration of the polarity reversal time scale by dated lavas on land, the dates of the event boundaries could not be clearly distinguished because of the large range of possible error in the dating method. For each sediment core in Figure 9-7a, the dates of the boundaries for each event can be determined from the corresponding straight line graphed in Figure 9-7b simply by reading the dates corresponding to the appropriate depths in the core. The dates assigned to polarity event boundaries are based on detailed studies of · this kind for many cores.

The ability to use fossil magnetization of deep-sea sediments in this way for dating purposes is a major development in marine geology. Sediment layers in widely spaced regions can be correlated. Any fossil, any unusual kind of sediment, any layer of volcanic ash from a nearby volcanic eruption, or any other feature of the sediment that may be of interest can be dated. Many events are recorded in the minerals and fossils of deep-sea sediments. One event that can be traced in detail is a polarity transition.

THE MAGNETIC FIELD DURING A POLARITY TRANSITION

The transition from one polarity to another in a sediment core is completed within a sediment thickness of a centimeter or two, corresponding to a thousand years or so for sedimentation rates illustrated in Figure 9-7. A core from a location with fast sedimentation rate is selected for a detailed study of the history of the earth's magnetic transition from one polarity to another. Each thin slice of sediment studied then represents a smaller number of years than a slice of the same thickness from a core with slower sedimentation rate. This can be seen from the lines for sediment cores A and D in Figure 9-7b.

Figure 9-8a shows a portion of a sediment core including the boundary between a length of normal magnetization and one of reverse magnetization. This section of the core is divided into slices 1 cm thick, as illustrated in Figure 9-5d, each slice corresponding to a time interval of perhaps 1000 years. For each slice the direction and the intensity of magnetization is measured. The direction is recorded as the angle above or below the horizontal plane, which corresponds to reversed or normal polarity, respectively, for a core from the northern hemisphere (Figures 9-1i and 9-1g). The results for the series of slices are plotted as points in Figure 9-8.

The line connecting the measured points for the magnetization direction in Figure 9-8b shows that the magnetic field reverses completely in a time interval of about 2000 years. The angle of dip decreases to zero and then increases again to its former value, with no significant change in angle of dip before or after this rather rapid reversal.

The line connecting measured points for the intensity of magnetization in Figure 9-8c is more irregular. The property varies with the type of sediment present, but it also shows systematic changes near a polarity transition boundary. The intensity of magnetization decreases for samples taken from a core length corresponding to a time of about 10,000 years on either side of the boundary indicated by the change in direction of the field. The precise correspondence of decreased intensity and of change in magnetization direction is apparent in Figures 9-8b and 9-8c. The intensity of magnetization of the sediments provides a measure of the magnetizing field at the time the sediments were deposited.

Data of the kind illustrated in Figure 9-8 measured for many polarity transition boundaries in many sediment cores gives the general picture shown in Figure 9-9. At a given point on the earth's surface, the magnetic field intensity remains fairly constant, with small increases and decreases in a cycle of about 10^4 years. In a period of 10,000 years before a polarity reversal the field intensity decreases by 60 to 80% with no change in the angle of dip. Then, within a period of 1000 to 2000 years, the field intensity decreases toward zero and returns to 20 to 40% of its maximum strength in the opposite direction, while the angle of dip decreases to zero and then increases to its

former value (but with the directional arrow reversed). The polarity transition is completed during the next 10,000 years as the magnetic field intensity builds up to its former maximum value. Note that the changes plotted in Figures 9-8 and 9-9 would appear much simpler if plotted on the time scale of Figure 9-7 or 9-3. The time interval of 20,000 years for the occurrence of a complete polarity transition would appear to be almost instantaneous on this time scale.

The dashed line in Figure 9-9 shows uncertainty about precisely what happens during the reversal. There is evidence from the magnetic properties of rocks in a large intrusive igneous mass (Figure 6-4) near Mount Rainier, which probably took about 1 million years to crystallize from a molten state. This indicates that the magnetic field intensity is reduced by a factor of about ten immediately before the reversal, the magnetic poles then swing right across the globe to their new positions, and the former

Fig. 9-8 Detailed results for polarity reversal from a portion of core C in Figure 9-7. Results of magnetization direction and intensity of magnetization are plotted against depth in core for a series of closely spaced sediment slices, as in Figure 9-5d.

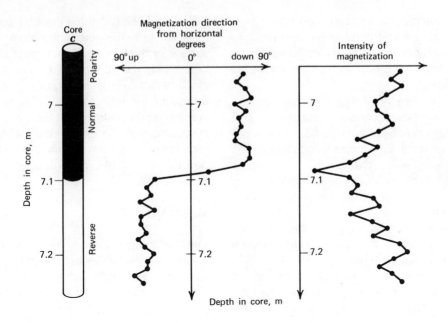

Depth in core, m

Fig. 9-9 Changes in intensity of magnetic field during the period of 40,000 years including the polarity reversal between epochs 1 and 2 (Figure 9-7), based on results for core C given in Figure 9-8.

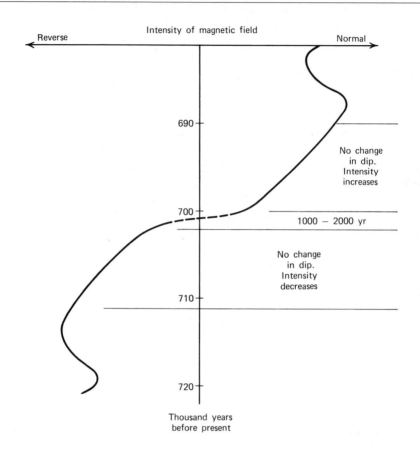

field intensity is restored. The whole process takes about 2000 years, according to this study by J. R. Dunn, M. Fuller, H. Ito and V. A. Schmidt of the University of Pittsburgh.

It is known that the earth's magnetic field intensity has decreased since 1835, at a uniform rate, by about 6%. If the rate of decrease were to continue unchanged, the magnetic field would be reduced to zero by about 2000 years from now. This change is considered to be part of the cyclic variation shown in Figure 9-9 for the period of maximum field intensity, and there is only a 5% probability that the present decrease will continue to zero.

There has been much speculation about what happens during the interval when the earth's magnetic field is of very low inten-

sity. It has been suggested that charged particles fall from the outer atmosphere to the surface of the earth and that more cosmic rays reach the surface. It has also been claimed that these effects could cause transformations or extinctions of living species. Some scientists believe that the evidence is mounting that the earth's magnetic field may have played an important role in the development of life. We will return to this topic in Chapter 15.

SUMMARY

The fossil magnetization directions of rocks may correspond either to the earth's present magnetic field or they may be reversed. It has been established that this is because the earth's magnetic field has reversed polarity completely at intervals through at least the past 1,000 million years of geological history.

The polarity reversal time scale has been calibrated back to 4.5 million years ago by measuring the age and magnetization direction of volcanic lavas on land. The results reveal a sequence of polarity reversal epochs at intervals of less than 1 million years, with shorter events of opposite polarity within them. For older rocks, the error in dating becomes larger than the polarity reversal intervals.

The pattern of frequent reversals extends back to about 70 million years but, from 70 to 120 million years ago, most rocks have normal polarity. For at least 600 million years, intervals of mixed polarity have alternated with intervals of dominantly normal or reversed polarity.

Long cores recovered from deep-sea sediments provide a continuous history of the earth's magnetic field. The age of the sediments at each depth where the direction of magnetization is reversed is obtained from the calibrated polarity reversal time scale. Rates of sedimentation can be calculated from sediment thicknesses and the magnetic dates, or time intervals. Similarly, any event recorded in a sediment layer can be dated. Measurement of intensity of magnetization in samples across a reversal boundary shows that for 10,000 years before a reversal the magnetic field strength decreases by 60 to 80%, then reverses completely in about 2000 years before increasing again in the opposite direction during the next 10,000 years.

SUGGESTED READINGS are at the end of the book.

chapter 10

Magnetic Anomalies and the Sea-floor Spreading Model

In Chapter 8 we reviewed the effect of magnetized rocks in the earth's crust on the magnetic field intensity. The presence of magnetized rocks either increases or decreases the earth's magnetic field intensity compared with the average values that might be expected if no magnetized rocks were present. The deviations from the average values are called magnetic anomalies, and they can be illustrated in magnetic field intensity maps or magnetic anomaly maps, or in cross sections through either, as illustrated in Figures 8-12 and 8-13.

Magnetic anomalies measured on continents vary in shape and size and, in the vicinity of some rock types, the anomalies may be very large. Those measured at sea level in oceanic regions, in contrast, are very small. At an influential symposium on "The Crust of the Earth" held at Columbia University in 1954, M. Ewing and F. Press stated that: "Magnetic measurements are as yet meager, but those available indicate that over large oceanic areas the magnetic field is unusually smooth." There did not seem to be much scientific future for magnetic surveys over the deep oceans.

Just four years later, however, it was discovered that, in the northeast Pacific Ocean, the very small magnetic anomalies were distributed in remarkable linear patterns, like a series of stripes. An interpretation for these in terms of polarity reversals and sea-floor spreading, proposed by F. J. Vine and D. H. Matthews in 1963, gave to Hess's geopoetry of sea-floor spreading the respectability of specific numbers and rates of spreading. The convection cell and conveyor belt of the sea-floor spreading model were joined by a magnetic tape recorder.

General acceptance of this model in the winter of 1966 and 1967 was the turning point in the revolution in earth sciences. In this chapter we review the model, how it works, and how it was extrapolated. In Chapter 13, we will examine the successive events that led to proclamation of revolution and then take a closer, more critical look at the magnetic anomalies that played such a significant role in persuading geologists and geophysicists that the sea floor had spread, that the plates had moved, and that the continents had drifted with them.

DISCOVERY OF MAGNETIC STRIPES

Ever since the Lamont Geological Observatory of Columbia University was founded in 1949, the oceanographic research vessels sailing from its dock have towed behind them a special instrument that measures continuously the intensity of the earth's magnetic field. A series of long magnetic profiles across the oceans has been obtained in this way. By 1957, nearly 20 crossings had been made over the mid-Atlantic ridge. These revealed a characteristic pattern of magnetic anomalies associated with the submarine ridge crest.

Figure 10-1a shows the map of magnetic field intensity for the Atlantic Ocean and adjacent continents, transferred from Figure 8-9b. A cross section through the magnetic field at a latitude near 50°N is illustrated in Figure 10-1b. The Atlantic Ocean there is about 3000 km wide. The diagram shows the trend of the average regional magnetic field along this line of latitude, decreasing gradually from about 55,000 γ in the west to about 45,000 γ in the east. The profile shows a remarkably smooth variation in magnetic intensity. In fact, there are minor deviations from the average regional magnetic field along its whole length, but these anomalies can barely be distinguished from the thickness

of the line in Figure 10-1b, except for those occurring directly above the mid-Atlantic ridge.

Figure 10-2a shows the magnetic field for the central portion of the profile in Figure 10-1b, with a greatly expanded vertical scale. It extends across 200 km on either side of the ridge crest. On this diagram, the small variations in the actual magnetic field intensity recorded by the instrument towed behind the cruising research vessel are compared with the dashed line representing the average regional magnetic field.

The magnetic anomaly profile in Figure 10-2b was constructed from the magnetic intensity profile of Figure 10-2a in the manner outlined in Chapter 8 (Figure 8-12) by subtracting the value of the average regional magnetic field at each point from the actual measured intensity. The magnetic anomaly pattern is characteristic for most crossings of the mid-Atlantic ridge. There is a large positive anomaly of more than 500 γ with a width of about 50 km directly over the ridge crest and a pair of negative anomalies of 300 to 500 γ over the adjacent parts of the ridge. Further away from the ridge crest the

Fig. 10-1 (a) Portion of Figure 8-9b showing lines for constant values of magnetic field intensity. (b) Graph showing the magnetic field intensity measured at sea level along line *AB* across the Atlantic Ocean. The intensity decreases regularly from west to east along this line and, plotted on this scale, there are few anomalies visible. The shaded area is enlarged in Figure 10-2.

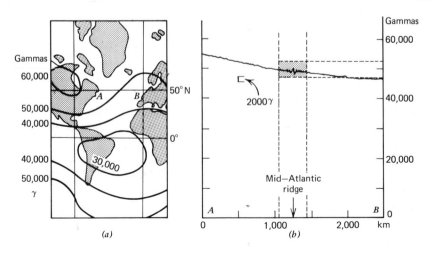

Fig. 10-2 *(a)* Enlargement of the shaded portion of Figure 10-1*b*. The dashed line shows the averaged regional magnetic intensity trend, a part of the general variation seen in Figure 10-1*b*. *(b)* Magnetic anomaly profile across the mid-Atlantic ridge, obtained by subtracting the average values in part *a* from the measured values.

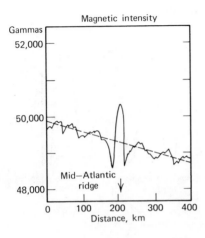
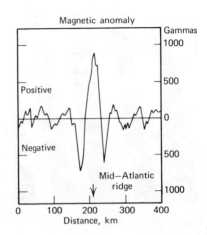

anomalies decrease in size to about 100 or 200 γ, with variable widths between 10 and 30 km.

The isolated, long magnetic profiles like that in Figures 10-1 and 10-2 provided no basis for quantitative examination or interpretation of the oceanic magnetic anomalies. The persistence of the central positive anomaly on all crossings of the ridge suggests that the anomaly is continuous along the ridge crest, but there was nothing to suggest that any of the flanking anomalies seen in different profiles could be correlated with each other. Indeed, there appears to have been no proposal that they were related in any way until after the discovery of the remarkable system of linear magnetic anomalies in the northeastern Pacific Ocean, which was reported in 1958.

Between 1955 and 1960, research vessels from the U.S. Coast and Geodetic Survey and from the Scripps Oceanographic Institution at the University of California, San Diego measured many magnetic profiles in an extensive area of the Pacific Ocean off the coast of North America. The profiles were spaced closely enough so that the magnetic intensity could be mapped with contour lines of equal intensity.

The regional magnetic field variation was determined by averaging the measured values over selected areas. At any point, subtraction of the average regional value from the measured intensity value gave the magnetic anomaly and, from these values, maps were constructed with contours for equal magnetic anomalies, as in Figure 8-13*a*. Simplified maps showed only the areas of positive anomaly (in black) and negative anomaly (in white). The results obtained are illustrated in Figure 10-3 (compare Figures 8-13 and 8-12).

The magnetic survey revealed a pattern of narrow, parallel anomalies, each about 30 km wide, extending approximately north-south for about 1000 km. The pattern is illustrated in Figure 10-3*a*. The total field intensity in this region ranges from about 45,000 γ to 55,000 γ (Figure 8-9*b*). The magnetic anomaly profile in Figure 10-3*b* shows positive and negative anomalies of about 400 γ, which is not quite 1% of the earth's field in this region. The remarkable regularity of this pattern of magnetic stripes points to a simple cause but, although many ideas were generated, no satisfactory explanation was forthcoming at this time.

Figures 8-12 and 8-13 illustrate one type of explanation proposed. Recent lava flows

Fig. 10-3 *(a)* Map illustrating the linear magnetic anomalies discovered in the Pacific Ocean, plotted as shown in Figure 8-13*b*. Details along line *AB* are shown by the magnetic anomaly profile in part *b*.

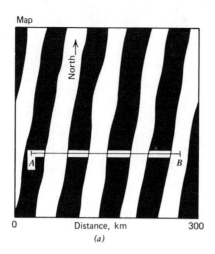

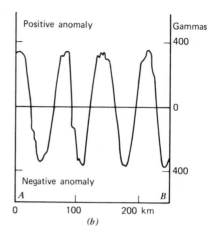

(a)   *(b)*

on the ocean floor would be magnetized with normal polarity, and these could produce positive magnetic anomalies. It was argued that the positive anomalies in Figure 10-3*a* were caused by magnetized lava flows occupying the valleys between a series of submarine, parallel ridges. The magnetic field above the ridges then corresponds to a negative anomaly, simply because of the averaging method used to obtain the regional field.

One reason for doubting this explanation is that the occurrence of a series of submarine ridges and valleys with the regularity implied by the anomaly pattern in Figure 10-3*a* is just as difficult to explain as the anomalies themselves. In fact, it has been clearly established that the magnetic anomaly pattern shows little correlation with the sea-floor topography in this region. It was not until 1963 that a satisfactory explanation was proposed.

THE LITHOSPHERE CONVEYOR BELT AND THE MAGNETIC TAPE RECORDER

In 1963, when F. J. Vine was a graduate student at Cambridge University, England, he sailed with a British oceanographic vessel

participating in the ·International Indian Ocean Expedition, and learned first hand about the existence of magnetic anomalies. Vine and his research supervisor, D. H. Matthews, published a three-page paper in *Nature* in 1963 with a novel explanation for the magnetic stripes. Very few scientists paid much attention to yet another attempt to explain the peculiar magnetic anomalies, for reasons that we will review below. Three years later, however, it was recognized that this short paper represented a most significant step in the formulation of the new global theories in earth sciences.

Figure 10-4 shows the first part of the Vine-Matthews proposal. They reasoned that the linear magnetic anomalies were caused by strips of the oceanic crust magnetized alternately in opposite directions, as shown in Figure 10-4. One set of crustal strips would reinforce the earth's magnetic field producing positive anomalies; the alternate set would decrease the field producing negative anomalies. The anomaly boundaries, as indicated in Figure 10-4, would correspond to the boundaries between the strips of magnetized rocks.

In order to explain the existence of ocean crust magnetized in alternating directions, Vine and Matthews appealed to reversals of the earth's magnetic field. Rocks magnetized during a period of normal polarity would be magnetized in the same direction as the present magnetic field; rocks magnetized during a period of reversed polarity would be magnetized in the opposite direction to the present magnetic field. Figure 9-2 shows that a series of alternately magnetized lavas can be produced one above the other in a growing volcano, but it is harder to envisage a mechanism for producing long, narrow strips of alternately magnetized lava side by side. Harder, that is, unless you happen to be receptive to geopoetry; in 1963 many geologists had not even heard of the geopoetry of sea-floor spreading.

Vine and Matthews combined the linear anomalies, polarity reversals, and sea-floor spreading into the following model. Let us assume that an area of the ocean floor had remained undisturbed by sea-floor spreading for a long period. Intermittent submarine lava flows would be magnetized but, because the polarity reversed from time to time, the lavas would be magnetized in alternate directions, and their effects would more or less cancel each other out. Through a long period of time, we would not expect any organized pattern of magnetization or of magnetic anomalies. The influence of the oceanic crust on the average magnetic field intensity measured at sea level would be slight.

Let us next assume, quite arbitrarily, that sea-floor spreading began 3 million years ago, producing an ocean ridge. This requires a convective upcurrent in the mantle. Figure 10-5a shows the effect produced during a portion of normal polarity epoch 3, between 3 and 2.5 million years ago (Figure 9-3). The diagram shows a schema-

Fig. 10-4 Interpretation of the linear magnetic anomalies in Figure 10-3a, illustrated here by the magnetic anomaly profile corresponding to Figure 10-3b, in terms of strips of ocean floor lavas magnetized in opposite directions. The magnetized rock strips produce the magnetic anomaly stripes. Vine and Matthews suggested that this alternating magnetization direction was caused by polarity reversals and sea-floor spreading.

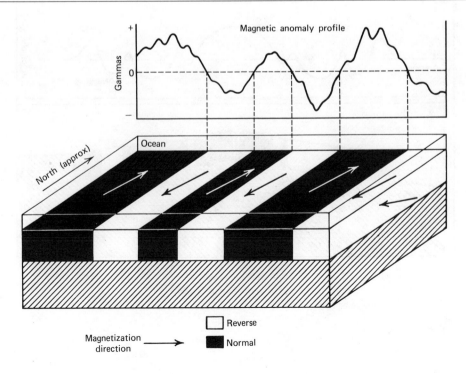

Fig. 10-5 Vine and Matthews' magnetic tape recorder, in stereo. Schematic representation of the process whereby sea-floor spreading and polarity reversals produce a series of magnetized lava strips parallel to an ocean ridge crest which, in turn, produce symmetrical magnetic anomalies in stripes parallel to the ridge crest. Compare Figures 10-4 and 10-3.

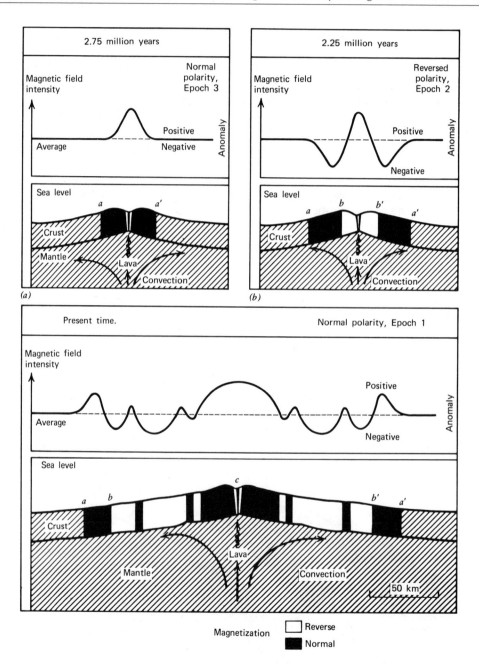

tic cross section through the crust and mantle at an ocean ridge, with convection in the mantle producing lava that generates new ocean crust as the conveyor-belt lithosphere plates move away from the ridge crest. At time 3 million years ago, the magnetic field intensity profile above the cross section, as plotted in the upper part of the diagram, would be approximate to the average regional field value, without significant anomalies. Points a and a' would have been coincident at the ridge crest.

During the period between 3 and 2.5 million years ago, the lava of the new ocean crust became magnetized in the direction of the earth's normal magnetic field as it cooled near the crest of the ridge. The magnetized lava was then carried away from the ridge crest, away from the region where additional new lava was being added to the lithosphere plate. At time 2.75 million years ago, as illustrated in Figure 10-5a, the plates had separated by a distance a-a', and new crust had formed between a and a'. This crust was magnetized in a normal direction, so it caused a positive anomaly in the magnetic profile illustrated in the upper part of the diagram.

Reversed polarity epoch 2 began 2.5 million years ago (Figure 9-3). Continued spreading for 0.25 million years produced the situation in Figure 10-5b, with the block of new crust b-b' magnetized in the opposite direction to the original block a-a'. This block had now become separated into two blocks, a-b and b'-a', which produced negative anomalies in the earth's reversed magnetic field. The block of crust b-b', which was magnetized during the reversed polarity epoch, produced a positive anomaly above it. When the earth's polarity changed again, the positive anomalies changed to negative, and vice versa.

In this way, the complete sequence of polarity reversals (Figure 9-3) becomes imprinted on the oceanic crust as if it were a tape recorder. Figure 10-5c is a cross section of the sequence of magnetized strips resulting from the polarity reversals of the past 3 million years. If we visualize this in

three dimensions, we see that the cross section of magnetized crust corresponds to the sequence of alternately magnetized strips depicted in Figure 10-4. The alternating directions of magnetization of the rocks produce the sequence of positive and negative anomalies, as depicted by the anomaly profiles in Figures 10-5c and 10-4 and, in three dimensions, these correspond to a series of linear magnetic anomalies, as shown in Figure 10-3a. Be sure to distinguish between the magnetized strips in Figure 10-4 and the magnetic anomaly stripes in Figure 10-3a.

Why is the central anomaly so much larger than those further away from the ridge crest (Figure 10-2)? The observation shows that the intensity of magnetization of the crust is greatest at the ridge crest and less further away. Volcanic activity occurring at places away from the ridge crest heats the crust and destroys part of its magnetism. It also adds new rock that has a 50% chance of being magnetized in the direction opposite to the crust, which would reduce the magnetic anomaly. Seawater percolating downward into the crust and solutions given off by rising lavas can react with the minerals; if the magnetic minerals are changed by reaction, the rocks lose magnetization. In fact, when the crust moves away from the ridge crest where it was formed, almost anything that happens to it causes a decrease in its intensity of magnetization.

Figures 10-5c and 10-3a illustrate several essential features of the Vine-Matthews model: (1) the anomaly profile is symmetrical about the ridge spreading center; the polarity reversals are recorded on the conveyor belt in stereo; (2) the linear anomalies on a map are parallel to the spreading center; and (3) the size and shape of the individual anomalies vary according to the width of magnetized blocks in the crust below, which relates, in turn, to the duration of the polarity interval during which the block was magnetized as it cooled near the ocean ridge crest; these shapes should permit identification and correlation of specific anomalies in different parts of the world. In fact, easily identifiable anomalies have been numbered for reference purposes and for correlation; the numbered anomalies, starting with number 1 at the ridge crest, are said to have characteristic magnetic signatures. Thirty anomalies have been numbered in this way, with estimated ages ranging to 70 million years on the extrapolated time scale given in Figure 10-9*b*. Some of these are shown in Figure 1-3 on the geomagnetic time scale.

Figure 10-5*c* is a schematic, idealized model. Figure 10-6 shows selected magnetic profiles across ocean ridges. These include some of the best examples that have been cited as compelling proof for the sea-floor spreading model. For each profile, the horizontal scale is given in kilometers and the vertical scale is given in gammas. Notice that the scales have been changed from one profile to another to emphasize the correlations.

In order to illustrate the degree of symmetry of the anomalies about the ridge crest, each profile is printed in two ways. The lower profile is reversed compared with the upper profile. If symmetry were perfect, each pair of profiles would appear identical, like mirror images. In some profiles, the Reykjanes Ridge for example, the symmetry is striking but, in some of the others, it is less obvious to the untutored eye. In many other profiles, as we will see in Chapter 13 (Figure 13-1), it is difficult to detect any symmetry.

Figure 10-6 also illustrates the magnetic signatures of anomalies numbered 1 through 5. These anomalies have shapes that supposedly make them easily identifiable in oceans throughout the world. The dashed lines show the correlations that have been made on the basis of counting the numbers of anomalies from the ridge crest and identifying the characteristic signatures. Some of these anomalies do show remarkably similar shapes but, for others, the similarity is less obvious. Additional examples of correlations of numbered anomalies are shown in Figure 13-2.

MAGNETIZATION OF OCEANIC SEDIMENTS

Figure 10-5 shows schematically the magnetization of the lavas that constitute the bulk of the ocean crust, but the sediments deposited above the lavas as the lithosphere plate moves away from the ridge boundary are not even represented. The sedimentary layer varies in thickness from 500 m to 1 km in the deep oceans, but it is much thinner near the ocean ridges, where new crust is constantly generated and only recently exposed to sedimentation.

Figure 10-7*a* shows a schematic cross section through the oceanic crust in a region where there is no sea-floor spreading and where the sedimentary layer overlies crust with no organized pattern of magnetization. Sediments accumulate on the ocean floor very slowly, becoming magnetized as they settle down, with the direction of magnetization reversing every time the earth's polarity reverses, as shown in Figure 9-7. This produces horizontal layers of alternately magnetized sediments, like a series of blankets on the ocean floor. The thickness of each blanket is measured in centimeters or meters.

The effect of a single magnetized sedimentary blanket of this sort on the magnetic field intensity recorded at sea level would be very slight, because the magnetization is weak and the layer is thin. If the effect could be detected at all, it would not produce a magnetic anomaly, it would simply increase or decrease the field intensity uniformly over the total width of the ocean floor receiving sedimentary deposits. The effects of a series of alternately magnetized sedimentary blankets would certainly be negligible, because they would cancel each other out.

Figure 10-7*b* illustrates how the simple sedimentation picture is modified by sea-floor spreading. The slope of the sea floor is exaggerated in order to show schematically the sedimentary layers. The crust is divided into blocks of normally and reversely mag-

Fig. 10-6 Examples of magnetic profiles measured across oceanic ridges, illustrating the degree of symmetry across the ridges, and the correlations of specific anomalies from one ridge to another. *(a)* North Pacific. *(b)* Pacific-Antarctic. *(c)* South Atlantic. *(d)* Reykjanes. (Examples selected for comparison by A. A. Meyerhoff and H. A. Meyerhoff, 1972, *Amer. Assoc. Petroleum Geologists Bull., 56,* 337-359.)

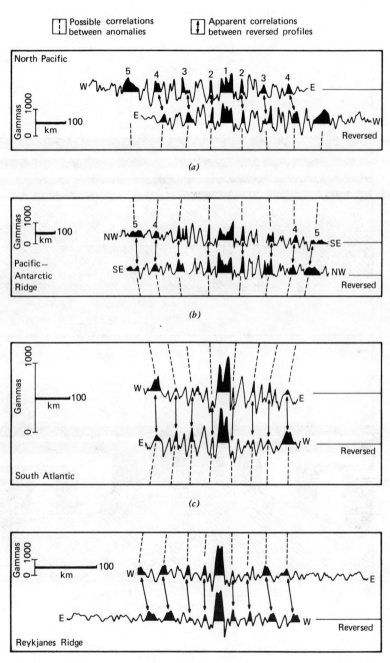

Fig. 10-7 Effect of polarity reversals on magnetization of deep-sea sediments *(a)* with no sea-floor spreading, a series of sedimentary blankets with alternating polarity is deposited above the crust, and *(b)* with sea-floor spreading, each magnetized sedimentary layer over-lies the older sediments and rests on a portion of the lava crust that was generated at the ridge crest during the same time interval, and subsequently migrated away from the ridge.

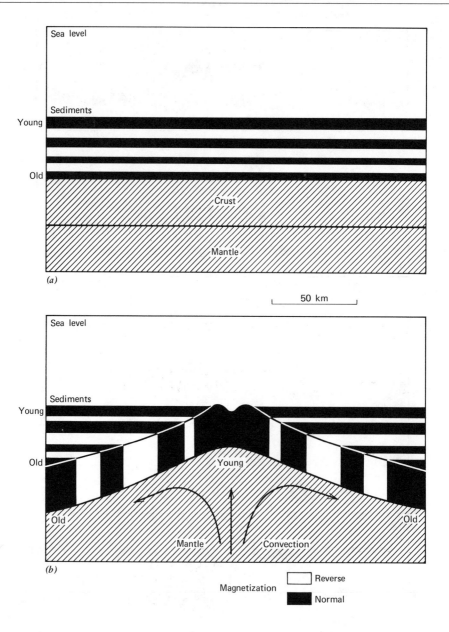

netized rock, with the youngest rock at the ridge crest. At the edges of the cross section, we see a sequence of magnetized sedimentary blankets overlaying the crust, similar to that in Figure 10-7a.

Consider the processes operating during a period of normal polarity. New ocean crust is being generated at the ridge crest with normal magnetization. Sediments are accumulating directly on this new crust and on top of the sedimentary layer on either side. The sediments deposited during the time interval for the formation of the central normally magnetized block are also normally magnetized.

During the next reversed polarity epoch, a younger sedimentary layer with reversed magnetization will be deposited above the normal layer and directly onto the new reversely magnetized section of crust generated from lava at the ridge crest. In this way, the pattern illustrated in Figure 10-7b is built up. Each magnetized sedimentary blanket extends from a section of the ocean crust that is of the same age interval and that has the same direction of magnetization. The total thickness of the sediments increases regularly with distance from the ridge crest.

The polarity reversal time scale is represented in the sediments in a vertical direction from the top downward and, in the crust generated from lavas, it is represented in a horizontal direction from the ridge crest outward. The magnetic anomalies recorded at sea level are caused by the lava blocks, with negligible contribution from the sediments.

RATES OF SEA-FLOOR SPREADING

If the observed sequences of linear anomalies associated with midoceanic ridges can be correlated with the polarity reversal time scale, we have a means for calculating rates of lateral movement during sea-floor spreading. The average velocity of movement is given by the distance of a specific anomaly boundary from the ridge crest divided by the age of the correlated

polarity epoch boundary. This age is the time taken for that part of the crust to be conveyed to its present position from the ridge crest where it was generated. The spreading rate determined in this way for a single plate is only one half of the separation rate, because the two plates are moving away from each other.

Figure 10-8 shows how spreading rates can be determined graphically from the magnetic anomalies. From Figure 10-5c, we can measure the distance from the ridge crest to each anomaly boundary: the distances are 30, 38, 45, 73, 80, 95, and 115 km. Each distance corresponds to a change in the polarity of magnetization of the underlying oceanic crust, and this correlates with the polarity reversal time scale in Figure 9-3. Therefore, in Figure 10-8a, for each distance a point is plotted at the corresponding age. An additional point, c, is 0 km at the ridge crest, where the crust was formed very recently at a time indistinguishable from 0 years on this scale.

The average rate of spreading is given by the best straight line that can be drawn through the plotted points. But, as shown in the figure, the points do not fit well onto a single straight line; they do fit well onto a line with two straight segments, with the change in slope occurring at d, a distance corresponding to an age of 1.1 million years. The interpretation of a result like this is that the spreading rate changed 1.1 million years ago.

Line cd represents 50 km of crust generated in 1.1 million years; point d moved through 50 km from ridge crest c, where it was generated, in 1.1 million years. This gives an average spreading rate of (50/1.1)km/ million years, or 4.5 cm/year. Similarly, the slope of line da shows that the crust at a moved (115 − 50) km from position d in (3 − 1.1) million years, which gives an average spreading rate of 3.4 cm/year.

Figure 10-8 b compares the spreading rates determined in this way from anomalies measured in the Pacific, the South Atlantic, and the North Atlantic Oceans. Each set of points is readily fitted

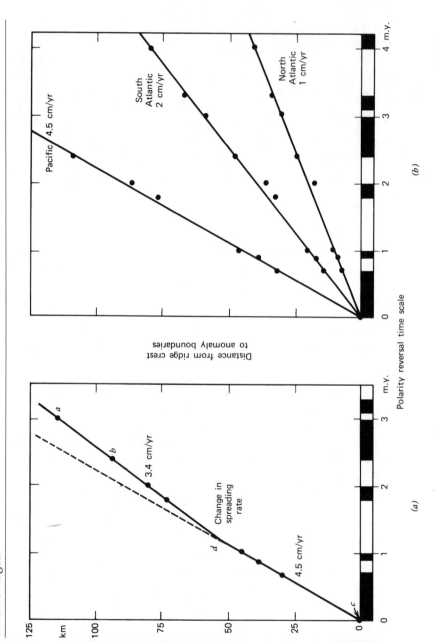

Fig. 10-8 Rates of sea-floor spreading estimated by plotting the distances of anomaly boundaries from the ridge crest against the calibrated ages of each polarity reversal that is correlated with each anomaly boundary. (a) Graph produced from anomalies in Figure 10-5c. There are two linear portions, indicating a change in spreading rate about 50 million years ago. Average spreading rate is given by the slope of the lines, distance per time interval. (b) Different spreading rates have been determined for different portions of the oceanic ridges.

to a straight line, as shown, and the slopes of the lines give average spreading rates of 4.5, 2, and 1 cm/year, respectively. Graphs like this for most ocean ridges give a linear relationship, indicating constant spreading rates from specific ridges for the past 4 million years. Examples with changes in spreading rate like the arbitrary example in Figure 10-8a seem to occur only where there are complications in the simple sea-floor spreading model, such as overriding of an ocean ridge by a continent.

We noted in connection with Figure 9-3 that the ages of polarity event boundaries could not be precisely established by dating lavas on land because of the range of error in the measurements. We also noted in connection with Figure 9-7 that the dates assigned to polarity events could be based on detailed studies of magnetized ocean sediment cores. Figure 10-8b shows that spreading rate graphs can also be used to adjust the polarity reversal time scale. The polarity event 2b near 2 million years ago gives two points that lie just below each straight line. If these points, for each location, were shifted down in time by a few thousand years, they would fit perfectly. Vine modified the polarity reversal time scale slightly by making minor adjustments of this sort.

Figure 10-8b shows that the spreading rate in the North Atlantic Ocean is slower than that in the South Atlantic Ocean, despite the fact that the mid-Atlantic ridge appears to be a single divergent plate boundary (Figure 5-1). This is consistent with the theory of plate tectonics.

Figure 5-7 shows how the separation of two plates is related to a pole of spreading. For plates separating by rotation about a fixed pole of rotation with constant angular velocity, Figures 5-6 and 5-7 show that the rate of spreading, measured as linear distance across the surface, should increase with distance from the pole, at least for one quarter of the circumference of the sphere. Results obtained for rates of sea-floor spreading by using the magnetic anomalies as in Figure 10-8b have confirmed that the rates along some ridge crests do vary ap-

proximately as expected according to the requirements of plate tectonics. Indeed, the variation in rates of spreading along a particular boundary has been used to estimate the position of the pole of spreading for a pair of plates.

EXTRAPOLATION OF THE POLARITY REVERSAL TIME SCALE

Figure 10-8b illustrates the kind of results obtained from linear anomalies in three oceans, which covers the time interval corresponding to the calibrated polarity reversal time scale, back to 4 million years. The distances from ridge crests to the oldest dated anomalies range from about 40 to 180 km. The pattern of magnetic anomalies extends much farther than this, and almost completely from ridge to continental slope in some oceans. If we assume that the spreading rates have remained reasonably constant, which appears to have been true for the past 4 million years (Figure 10-8b), then the spacing of the linear anomalies from the ridge axis can be used to extrapolate the polarity reversal time scale beyond the 4 million year limit imposed by the error in the rock-dating method.

One way to use a graph is shown in Figure 10-8b. Data points are plotted, a line is drawn through them, and a value for the spreading rate is determined from the line. Figure 10-9a shows a different use for a graph, with extrapolation of the data. We transfer the line for the South Atlantic Ocean from Figure 10-8b, assume that the spreading rate remained constant, extrapolate the line with constant slope through the total distance within which linear magnetic anomalies are recorded in the South Atlantic, and plot the distances of anomaly boundaries on the line. Each distance corresponds to a date that, according to the Vine-Matthews model, is the date of a polarity reversal.

In 1968, J. R. Heirtzler and his associates from Lamont Geological Observatory published the results of a detailed comparison and correlation of the linear magnetic

Fig. 10-9 Extrapolation of the polarity reversal time scale. *(a)* The spreading curve for the South Atlantic Ocean is reproduced from Figure 10-8*b* near the origin of the graph. Magnetic anomalies can be measured as far as 1500 km from the ridge crest in this region and the axes of the graph, distance and time, have been extended accordingly. Making the assumption that the spreading rate has remained constant, the spreading rate curve has been extrapolated as the dashed straight line. This line is then used to assign a date to the polarity reversal corresponding to each anomaly boundary out to 1500 km from the ridge crest. *(b)* The dates assigned to each anomaly boundary provide the extrapolated polarity reversal time scale; compare Figures 9-3, 9-4, and 1-3. *(c)* Using the extrapolated polarity reversal time scale, the dates assigned to anomaly boundaries in the Pacific Ocean give points that lie on a line with three distinct portions, each with a different spreading rate.

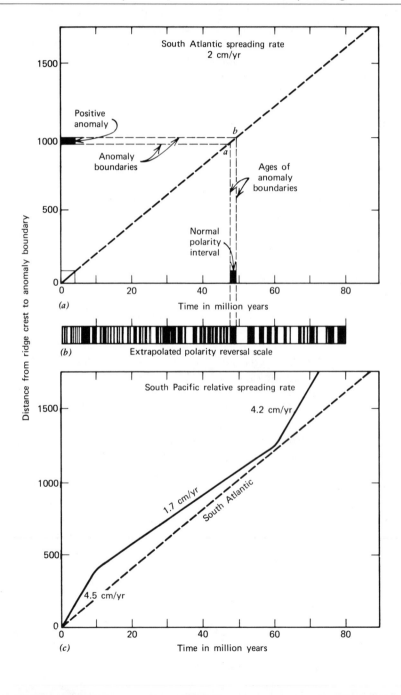

anomaly patterns that had been mapped over extensive regions of the North Pacific, South Pacific, South Atlantic, and Indian Oceans. They concluded that the pattern was bilaterally symmetrical about the crest of the ridge in each ocean, and that the pattern was the same in all oceans. They assigned a provisional age scale to the sequence of anomalies up to 80 million years (Figure 1-3), by extrapolation of the calibrated results for the South Atlantic Ocean, as shown in Figure 10-9.

The width of the South Atlantic Ocean is about 5000 km, becoming wider with distance from the equator, and magnetic anomalies have been mapped for distances of 1500 km from the ridge crest. The small box near the origin in Figure 10-9a shows the magnitude of the extrapolation involved in extending the spreading rate line through a distance of 1500 km.

The procedure is as follows. A positive anomaly is located between 950 and 990 km from the ridge crest, and its boundaries are plotted as points a and b on the extrapolated line. The graph gives ages of 48 and 49.5 million years for points a and b, and these ages give the times of polarity reversals. The time interval between them is therefore assigned to a normal polarity epoch in Figure 10-9b. The other polarity epochs and events shown were dated in the same way, and this provides a polarity reversal time scale extrapolated to 80 million years, well beyond the 4 million year calibrated scale. This mixed polarity interval is shown on Figures 9-4 and 1-3.

This extrapolated time scale can then be used to estimate the rates of spreading in other oceans by plotting anomaly boundaries against the extrapolated polarity reversal boundaries. The results obtained are not absolute values; they are rates relative to the assumed constant spreading rate for the South Atlantic Ocean. The line obtained for the South Pacific Ocean is given in Figure 10-9c. It consists of three fairly distinct parts, each with a different slope corresponding to a different spreading rate. For the past 10 million years, the slope gives about 4.5 cm/year, similar to that for another part of the Pacific plotted in Figure

10-8b. For the period between 10 and 60 million years ago, the spreading rate was much slower, 1.7 cm/year but, before 60 million years ago, it was faster (4.2 cm/year).

An alternative approach would be to assume that the spreading rate from the South Pacific ridge had remained constant and to extend the recent calibrated rate of 4.5 cm/year as a straight line out to the most distant anomaly, at about 1700 km from the ridge crest. Then the distances to anomaly boundaries plotted on the extrapolated line would give the dates of polarity reversals and, from these, we would obtain a different extrapolated polarity reversal time scale. It would be similar to that in Figure 10-9b, but more compressed. This time scale could then be used as a basis for plotting a spreading line from the anomaly boundaries in the South Atlantic Ocean, and the result would be a line with three portions, each with a different slope. If the South Pacific spreading rate remained constant, then the South Atlantic spreading rate changed twice during the past 70 million years. This transformation can be made by scaling numbers from Figure 10-9c and replotting them as outlined above.

Whatever we use as a basis for comparison, an assumed constant spreading rate for one ocean or another, we find, as in Figure 10-9c, that the rates of spreading of one ridge relative to another may change with time. The extrapolated polarity reversal time scale given in Figure 10-9b is obviously only provisional. The scale will require modification if the spreading rate from the South Atlantic ridge has not remained constant, if there have been worldwide increases or decreases in spreading rates from all ridges, or if spreading has been episodic, with all movement stopping for periods of unknown duration.

DATING THE OCEAN FLOOR

We noted in connection with Figures 10-5c and 10-6 that many anomalies had characteristic shapes, reflecting the distribution of

Fig. 10-10 Schematic illustration of the distribution of recorded linear magnetic anomalies. Dates for these, estimated as shown by extrapolation in Figure 10-9, are indicated up to 80 million years. These lines give isochron maps, showing the age of the ocean floor. (Based on data from Heirtzler and others, 1968, *Jour. Geophys. Res., 73*, 2119-2136, and W. C. Pitman and M. Talwani, 1972, *Bull. Geol. Soc. Amer., 83*, 619.)

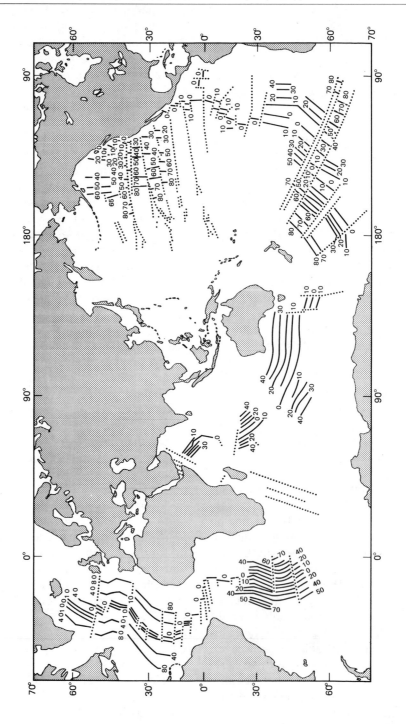

magnetized blocks on the ocean floor. These anomalies, with magnetic signatures, can be assigned an age, using the provisional time scale in Figure 10-9*b*. The distribution of the anomalies in the ocean basins therefore corresponds to the distribution of isochrons for the ocean basin floor. The locus of a specific magnetic anomaly is a line of constant age, where the age corresponds to the time of magnetization. This is interpreted as the time that this part of the ocean floor was brought to the surface as lava at an active oceanic ridge. Despite the provisional nature of the ages in the extrapolated time scale, the ability to contour the ocean basins with isochrons provides the prospect of unraveling the history of the ocean basins and the resultant movement of the continents, with a precision of detail inconceivable during the debate about continental drift that occupied the first half of the century.

Figure 10-10 shows some of the ocean floor isochrons. These may be considered as growth lines from the ridge crests, which are indicated by the 0 isochrons. The dotted lines show the fracture zones that have displaced the isochrons. A large, colored version of this map, including additional data, was prepared in 1974 by W. C. Pitman, R. L. Larson, and E. Herron of the Lamont-Doherty Geological Observatory, for the Geological Society of America. It is available from the Society for a small price. Using these data, together with all available information from sea-floor topography and earthquakes, several papers dealing with the history of the Atlantic, Indian, and Pacific Oceans have already been published.

Remember that Figure 10-10 is based on acceptance of the Vine-Matthews model idealized in Figure 10-5 and, on a very long extrapolation of the model and the polarity reversal time scale, like the 20-fold extrapolation from Figure 10-8*b* to 10-9*a*. As we will see in Chapter 13, these conditions appear to be acceptable to most geologists and geophysicists, although a few have raised some questions.

SUMMARY

Anomalies in the earth's magnetic field measured in oceanic regions are very small, rarely exceeding 2% of the average field strength. By 1957 it had been established that there is a distinctive positive anomaly over the crest of the ocean ridge system. Magnetic surveys in the northwest Pacific Ocean published in 1958 revealed a pattern of narrow, parallel anomalies, each about 30 km wide, extending for about 1000 km.

Vine and Matthews, in 1963, reasoned that the linear anomalies were caused by strips of oceanic crust magnetized alternately in opposite directions, and they explained this arrangement by considering the effect of polarity reversals on the direction of magnetization of the new ocean crust generated at the crest of the ocean ridges by sea-floor spreading. The complete sequence of polarity reversals becomes imprinted on the oceanic crust as if it were a tape recorder in stereo, and the strips of magnetized rocks produce the striped magnetic anomalies, symmetrical about the spreading center.

To each magnetic anomaly boundary at a measured distance from the spreading center, a date can be assigned from the calibrated polarity reversal time scale. The distances and times so determined give rates of sea-floor spreading of 4.5, 2, and 1 cm/year, respectively, for the Pacific, South Atlantic, and North Atlantic Oceans.

If constant spreading rate is assumed, anomaly boundaries further away from the spreading center than the calibrated distance can be assigned provisional ages. In this way, the polarity reversal time scale has been extrapolated to 80 million years ago. This distribution of magnetic anomalies corresponds to the distribution of isochrons for the ocean basin floor. A specific magnetic anomaly is a line of constant age, giving the time that this part of the ocean floor was formed at an active ridge crest.

SUGGESTED READINGS are at the end of the book.

chapter 11

Continental Drift and Paleomagnetism

From the evidence for mobile plates and for spreading sea floors comes the conclusion that the continents drift. The continents drift away from divergent plate boundaries as fast as the sea floor spreads, and we can estimate their drifting rates for the past 80 million years from the linear magnetic anomalies by using the extrapolated polarity reversal scale (Figure 10-9b). We conclude that they have been drifting for longer than 80 million years because the magnetic anomalies occur further away from ocean ridges than distances correlated with ages of 80 million years.

The spreading rate for each side of the South Atlantic Ocean is 2 cm/year, which means that each year, 4 cm of new ocean floor is generated at the mid-Atlantic ridge, as depicted by Figure 2-3. If we could reverse the process of sea-floor spreading at the same rate, we would move the lithosphere conveyor belt back into the earth's interior down the mid-Atlantic rift valley, decreasing the width of the Atlantic Ocean and bringing Africa and South America back toward each other.

At the latitude of Figure 3-10, the present width of the Atlantic Ocean is about 5500 km. Reversal of sea-floor spreading for 80 million years would remove 3200 km from

the center of the Atlantic Ocean, narrowing the gap between Africa and South America to 2300 km. If the spreading process could be reversed at a constant rate, it would take about 138 million years to close the Atlantic Ocean basin completely and to bring the continental slopes of Africa and South America into contact. From this kind of argument, we could infer that the South Atlantic Ocean began to open up by rifting and drifting apart of a supercontinent about 138 million years ago.

But our time scale is calibrated only to 4.5 million years (Figure 9-3) and extrapolated only to 80 million years (Figure 10-9b). Is there any independent evidence that the sea floor spreads, that the plates move, and that the continents drift for times further back than this? Yes, there is. Once again, we must turn to the earth's magnetic field and to the fossil magnetism produced in rocks at their times of formation.

Instead of mapping the estimated ages of magnetized rocks on the ocean floor, as we did in Chapter 10 (Figure 10-10), we now map the former positions of magnetized rocks on the continents and track the movement of the continents over the surface of the earth. This was the new evidence of the 1950s that began the revival of interest in the old debate (Table 2-1). Before examining the new evidence, let us briefly review the various lines of evidence that were argued in favor of continental drift before this.

EVIDENCE CITED FOR CONTINENTAL DRIFT EARLY IN THE DEBATE

Early protagonists of continental drift developed four main lines of evidence supporting their contention that the continents were once joined together: (1) paleoclimatology; (2) paleontology (the study of fossils); (3) the geometrical fit of the continents; and (4) the matching of rocks and major geological structures across oceans. As we noted in Chapter 2, the evidence available during the first half of the century was not regarded as conclusive, and we need not cover it in any detail here. A lighthearted review of the debated evidence published by R. S. Dietz in 1967 includes several expressive illustrations by his artistic colleague, J. C. Holden. Dietz and Holden have kindly granted permission for reproduction of some of these pictures here.

Paleoclimatology is concerned with the reconstruction of ancient climatic zones from evidence stored within rocks of known ages. Several kinds of sedimentary rocks are indicative of climate. Desert sands, for example, produce sandstones with distinctive characteristics. Rocks formed around the margins of glaciers and ice sheets are easily recognized. Fossil coral reefs that once flourished in warm waters are incorporated into many sedimentary rocks. We find warm-weather fossil coral reefs in rocks in Greenland and cold-weather glacial deposits in subtropical latitudes. From detailed studies of these and other sedimentary rocks formed within specific time intervals, geologists produced maps showing the positions of ancient climatic zones. Some interpreted these maps as evidence that the continents had moved relative to their present latitudes, but others interpreted them as evidence that there have been major climatic fluctuations in the past, even at specific latitudes.

There was a lively debate about the distribution of certain fossil animals and plants that are believed to be incapable of crossing deep water. Some species of these ancient creatures and plants occur in limited geographic regions on either side of oceans. The paleontological evidence indicates that land connections must have existed between the now separated continents until about 200 million years ago, but not more recently than this. Figure 11-1 illustrates some of the hypotheses invoked to explain this. One idea was that these connections were made by temporary land bridges that permitted the animals to island-hop from one continent to another. Another idea was that the connection was real when the continents were together, before continental drift severed the connection.

Attempting to fit together the continents marginal to the Atlantic Ocean has been a popular pastime since the last century. A glance at the map of the Atlantic Ocean shows why (Figure 3-7). South America so obviously fits snugly against Africa, and the mid-Atlantic ridge divides the ocean so symmetrically. In other places, however, a certain amount of scientific license was required and was taken in order to make continental margins fit across an ocean basin. If you use a very small globe or a map that can be stretched, fits are more easily made, as shown in Figure 11-2.

If two continents were once joined together, then the rock types and the geological structures of the continents near opposite shorelines should match each other closely. There are remarkable similarities between rocks of various ages from Brazil and South Africa, but not all geologists were convinced that this required that Brazil and South Africa were once adjacent.

REVIVAL OF OLD ARGUMENTS

Since 1965, there have been many publications describing geometrical fits between continental masses. The computer has been put to work to move the continents over the

Fig. 11-1 The occurrence of similar species on continents widely separated by the ocean have been explained as shown in the four sketches by J. C. Holden. (Reproduced by kind permission of J. C. Holden; published by R. S. Dietz, 1967, *Sea Frontiers, 13,* 66-82.)

surface as shells on a sphere, with the geometrical constraints illustrated in Figure 5-6. This procedure gives a less subjective fit than the sketchbook procedures adopted by many in earlier years (Figure 11-2). One recent reconstruction of the single super-continent of Pangaea is shown in Figure 11-3.

Of course, as often pointed out by adversaries of continental drift, the mere fact that continental margins can be fitted against each other is no proof that they ever were in contact. Figure 11-4 was prepared by Ye. N. Lyustikh to demonstrate the amazingly precise fits of 15 shorelines, drawn to the same scale, none of which could have anything to do with continental drift. But the computer fits do provide frameworks for the comparison of the geology at continental margins across ocean basins. Continental

reconstructions such as that in Figure 11-3 have to satisfy both the requirements of geometry, which is taken care of by the computer, and the requirement that the geology matches across the boundaries. This principle is illustrated in Figure 11-5. Recent work involving age measurements is beginning to provide more precise data.

Figure 11-6 shows Africa and South America fitted together, as in Figures 11-3 and 11-5a. There is a large area in northwest Africa where the metamorphic rocks beneath the cover of recent sediments, forest, and desert have similar ages, near 2,000 million years. To the west and east of this province, the rocks are only 600 million years old. The boundaries between these age provinces head out to the sea. If Africa and South America had been part of a single continent when these rocks were

Fig. 11-2 Fitting the continents together has occupied the attention of many geologists. As illustrated in these sketches by J. C. Holden, (a) good fits are more easily achieved on a ping-pong ball-sized globe; (b) many drifters were able to force continental reconstruction by considering the continents to be as pliable as a rubber sheet. (Reproduced by kind permission of J. C. Holden; published by R. S. Dietz, 1967, see Figure 11-1.)

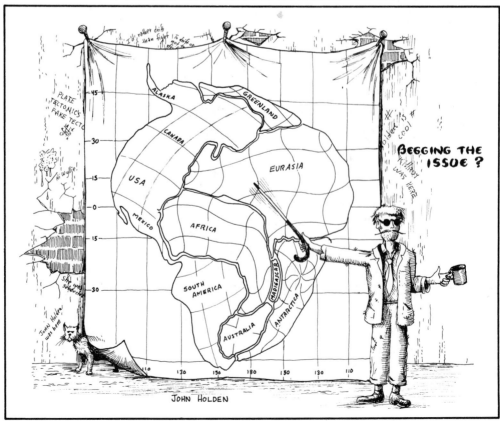

metamorphosed 600 and 2,000 million years ago, as indicated in Figure 11-6, the boundaries between the two age provinces should extend across the contiguous continental shelves and continue into South America.

A research program planned in 1966 by P. M. Hurley and R. J. Rand to test this discovered that the same two age provinces do exist in Brazil, and confirmed that the boundary lines between them lay exactly where predicted. The small area of very old rocks shown in Brazil is believed to represent a fragment of the African province, carried away on South America when the continents drifted apart. In central Africa, there is another region of rocks with ages greater than 2,000 million years that is matched across the Atlantic Ocean by a similar region.

This kind of evidence, the matching of age provinces and other geological features across ocean basins, is necessary but not sufficient evidence for continental drift. If the continents had drifted apart, then it is necessary that the geological features match. But observations that geological features match across ocean basins are not in themselves proof that the continents have drifted. The evidence from fossil magnetism is more direct because it shows the relative positions of the continents through time and makes it possible to trace the paths followed by the continents.

INTERPRETATION OF FOSSIL MAGNETISM

The methods of paleomagnetism depend on the fact that some rocks become magnetized when they are formed (Figures 8-10 and 8-11) and they preserve a fossil record of the direction of the earth's magnetic field at their time and place of origin. If this direction is measured, it shows the attitude in the rock of the lines of force at the time that the rock was magnetized. For young rocks, the direction of magnetization approximates closely the lines of force of the present magnetic field but, for many old rocks, the direction may differ considerably from this. If so, it indicates that something has moved. Either the rock has moved since it was formed, or the magnetic pole has moved, or both have moved.

Fig. 11-3 Reconstruction of the supercontinent Pangaea 225 million years ago. (According to R. S. Dietz and J. C. Holden, 1970, *Jour. Geophy. Res., 75,* 4939-4956.)

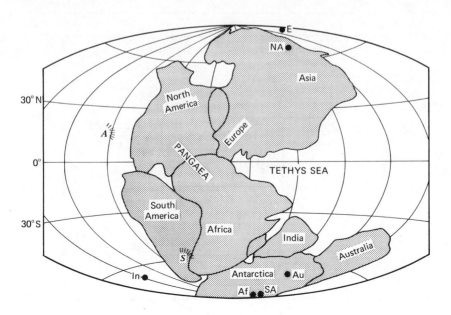

● Paleomagnetic pole positions: NA-North America E-Eurasia In-India SA-South America AF-Africa AU-Australia

The fossil magnetization provides information about the direction to the north magnetic pole and the distance of the rock from the magnetic pole (Figure 8-8) at the time of magnetization. This information for a series of rocks with different ages on a continent is usually summarized in a *path of polar wandering,* as illustrated in Figure 12-1. This shows the relative positions of a continent and the magnetic pole as a function of time, with the arbitrary stipulation that the continent is fixed in position with respect to the present geographic coordinates. This concept usually gives students trouble, so let us examine it in more detail.

In the first place, the path of polar wandering should properly be called a *path of apparent polar wandering* or a *path of polar wandering relative to a continent* because it definitely does not show how the position of the magnetic pole has changed with time. In fact, the preferred interpretation is that the magnetic pole position has not changed significantly through time.

We make three assumptions. First, that the earth's magnetic field has always been equivalent to that illustrated in Figure 8-5 with only two magnetic poles; more complex systems with several poles are conceivable. Second, that the earth's magnetic

Fig. 11-4 "Fits" of similar coastlines and Japanese Islands. Try to locate these on Figure 3-1. All coastlines drawn at same scale. 1, western coast of South America from 15 to 40°S lat.; land is on right. 2, western and northwestern coast of Australia from 35°S lat. to 133°E long.; south is on top and land is on left. 3, western coast of North America from 175°W long. to 35°N lat.; land is on right. 4, southeastern coast of South America from 5 to 35°S lat.; land is on left. 5, northeastern coast of South America from 35 to 60°W long.; east is at top and land is on right. 6, western coast of Africa from 4°N to 30°S lat.; land is on right. 7, eastern coast of Africa from 0 to 24°S lat.; land is on left. 8, Japanese Islands. 9, western coast of South America from 5°N to 18°S lat.; land is on left. 10, southern coast of Australia from 117 to 144°E long.; west is at top and land is on right. 11, eastern coast of Australia from 17 to 38°S lat.; land is on left. 12, eastern coast of Greenland from 84 to 60°N lat.; land is on left. 13, eastern coast of Africa from 15°S to 12°N lat.; south is at top and land is on right. 14, western coast of Africa from 5 to 25°N lat.; south is at top and land is on left. 15, eastern coast of India from 8 to 22°N. lat.; south is at top and land is on right. (Published with permission of Royal Astronomical Society. From Ye. N. Lyustikh, 1967, *Roy. Astron. Soc. Geophys. Jour.,* 14, 347-352.)

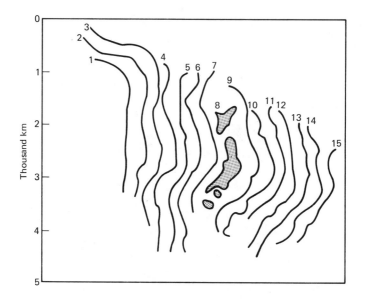

Fig. 11-5 With continental jigsaw puzzles, the clues for fitting the pieces include their shapes and design of the picture. In these sketches by J. C. Holden, (a) the pieces fit and the picture matches across the coast lines, but (b) although the pieces fit, the picture does not match. (Reproduced by kind permission of J. C. Holden. Published by R. S. Dietz, 1967, see Figure 11-1.)

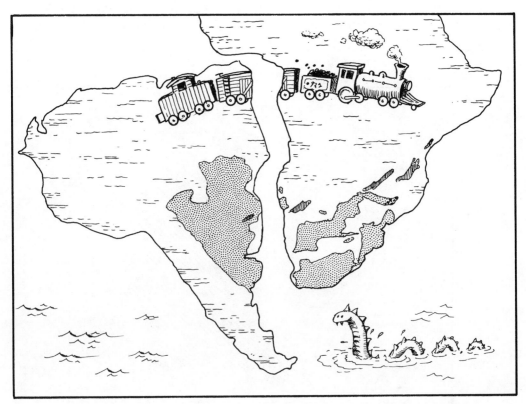

Fig. 11-6 Continental rocks can be dated, and areas of continent with rocks of similar ages can be mapped. Boundaries between areas with different rock ages have been mapped in Africa and South America. When these continental pieces are fitted together as in Figure 11-3, the pictures also fit, as in Figure 11-5(b), because the boundaries can be traced from Africa to America. (Maps based on data from P. M. Hurley and others, 1967, Science, *157*, 495-500, and P. M. Hurley, 1973, in *Implications of continental drift to the earth sciences,* Vol. 2, D. H. Tarling and S. K. Runcorn, Academic Press.)

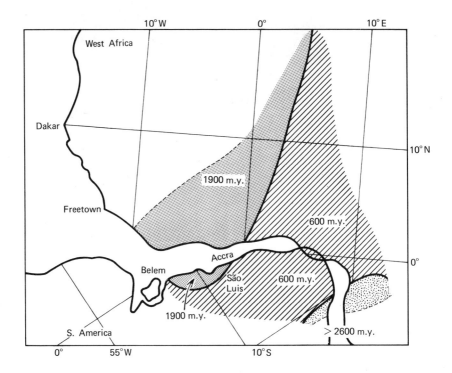

axis remains close to the earth's rotation axis, which means that the magnetic poles remain close to the geographic poles. Evidence supporting this assumption is given in Figure 12-1a for the past 20 million years, at least. Third, that the rotation axis has remained in essentially the same position in space as the main mass of the earth spins around it, with the outer shell, the lithosphere plates, sliding over the interior.

With these assumptions, it is clear that when we examine the relative movement of poles and continents, it is the continents that are moving with respect to a fixed pole. Why then, do we not plot the paleomagnetic data in terms of *paths of continental wandering?* There would be no need to qualify the phrase with the words "appar-

ent" or "relative." The reason is that a path of continental wandering plotted on the basis of paleomagnetic data does not display all of the information available from the data. We would need to plot a series of continent outlines rather than a path. We can see this if we follow the procedures actually used to plot the results of measurements.

MEASUREMENT AND REPRESENTATION OF FOSSIL MAGNETISM

The first step is to collect suitable specimens of rock. These have to be carefully oriented with respect to the present geographic coordinates and with respect to the original horizontal plane at their time of formation.

The directions of the present geographic north and of the horizontal plane are marked on a rock specimen before it is broken loose by a hammer-wielding geologist. If the rock has not been tilted or folded since it was formed, the mark shows also the original horizontal plane. But, if tilting has occurred, it may be necessary to work out the geological structure of the region so that the original attitude of the specimen (specifically, the horizontal plane) can be located. This is usually a simple matter for sedimentary rocks and for many lavas because they are often formed in clearly layered sequences, with each layer originally horizontal.

The oriented specimens are then brought back to the laboratory and, with very sensitive instruments, the direction of the very weak magnetization of each specimen is determined. The age of formation of the rock also is measured in the laboratory.

Figure 11-7 illustrates results for two specimens. Each rock is sitting on a table with its original horizontal plane parallel to the present horizontal surface of the table. The direction to the present geographic north, as recorded when the specimen was collected, is represented by vertical planes.

The measured direction of magnetization for each specimen is shown by an arrow, and the attitudes of these arrows in space are shown by the lines drawn on the vertical planes through the arrows. The direction in space of each arrow is defined by two angles, which are illustrated in three dimensions in Figures 11-7a and 11-7d, and on maps in Figures 11-7b and 11-7e.

For the specimen in Figure 11-7a, the direction is defined by the angle between the two vertical planes, i, measured in a horizontal plane to the west of north and by the angle of dip, α, in a northerly direction, measured downward in the vertical plane through the arrow. The northerly angle of dip shows that the rock was magnetized in the northern hemisphere, and the direction of magnetization shows that it was during a period of normal polarity (Figure 9-1).

For the specimen in Figure 11-7d, the direction of the arrow is defined by the angle j measured in a horizontal plane to the east of south and by the angle of dip, β, in a southerly direction. This result is represented on a map as in Figure 11-7e. The southerly angle of dip shows that the rock was magnetized in the southern hemisphere, and the direction of magnetization shows that it was during a period of reversed polarity (Figure 9-1).

Does it make any difference if the rocks were formed and magnetized during a period of reversed polarity? No, because the only thing that changes is the direction of the arrows. The position of each line of magnetization in space is still defined by the same two angles. This is shown by comparison of the maps in Figures 11-7b and 11-7e with those in Figures 11-7c and 11-7f. By convention, the positions of ancient magnetic poles are plotted in the northern hemisphere for all specimens, whether their polarity is normal or reversed.

Each map in Figures 11-7b and 11-7e shows the direction to the north magnetic pole at the time of formation of the rock relative to the position of the rock in the field where it was collected. We know that when the rock was magnetized, the angle between the magnetic and rotation axes was just a few degrees. Therefore, we can deduce that since each rock was formed, it has been rotated through an angle: the angle i for the specimen in Figures 11-7a and 11-7b, and the angle j for the specimen in Figures 11-7d and 11-7e.

For each rock, the angle of dip gives approximately its latitude at the time of formation and magnetization, as shown by Figures 8-8 and 8-9a. Figure 8-9a shows that for a given angle of dip, the present latitude can vary quite significantly. We have no way of knowing the extent of similar irregularities tens or hundreds of millions of years ago, so we assume an ideal world with ancient lines of constant dip parallel with the lines of latitude. This introduces a fairly large error, as we will see, but this is minimized by taking the average of many measurements. The relationship between

Fig. 11-7 Rock specimens placed on table, in the same orientation with respect to geographic coordinates that they occupied when they were in the field (see Figures 8-10 and 8-11). The maps *(b)* and *(e)* show the orientation of the magnetization direction in each specimen. The maps *(c)* and *(f)* show that the same information would be available if the rocks had been magnetized with opposite polarities.

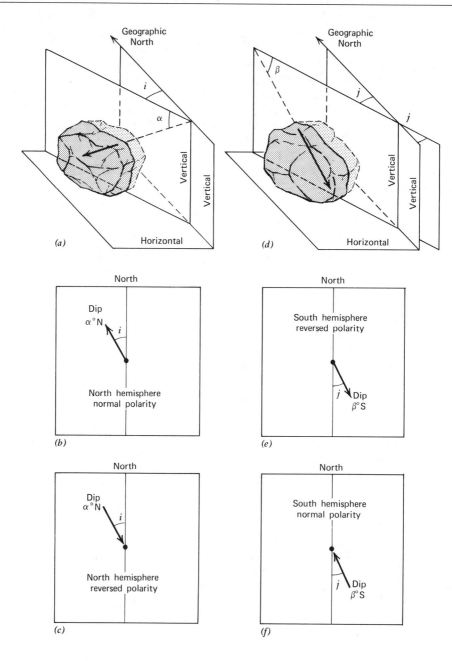

angle of dip, paleolatitude, and distance is shown in Figure 8-8c.

If we know the paleolatitude of the rock at its time of formation, we know from Figure 8-8c how far away from the magnetic pole it was at that time. This information permits us to plot the position of the ancient magnetic pole on the map relative to the rock specimen, with the present position of the rock defined in terms of the geographic coordinates (back in the field where it came from, as in Figures 11-7b and 11-7e, and not on the laboratory table, as in Figures 11-7a and 11-7d). This is illustrated in Figure 11-8a, which is derived from Figure 11-7b. The same relationships are compared on a globe in Figure 11-8d.

In Figure 11-8, the present positions of rock specimen and magnetic pole are indicated by R_N and P_N (N for now at time T_N), and the original position of the rock when it was formed and magnetized (at original time T_O) is indicated by R_O. P_O is the position of the original pole relative to the magnetized rock specimen, wherever the rock may wander over the surface of the earth.

The distance between P_O and the rock specimen is constant, wherever the rock may be. If the rock rotates, line $R_N P_O$ rotates with it. The rock is represented by a square so that its rotation can be depicted.

Figures 11-8a and 11-8d show R_N and P_N, and the position of P_O relative to R_N at time T_N, now. We want to find R_O, the original position of the rock at time T_O. We can do this by putting P_O back where it belongs, at the north pole, so that $P_N = P_O$.

First, we rotate the rock around a vertical axis so that its direction of magnetization is pointing in the proper direction, toward P_N, as in Figures 11-8b and 11-8e. The second operation is to move P_O north to coincide with P_N. The distance $P_O - R_O$ is fixed, by definition, and equal to the distance $P_O - R_N$ in Figures 11-8a and 11-8b, so this gives the position of R_O in Figure 11-8c. At R_O, the angle of dip of the earth's magnetic field is the same as the angle of dip of the fossil magnetization, $\alpha°N$.

Figures 11-8c, 11-8f, 11-8a, and 11-8d show that since the rock was formed and magnetized, in time $T_O - T_N$, it has drifted

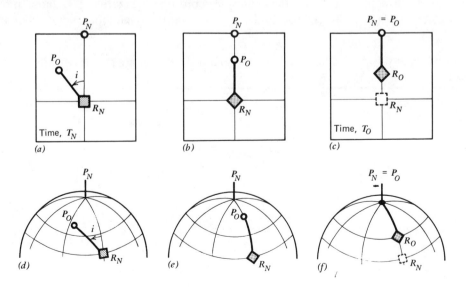

Fig. 11-8 (a and d) Map and globe showing information of Figures 11-7a and 11-7b. P_O is the position of the fossil magnetic pole relative to the position of the rock where it was collected, R_N. (b and c) show two steps that return the fossil pole position to the present north geographic pole, producing a rotation and a northward movement of the rock to its original position when it was formed and magnetized, R_O.

southward from the latitude of R_O to R_N; and it has also rotated through angle i in an anticlockwise direction. This is all of the information that can be deduced from the measured magnetic properties of the rock specimen illustrated in Figure 11-7a. We know nothing about possible movements in east-west or west-east directions.

The angle of dip, α (Figure 11-7a), tells us only that R_O was situated on a particular line of latitude (Figure 8-8c) and the direction of magnetization depicted in Figure 11-7a would be the same wherever R_O was on that line of latitude, as illustrated in Figure 11-9. The original position of the rock specimen could be R_O', R_O'', R_O''', or any other point on the latitude line, a very large distance to east or west of R_N. We have no way of determining this from measurements on rocks from a single place or a single continent.

The two ways of summarizing the data from study of the single specimen shown in Figure 11-7a are illustrated in Figure 11-10. Figure 11-10a is a Path of Apparent Polar Wandering and Figure 11-10b is a path of apparent continental wandering. Figure 11-10b does not show the rotation of the rock.

In Figure 11-10a we hold the rock specimen fixed in position R_N, where it was collected, and we plot successive relative pole positions determined at different times. In this example, we have data for only two times. At time T_O, relative to the present position of the rock, the pole was in position P_O. If the rock had been formed and magnetized in its present position at the present time, T_N, the direction of magnetization would show that the pole was in position P_N. Thus we conclude that in the time interval $T_O - T_N$, the pole has moved along the path $P_O - P_N$ relative to the present position of the rock, R_N. This way of representing the data includes the northerly component of the movement ($P_O - P_N$ in Figure 11-8b equals $R_N - R_C$ in Figure 11-8c), as well as the rotation.

In Figure 11-10b, we assume that the pole position $P_O = P_N$ remains unchanged through time interval $T_O - T_N$, while the rock specimen (situated on a continent) moves south along the path $R_O - R_N$. This path of continental wandering relative to the pole does not display the rotation. This path shows no component of movement to east or west during the time interval $T_O - T_N$. With our assumption that the magnetic pole remained essentially coincident with the geographic pole through the time $T_O - T_N$, the true path of continental wandering could be any one of an infinite number of paths in Figure 11-9, such as $R_O' - R_N$, $R_O'' - R_N$, $R_O - N$, and $R_O''' - N$.

THE NORTHWARD DRIFT OF INDIA

According to Figure 12-3, India moved northward through thousands of kilometers

Fig. 11-9 Other positions of R_O in addition to Figure 11-8f that are consistent with the magnetization data.

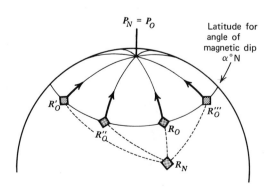

Fig. 11-10 (a) P_O-P_N is path of apparent polar wandering in time interval T_O-T_N from Figure 11-8a to 11-8c. (b) R_O-R_N is the path of continental wandering in the same time interval, with neither rotation nor latitudinal movement (Figure 11-9) indicated.

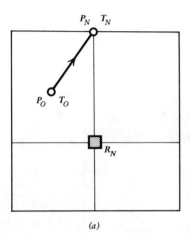

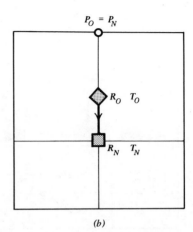

(a)   (b)

with little rotation and little change in longitude. The path followed by India can therefore be tracked from paleomagnetic data by determining successive paleolatitudes as a function of the angle of dip of the fossil magnetism, without the complications of rotation and possible longitudinal drift illustrated in Figures 11-8 and 11-9. Figure 11-11 uses estimated rather than real data, simply to show how this is done.

The Deccan Plateau of India is built up of successive lava flows with ages ranging from more than 180 million years to just a few million years. Measurements of the ages and fossil magnetization of these rocks reveals a systematic picture, which is illustrated schematically in Figure 11-11a. A series of rock specimens is shown sitting on laboratory benches, arranged in order of decreasing age. It turns out that the magnetization direction of the rocks is approximately north-south, and all of the specimens are oriented in this direction.

The arrow in each rock gives the orientation of the lines of force when the rock was magnetized, and this gives the angle of dip. For convenience of representation, the magnetization for normal polarity is assumed, but the same information is available from rocks magnetized with reversed polarity (Figures 9-1, 11-7b and 11-7c). The

angles of dip of the ancient magnetic field are compared with that of the present field at the location of the specimens, near 18°N latitude.

From the angles of dip we can see that the 20 million-year-old rock was magnetized in the northern hemisphere, but the rocks of 65 million years and older must have been in the southern hemisphere when they were magnetized. Match the magnetized directions with the lines of magnetic force in Figures 8-5 and 9-1. The paleolatitude for each rock when it was magnetized can be read from the graph in Figure 8-8c, and the values are given in Figure 11-11a. According to this graph for an idealized magnetic field, the present latitude of the location with dip of 20°N should be 12°N instead of the actual value of 18°N. This difference gives some idea of the errors involved in this method using the idealized graph.

The successive paleolatitudes give us a path of apparent continental wandering corresponding to $R_O - R_N$ in Figures 11-9 and 11-10b. In this example, the direction is from south to north. The fact that the magnetization directions deviate only slightly from north shows that the land mass has not rotated much during the time of drift.

Fig. 11-11 Fossil magnetic data from lavas in India. *(a)* Magnetization data for rocks of various ages, corresponding to Figure 11-7. (These are not actual data, but simplified examples to illustrate the method.) Paleolatitude can be determined from these data using Figure 8-8*c*. *(b)* Graph showing the paleolatitude of each rock at the time it was formed and magnetized. The curve through the data points shows that India has drifted from south to north, with three periods of drifting distinguished by different rates of movement.

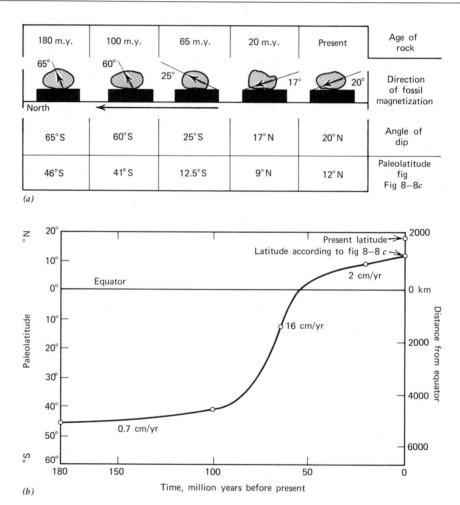

180 m.y.	100 m.y.	65 m.y.	20 m.y.	Present	Age of rock
65°	60°	25°	17°	20°	Direction of fossil magnetization
65°S	60°S	25°S	17°N	20°N	Angle of dip
46°S	41°S	12.5°S	9°N	12°N	Paleolatitude fig Fig 8–8*c*

(a)

(b)

The paleolatitude of each specimen is plotted as a function of its age of formation or time of magnetization in Figure 11-11*b*. The curve drawn through the points then represents the rate of movement of the land mass in a northerly direction, as mapped in Figure 12-3. With only five points for a period of 180 million years, there is not much control on the position of the line. Any reasonable curve through the points, however, shows at least three periods with different rates of movement. For the curve

plotted, there is a period between 180 and 100 million years ago with slow movement averaging 0.7 cm/year. Between 100 and 50 million years ago the rate of movement increased markedly, reaching 16 cm/year and, for the past 50 million years, movement slowed down to about 2 cm/year.

The graph in Figure 11-11*b* has been constructed to show a technique. It gives an indication of the rate of movement of India from a position in the south to its present position just north of the equator (Figure

12-3), but it could be improved by the inclusion of more paleomagnetic points and of spreading rates from sea-floor anomalies. There is evidence, for example, that movement of India practically ceased between about 55 and 35 million years, but it would require many paleomagnetic measurements of rocks in this age interval to establish this, using the method shown in Figure 11-11.

SUMMARY

Early arguments for continental drift were based on paleoclimatology, paleontology, the geometrical fit of the continents, and the matching of rocks and major geological structures across oceans. Recently, continental fits have been revised using computers and the geometrical constraints of plate tectonics, and radiometric dating of rocks has been used in matching portions of continents.

The direction of fossil magnetization in a rock gives an estimate of the direction and distance to the magnetic pole, relative to the rock, at its time of formation. For many old rocks, the fossil magnetic pole does not coincide with the present magnetic pole. Either the rock and its continental host have moved since the rock was formed, or the magnetic pole has moved, or both have moved. For a given continent, by measuring the ages and fossil magnetic poles for many rocks, the relative positions of continent and magnetic pole as a function of time can be shown in a *path of apparent polar wandering*. The preferred interpretation is that the magnetic pole position has remained close to the earth's rotation axis, and the continent has moved relative to the pole. The paths of polar wandering give information about north-south movement and the rotation of continents.

The Deccan Plateau of India is built of lava flows with ages ranging from 180 million years to a few million years. The record of fossil magnetization in these lavas gives the rate of movement of India from a position near latitude 45°S to its present position in the northern hemisphere. India moved with a velocity near 1 cm/year for 80 million years, then moved considerably faster until it crossed the equator about 50 million years ago before colliding with Asia.

SUGGESTED READINGS are at the end of the book.

chapter 12

Paths of Polar Wandering and Migration of the Continents

In the preceding chapter we examined the principles of paleomagnetism, fossil poles, magnetic paleolatitudes, and the difference between paths of polar wandering and continental wandering and rotation. In this chapter, we will first examine how specific paths of polar wandering are constructed from the paleomagnetic data and then track the movements of continental masses back through time to 500 million years ago.

It appears that the continental fragments have collided and combined tõ form supercontinents, which have, in turn, been disrupted with subsequent dispersal of continental fragments with different boundaries. The dispersed continents must eventually collide with each other along convergent plate boundaries, producing mountain ranges with special characteristics (Chapter 6).

PATHS OF POLAR WANDERING

Because of the large error involved in converting the angle of dip of magnetization to paleolatitude (Figures 8-8c and 8-9a), individual measurements such as those shown in Figure 11-7 are of limited use for studies

of continental drift, although they do provide information about polarity reversals, as discussed in connection with Figure 9-4. Therefore we take the average of a large number of measurements of individual rocks that formed during a specific interval of time and refer to this as the mean paleomagnetic pole position relative to the continent for that interval of time.

The procedure is illustrated in Figure 12-1. The maps show the northern hemisphere of the world, looking directly down at the north pole, which is marked by a cross in the center. The equator is the outer circle. Suitable rocks are collected and their ages and directions of magnetization are measured. For each rock a result such as that in Figures 11-7a and 11-7b is obtained, and this gives the former pole position relative to the rock, such as P_O in Figures 11-8a and 11-8d.

Figure 12-1a shows the original pole positions for individual rocks collected from all over the world with ages of 0 to 20 million years. The pole positions are distributed over a considerable area, but they are centered around the geographic pole. The average position of the paleomagnetic pole during this interval of 20 million years, indicated by the square, is almost indistin-

Fig. 12-1 Method for determining apparent polar wandering path for North America. (a) Rocks with ages between 0-20 million years have fossil magnetic pole positions shown by large dots. The pole positions were determined as in Figures 11-7 and 11-8a and 11-8d. The average position of these fossil magnetic poles is shown by the large square. This is transferred to part d as the paleopole position for the time 10 ± 10 million years ago, which is almost identical with the present time. The other squares in d were determined similarly, as illustrated by diagrams b and c. The path is drawn through the average paleopole positions.

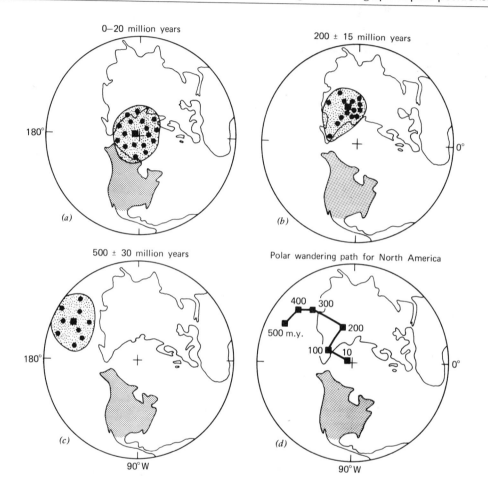

guishable from the geographic pole. This result is consistent with rocks from all continents, and it is considered good evidence that the magnetic axis remained close to the rotation axis during this period. One of our basic assumptions is that this continues to be true back through earth history.

For rocks older than 20 million years, the fossil pole positions are located further away from the present pole, and it is found that results from different continents diverge from each other. Figure 12-1 shows

the variation with the age of rocks for the North American continent.

Figure 12-1b shows schematic results for a group of rocks with ages within the interval 185 to 215 million years. This 30 million year period is listed as 200±15 million years. The former pole positions for individual rocks are distributed over a large area in north and east Asia. There is overlap with the distribution of ancient pole positions in Figure 12-1a, but the square, which is the mean paleomagnetic pole position relative to North America 200 million years ago, is well removed from the average for the interval 0 to 20 million years.

Figure 12-1c shows similar results for a group of older rocks from North America with ages in the interval 470 to 530 million years, which is listed as 500±30 million years. The fossil pole positions for the individual rocks are shown by the points distributed widely in the Pacific Ocean. The area covered by these points is separate from the area covered by the 200 million year old points. The square gives the mean paleomagnetic pole position relative to North America 500 million years ago.

The mean paleomagnetic poles from Figures 12-1b and 12-1c are plotted in Figure 12-1d, with additional poles, similarly determined, for every 100 million years. The line connecting these points gives the path of apparent polar wandering relative to North America for the past 500 million years. Working backward from this empirically determined path, for any time interval back to 500 million years ago, we can estimate two things: the angular rotation of North America, and the distance the continent has moved in a northerly direction.

Figure 12-2a shows the path of apparent polar wandering relative to Europe and northern Asia based on rock specimens collected from the shaded area of the map, with ages back to 530 million years. The mean paleomagnetic pole positions for each 100 million years back through earth history were located in the same way as those plotted in Figure 12-1.

The path for Eurasia differs from that for North America, as shown in Figure 12-2b, where both paths are plotted on the same map. They are barely distinguishable back through 100 million years but, between 100 and 400 million years ago, the paths are quite well separated. As we noted in connection with Figures 11-9 and 11-10, individual paths of apparent polar wandering tell us only about relative movements in northerly or southerly directions. Both continental masses have moved northward, with considerable rotation, during the past 500 million years. Similar paths of apparent polar wandering have been determined for every continent. Each path is different.

Two of our basic assumptions for interpreting fossil magnetism are that there is only one north magnetic pole, and that this coincides with the rotation axis (Figure 12-1a). How, then, can we explain two paths of apparent polar wandering in Figure 12-2b? One interpretation is that the conti-

Fig. 12-2 (a) Apparent polar wandering path for Eurasia, determined as shown in Figure 12-1, using rocks from Eurasia. (b) Comparison of apparent polar wandering paths for North America and Eurasia.

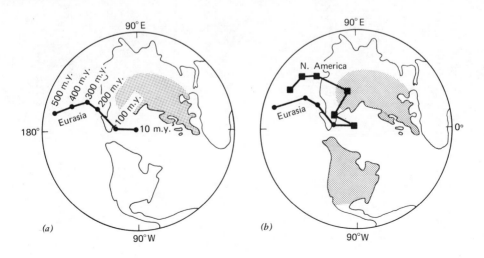

Fig. 12-3 Continental drift. The disruption and dispersal of the supercontinent of Pangaea during the past 200 million years. Pangaea is shown in Figures 11-3 and 12-4a. The hatched curves A and S show the present position of the Antilles arc in the Caribbean, and the Scotia arc in the South Atlantic, see part d. Compare part d with Figure 3-12. (Maps from R. S. Dietz and J. C. Holden, 1970, *Jour. Geophys. Res.*, 75, 4939-4956.)

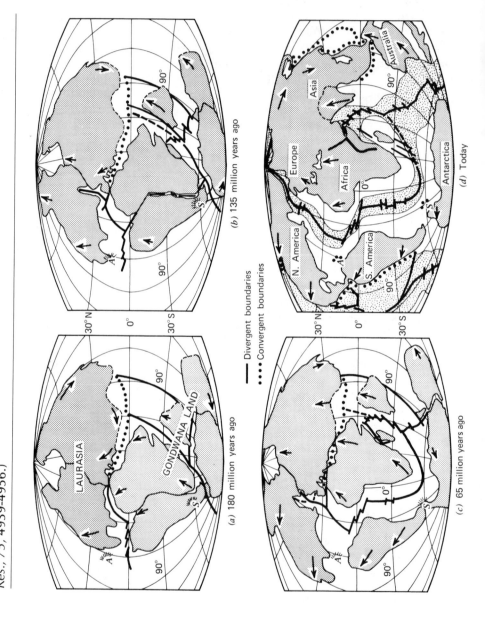

tions of fossil magnetic poles relative to the continents. (*b*, *c*, and *d*) show the relative positions of the continents at the times indicated, based on the interpretation of apparent polar wandering paths. This gives control in a north-south direction, but no direct information about continental drift in an east-west direction (see Figure 11-9). Note especially the ocean bounded by the lines *ab* that closed to form the Ural Mountains between Europe and Asia. Note also the transfer of Scotland and Northern Ireland from North America in part *d* to Europe in part *c*. (These sketch maps are based on results of J. C. Briden, A. G. Smith, and G. Drewry, 1973, in *Organisms and Continents through time*, editor N. F. Hughes, Paleontology Special Paper 12.)

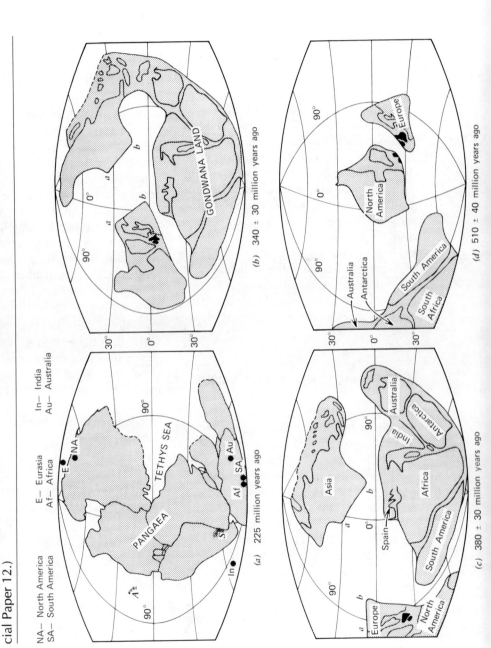

NA— North America   E— Eurasia   In— India
SA— South America   Af— Africa   Au— Australia

(*a*) 225 million years ago

(*b*) 340 ± 30 million years ago

(*c*) 380 ± 30 million years ago

(*d*) 510 ± 40 million years ago

nents have moved not only relative to the pole, but also relative to each other. It was suggested that where the paths are moving toward each other in the interval 200 to 100 million years, the two continents were moving away from each other with an east-west component of movement. However, others maintained that paths could be made to coincide by suitable changes in latitude and rotational movements of the continents, and that these two paths of apparent polar wandering do not provide unambiguous evidence for continental drift in east-west directions. However, if we put together the information summarized in many paths of apparent polar wandering with the plate movements deduced from sea-floor spreading and include restrictions such as not permitting the continents to overlap or to leap-frog each other, then it becomes possible to chart the relative movements of continents. The game of reconstruction of continents (Figure 11-2) can be conducted with a logical scientific basis, with results shown in Figures 12-3 and 12-4.

IS PALEOMAGNETISM REALLY VALID?

Some geologists are not altogether happy with the statistical treatment required to average a large number of widespread results such as those shown in Figure 12-1 in order to obtain a simple curve such as those in Figure 12-2*b*. They note that the spread of poles for one time interval may show considerable overlap with the spread for another interval, and that the actual results used to locate a polar wandering path for one continent may overlap those used for another continent. Not everyone is satisfied that the two paths shown in Figure 12-2*b*, for example, can really be distinguished from each other.

Another question is whether the hypothetical bar magnet model for the earth's magnetic field has been a valid model in the distant past. Apparently, it is theoretically conceivable for a magnetic field produced by a rotating fluid in the outer core to have more than two magnetic poles. If the field were multipolar, we would have to ascertain which pole path we were plotting and the position of this pole relative to the others. From the limited data we have available from old rocks we can not even be positive that there are only two poles.

Finally, a famous geophysicist, Sir Harold Jeffreys, had this to say about rock magnetism in 1970:

When I last did a magnetic experiment (about 1909) we were warned against careless handling of permanent magnets and the magnetism was liable to change without much carelessness. In studying the magnetism of rocks, the specimen has to be broken off with a geologic hammer and then carried into the laboratory. It is supposed that in the process, its magnetization does not change to any important extent, and though I have often asked how this comes to be the case, I have never received any answer.

Despite these doubts, many paleomagnetic data have been gathered, tested, and compiled into paths of polar wandering that have been modified, refined, and moved about quite a bit since the 1950s. These, together with magnetic anomaly data, have made it possible to trace the history of continental drift.

THE DISPERSAL OF LAURASIA AND GONDWANALAND

All theories of continental drift require that before about 200 million years ago the continents were grouped together into a single block or into two blocks. Alfred Wegener proposed the name Pangaea for a single supercontinent, which was partly covered by shallow seas and surrounded by a great ocean covering about 60% of the earth's surface (Figures 11-3 and 12-4*a*). The northern part of Pangaea, composed of North America and Eurasia, constitutes Laurasia, and the southern continents constitute Gondwanaland (Figure 12-3*a*).

In 1970, R. S. Dietz and J. C. Holden repeated Wegener's construction of Pangaea with cartographic precision. Instead of using trial and error methods (Figure 11-2), they used the new geometrical and geological fits (Figure 11-6) to position Pangaea for the first time in absolute coordinates on the globe. Their guiding rationale for reconstruction was the drift mechanism associated with plate tectonics and sea-floor spreading, and they also used paleomagnetic pole positions (Figures 11-11, 12-1, 12-2). Using this information and following the rules of plate tectonics, they prepared four maps illustrating the breakup and dispersion of the continents during the past 180 million years, as shown in Figures 11-3 and 12-3. Absolute geographic coordinates were assigned for the continents as well as for the active ocean rift zones and the ocean trenches as they migrated to their present positions.

The magnetic anomalies indicate the relative movements of the plates carrying the continents for about 100 million years, and paleomagnetism gives latitude but not longitude for several hundred million years. Dietz and Holden chose a center near the volcanic island of Tristan da Cunha in the South Atlantic Ocean, which they assumed had remained fixed in position while the adjacent plates spread away from it. Given a fixed reference point, the motions of the plates could then be plotted in terms of latitude and longitude.

The position of the assembled continents in Figures 11-3 and 12-4a, in terms of the present coordinates of latitude and longitude, was estimated by tracking the movements of the continents back through time from the present, with reference to the fixed reference point. The accuracy of the navigation decreased further back than 100 million years ago, when the magnetic anomaly record terminated, but dead-reckoning carried the continents back to the general region shown.

The next step in the reconstruction of Pangaea as it existed 225 million years ago was to fit the continents together, using computer methods and geological matching of structures across continent boundaries. The fit was made not along the shorelines, but along lines 2 km below sea level, about halfway down the continental slopes, which form the real boundaries to the continental masses (Figure 3-7).

A test was provided by checking the 225 million year paleomagnetic pole positions for each continent. Each pole position was plotted first in its present-day position, then the continent was moved and rotated into the position shown in Figure 12-4a; the position of the paleomagnetic pole moved simultaneously. If the reconstruction is correct, then the adjusted paleomagnetic poles should cluster around the present poles (see also Figure 11-3).

The procedure is easily visualized for North America. The position of the 225 million year paleomagnetic pole for North America can be seen in Figures 12-1b and 12-1d. Figures 11-8a and 11-8d give a reasonable representation of the positions of the continent (R_N) and the paleomagnetic pole (P_O). If the original position of the continent is known (R_O in Figures 11-8c and 11-8f) then the present continent can be moved and rotated into that position as shown. Ideally, this brings the paleomagnetic pole, P_O, into coincidence with the present magnetic pole, P_N. In Figures 11-3 and 12-4a, the adjusted paleomagnetic pole positions for the northern and southern continents do not coincide precisely with the present poles, but they do fall within the Arctic Circle or the Antarctic Circle.

In Figures 11-3 and 12-3 the two hatched crescents, A and S, show the present positions of the modern Antilles arc in the West Indies and the Scotia arc in the South Atlantic. These serve as geographic reference points. According to Figure 11-3, the site of New York City was on the equator 225 million years ago, and at longitude 10°E instead of its present 74°W. The Japanese Islands were in the Arctic, whereas India and Australia were in Antarctica. The present Mediterranean Ocean was represented by the large Tethys Sea, which was a branch of the enormous ocean that girdled the rest of the globe.

Pangaea began to break up along rifts about 200 million years ago. This date coincides with the age of large quantities of lava that were erupted in many of these locations. After 20 million years of rifting and drifting, Laurasia had separated from Gondwanaland and rotated clockwise, as shown in Figure 12-3a, and an ocean ridge developed between North and South America. Sea-floor spreading from this rift kept the continents moving apart as the plates diverged. Another rift had separated South America and Africa from the rest of Gondwanaland and, in the same time interval, India had also separated from Antarctica. The plates containing the eastern parts of Laurasia and Gondwanaland were converging, as indicated by the arrow showing the movement of the continents, and a convergent plate boundary is represented by an ocean trench along the long dotted line through the Tethys Sea, which here begins to look more like the Mediterranean than it did in Figure 11-3.

Figure 12-3b shows the situation 135 million years ago, after 65 million years of plate movement and continental drift. Continued sea-floor spreading has opened up the North Atlantic and the Indian Oceans. The Tethys Sea continued to close at its eastern end, and Eurasia slid westward with respect to Africa along the western part of the plate boundary. Sediments accumulating in the Tethys Sea were later folded and uplifted to form the Alpine and Himalayan mountain ranges. A rift began to split South America away from Africa, probably starting as a rift valley resembling the present Great Rift Valleys of Africa and changing into a narrow sea something like the present Red Sea.

After 135 million years of drift, 65 million years ago, as in Figure 12-3c, the South Atlantic Ocean had widened and become continuous with the North Atlantic Ocean, but the mid-Atlantic ridge had not penetrated into the Arctic Ocean. The two American continents had migrated far westward, as shown by the positions of the Antilles and Scotia arc reference lines. Laurasia had continued to rotate clockwise,

and Africa had moved counterclockwise while drifting northward, almost closing the eastern end of the Tethys Sea, causing great compression, shear stresses, and mountain building. Madagascar had separated from Africa along a new rift, and India had continued to move northward. The northward movement of India in Figure 12-3 can also be tracked in Figure 11-11b.

The system of ocean ridges and trenches shown in Figure 12-3c adjusted during the past 65 million years as the continents moved into their present positions, shown in Figure 12-3d. The mid-Atlantic ridge penetrated through to the Arctic Ocean, Australia broke away from Antarctica and drifted northward because of sea-floor spreading from the new ocean ridge, the two American continents were connected by an isthmus built largely of volcanic eruptions, the Mediterranean Ocean was formed by closure of the Tethys Sea at its eastern end, the Red Sea was produced by extension of a branch of the Indian Ocean ridge, and India came to the end of its journey northward by colliding with Asia, sliding partly below Asia along the former ocean trench and convergent plate boundary, then rising upward because of its low density and buoyancy compared with the mantle, thus causing the formation of the Himalayas and the Tibetan Plateau. The continents are now in their correct positions with respect to the Antilles and Scotia arcs.

The plate movements continue, causing the continents to drift in direction shown by the arrows. Dietz and Holden extended their analysis into the future and extrapolated the movements of plates in order to estimate the positions of the continents 50 million years from today. Among the predicted changes are continued opening of the Atlantic Ocean, movement of Australia to the Asian plate, virtual closure of the Mediterranean Sea by the northward drift of the African plate, creation of new land in the Caribbean by compression and resulting uplift, and some significant changes in the geography of California. The boundary between the Pacific plate and the North

American plate follows the San Andreas fault zone (Figure 5-1). Baja California and a sliver of California west of the fault zone are drifting to the northwest with the Pacific plate, while the North American plate drifts westward. Dietz and Holden estimated that in about 10 million years Los Angeles will be abreast of San Francisco, still attached to the mainland, and that in about 60 million years Los Angeles will start sliding into the Aleutian trench south of Alaska, as the Pacific plate sinks down into the earth's interior (compare Figure 6-8).

BEFORE PANGAEA

Figure 12-3 shows that the supercontinent of Pangaea was disrupted and dispersed into widely separated continents during a period of less than 200 million years. This is a minute fraction of geological time (Figure 1-3). If the present pattern of sea-floor spreading continues, the continents will continue to separate as new lithosphere is generated at ocean ridge crests, and the older ocean floors will disappear into the earth's interior along the lines of ocean trenches at convergent plate boundaries. The Atlantic Ocean will grow, and the Pacific Ocean will decrease in area. The chances are that the dispersing continents will collide with each other on the opposite side of the globe, forming a new supercontinent surrounded by a new global ocean. At present rates of movement, this could occur within a few hundred million years, another minute fraction of geological time.

It is possible that the supercontinent of Pangaea drifted as a unit over the surface of the earth through hundreds of millions of years before it was disrupted and dispersed. Indeed, there is some evidence to suggest that Pangaea existed as a single continental block between 1,700 and 200 million years ago. There is other evidence, however, to suggest that Pangaea was produced by the collision and combination of several separate continental masses some time between 340 and 250 million years ago.

Figure 12-4 illustrates schematically the possible positions of continental masses before they collided to form Pangaea. These are based on paleomagnetic data, and therefore have no longitudinal control. Figure 12-4a reproduces Figure 11-3, Pangaea 225 million years ago.

In Figure 12-4b, 340 million years ago, Laurasia has not yet formed from its two parts. The continental mass including North America and northern Europe (excluding the Mediterranean region) will collide with the Asian part along line ab, which is the line of the Ural Mountains, at some time between 340 and 225 million years ago. The southern continents are united in Gondwanaland, including also the Mediterranean part of Europe, but note that their estimated relative positions differ quite significantly from those in Pangaea. Gondwanaland and Asia are both rotated compared with 225 million years ago, and they may be connected in the east, as shown. Early versions of the North Atlantic Ocean and the Tethys Sea exist between Laurasia and Gondwanaland.

Going back another 40 million years to Figure 12-4c, Gondwanaland appears to have remained still, the Asian part and the American-European part of Laurasia are rotated anticlockwise compared with 340 million years ago, the eastern connection between Gondwanaland and Laurasia does not exist, and the distance between the boundaries ab is greater.

Figure 12-4d shows that the American-European part of Laurasia consisted of even smaller continental masses 510 million years ago. There is an American continent, separated by an ocean from the European continent. Note the extreme rotation indicated for Gondwanaland between 510 and 380 million years. Africa and South America are almost upside down compared with their later positions. South Africa and the tip of South America here point toward the north.

Let us now trace the history of the North American and European continental masses and the Atlantic-like ocean that appears between them from time to time. Starting with Figure 12-4d, 510 million years ago, we have two continents separated by an ocean. The northern parts of Scotland and Ireland form part of the American continent, and the southern parts are attached to the European continent. A part of present Newfoundland is attached to the European continent.

At some time between 510 and 380 million years ago, this ocean was closed as shown in Figure 12-4c, and the previously separated parts of Scotland, Ireland, and Newfoundland became joined. A high mountain range was produced by the collision. An ocean existed between the American-European part of Laurasia and Gondwanaland. Within 40 million years, rotation and northward movement of the American-European part of Laurasia had narrowed this ocean and, in Figure 12-4b, it lies between North America and Africa, with an extension through the lower part of Europe.

Continued movement of the American-European part of Laurasia between 340 and 225 million years ago brought about a continental collision along boundary ab, raising the Ural mountains and making Laurasia continuous. The ocean between North America and Africa was closed, forming Pangaea. Rotation of Gondwanaland opened up the Tethys Sea and moved Antarctica, India, and Australia down to the south, as shown in Figure 12-4a.

Pangaea began to split up again 25 million years later, and the North Atlantic Ocean opened up again along approximately the same line between North America and Africa, as shown in Figure 12-3a. As the ocean widened, the rift extended northward, made an abortive attempt to open up an ocean between Greenland and North America, as illustrated in Figure 12-3b for 135 million years ago, and then penetrated between Greenland and Europe, as shown in Figure 12-3c for 65 million years ago. The ocean eventually opened completely, as shown in Figure 12-3d, but along a slightly different line than that shown in Figure 12-4d for 510 million years ago.

This outline history shows that the conti-

nents have combined and dispersed in various ways at several times during the past 500 million years. It is fairly well established that when the continents collide, a mountain range is formed. Five hundred million years is less than 15% of the time since the formation of the earth, and there is geological evidence for the formation of mountain ranges much older than 500 million years. Many geologists are studying the history of ancient mountain ranges, attempting to explain them in terms of plate tectonics and continental collisions, and hence to unravel the history of plate movements and continental drift back through time.

World patterns of climate must have changed significantly if the continents migrated as illustrated in Figures 12-3 and 12-4. When the continental fragments are combined, as in Pangaea 225 million years ago, the storm winds were free to roar more than halfway around the world across the vast expanse of the ocean, unimpeded by continents or mountain ranges. Similarly, the system of ocean currents would be much simpler than at the present time, when the currents have to circulate around the seven seas, diverting wherever they impinge on a continent. The inner parts of the supercontinent were so far away from the ocean that they probably received little rain.

As a supercontinent fragmented and dispersed, the large ocean would become subdivided into smaller oceans whose shape and distribution would change as the continents moved (Figure 12-3). Ocean currents and wind patterns would also change with time.

Paleoclimatic studies have been applied in several ways in connection with continental drift in the past, either to prove it or to disprove it. A. A. Meyerhoff concluded that the planetary wind-current and ocean-current pattern has been essentially the same for the past 800 to 1,000 million years, which certainly does not permit the continental migration illustrated in Figures 12-3 and 12-4. We will return to Meyerhoff's interpretation in Chapter 13.

PRIORITY IS FOR THE BIRDS

One of the striking aspects of the present activity in geology and geophysics is that scientists with quite different interests are finding that their own specialties not only relate to the new global schemes but also may shed light on problems that previously appeared to be unrelated. There has been considerable discussion in the literature about which scientist or philosopher should have priority for proposing that the continents had drifted or for having expressed the germ of the idea. In an exchange of letters in *Geotimes*, R. S. Dietz (November 1972) was able to combine ornithological data with the continental reconstructions of Figures 11-3 and 12-3 to support his contention that priority for the idea did not strictly belong to any person, but to the bird illustrated in Figure 12-5. He wrote:

I failed to recognize a prior priority. Since long before the advent of man on this earth, let alone geologists, a doughty bird, the sooty hoodwink (Oceanites erraticus), already "knew" about continental drift.

Displaying true grit, this bird migrates each year from its feeding ground in the Antarctic peninsula to its nesting site in Spitzbergen. Even more remarkable, its course is not direct; instead, the sooty hoodwink flies a zigzag path, lacing back and forth across the Atlantic. This bird first touches down in Argentina, then it flies to southern Africa landing amongst the fever trees of the Limpopo River, then on to Brazil, then back to North Africa, to Newfoundland, and to Land's End and finally to Spitzbergen via Greenland. (Apparently blown off course by heavy storms, an errant flock was sighted this spring in the spaghetti fields of the Po Valley.) With timing equalled only by the renowned swallows of San Juan Capistrano, the sooty hoodwink arrives in Spitzbergen precisely on April 1.

The reason for this circuitour flight, at odds with the principle of least work, becomes apparent once we recognize that the Atlantic is a rift ocean split asunder by continental drift. Once this ocean is closed, this

zigzag course becomes the shortest distance between 2 points on the globe, the great circle route (see sketch by my colleague John Holden).

R. S. Dietz, a geologist with a fertile imagination, has been involved with other priority matters. Priority for an idea is usually assigned to one scientist or team in the history books, but it is not uncommon for several groups to reach similar conclusions almost simultaneously. Then there may be a hassle over who is the first to get the idea into print. Hess is credited with the idea of sea-floor spreading but, in the late 1950s, many geologists were developing conceptual models involving mantle convection beneath ocean ridges. One of these was H. W. Menard of the Scripps Institution of Oceanography.

In his 1971 book, *Science, Growth and Change*, Menard described some of the events related to the publication of Hess's geopoetry. Hess wrote the paper and mailed copies of the typed manuscript to many colleagues in 1960. This is common practice, the idea being to keep scientists informed about who is gathering what data and who is thinking about what. Menard reported that no one expressed much interest in the preprint. The paper was eventually published in 1962, with the title "History of the Ocean Basins". In the meantime, Dietz had published similar concepts in *Nature*, in a 1961 paper titled "Continent and Ocean Basin Evolution by Spreading of the Sea Floor." Menard added that in 1963, Dietz "tidied things up" by giving priority to Hess, which was a most unusual act considering the prevalence of multiple discovery at this time in the earth sciences.

This topic was also discussed by B. Heezen of Columbia University in a book review. Referring to the suggestion that Dietz got the idea for sea-floor spreading from Hess's preprint, he wrote in 1974: "But from whom had the then-landlubber Princetonian geologist (Hess) gotten it? Was it from his seagoing rivals at nearby Columbia University, who he often complained

Fig. 12-5 A doughty sea-bird. The tortuous migratory flight of this bird, shown in part *a*, apparently has its origin in the days when Pangaea existed, and the bird flew a direct route from south to north, as shown in part *b*. (Flight of fancy drawn by J. C. Holden, and reproduced with his kind permission. Published with letter to *Geotimes,* November 1972, by R. S. Dietz.)

(a)

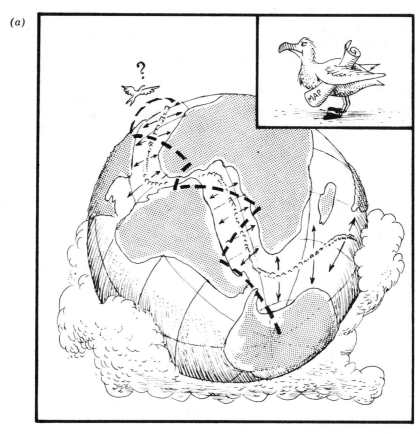

(b)

JOHN HOLDEN

were so busy making discoveries that they took insufficient time to publish their findings or to contemplate the ramifications of their observations?''

Wherever the idea of sea-floor spreading did originate, it was Hess who prepared the first full-length paper putting all the details together, and it is Hess who is credited with priority. We will see in the next chapter that the race for priority with the magnetic tape recorder, which dates the spreading, depended on factors over which authors have no control: the editor's choice of referees who advise him on acceptance or rejection of manuscripts. Chapter 18 includes an outline of another race for priority, with scientists from three universities grappling with the problem of earthquake prediction and theoretical models for earthquakes.

SUMMARY

The path of polar wandering for a continent is determined as follows. The fossil magnetic poles are measured for a large number of rocks with ages within a specific interval of time, covering tens of millions of years. The average position is the mean paleomagnetic pole position for that interval of time. Similar pole positions are located for a series of time intervals, and the line connecting these points gives the path of apparent polar wandering for hundreds of millions of years. Paths of polar wandering have been determined for every continent. Each path is different, suggesting that the continents have moved not only relative to the poles, but also relative to each other.

Dietz and Holden, using new geometrical fits of the continent, paths of polar wandering, and the constraints of plate tectonics and the oceanic magnetic anomalies have reconstructed the ancient supercontinent that existed about 200 million years ago, and traced the stages of breakup and dispersal of the present continents with cartographic precision. The site of New York City, for example, was situated on the equator 225 million years ago, and at longitude 10°E instead of its present 74°W.

There is evidence that the supercontinent was produced by the collision and combination of several separate continental masses between 340 and 250 million years ago. Paths of polar wandering are interpreted to sketch the movement of these continental masses starting 510 million years ago. The evidence suggests that ocean basins have opened, closed, and opened again. A mountain range is produced along the line where continents collide as an ocean basin is closed. Ocean currents, wind patterns, and world climates must have changed as the continents migrated.

SUGGESTED READINGS are at the end of the book.

chapter 13

A Revolution Proclaimed –
The Evidence Challenged

We outlined the development of the revolution in earth sciences in Chapter 2, and the sequence of major events was summarized in Table 2-1. This is a good time to return to Table 2-1 to review the positions in history of the topics and evidence discussed so far. Figure 13-9 is another historical summary. In the 1950s we had revival of interest in continental drift arising from the exploration of the ocean floor (Chapter 3) and from the development of paleomagnetism (Chapter 11). In 1960 H. H. Hess presented the hypothesis of sea-floor spreading. The Vine-Matthews model, combining sea-floor spreading with linear magnetic anomalies and polarity reversals (Chapter 10), was a key development in 1963, but it did not have much impact until the winter of 1966 and 1967. In this chapter, we will examine in detail how everything seemed to come together in this brief period, because this was when revolution was proclaimed. The general acceptance of sea-floor spreading was promptly followed by the formulation of plate tectonics and the incorporation of all of the earthquake evidence reviewed in Chapter 5.

Applications of plate tectonics have since been made to all kinds of problems. Enthusiastic geologists and geophysicists have enjoyed the euphoria of a great, new global theory that explains away their problems. The evidence described in Chapter 10 and the earthquake evidence in Chapter 5 is persuasive. Most geologists and geophysicists have become convinced of the essential validity of plate tectonics and the validity of the provisional dating methods involving the oceanic magnetic anomalies but, as we noted in Chapter 2, not everyone is satisfied. The dissident voices are heard less often these days. Let us hope, however, that they do not become muted, because someone has to ask those difficult questions that the conceptual models do not explain. After recounting the stages of development of the revolution, we will therefore examine some of the criticisms that have since been leveled at the Vine-Matthews model and its extrapolations.

THE VINE-MATTHEWS HYPOTHESIS AND HOW L. W. MORLEY WAS SCREWED BY THE ESTABLISHMENT

The Vine-Matthews model of Figures 10-4 and 10-5 was at first received with some scepticism because not one of the three basic assumptions that they invoked was generally accepted in 1963. These were (1) sea-floor spreading, (2) the contribution of magnetized crustal rocks to the production of the oceanic magnetic anomalies, and (3) polarity reversals of the earth's magnetic field. There were other difficulties. The best known examples of linear anomalies, in the northeast Pacific Ocean, did not appear to be parallel to any existing or preexisting oceanic ridges and, in surveys of known midoceanic ridges, the existence of linear anomalies paralleling the central anomaly had not been established.

The generally unfavorable climate for ideas of this sort is demonstrated by the fact that a Canadian, L. W. Morley, independently worked out the same model but could not get it accepted for publication in a scientific journal. His manuscript was first submitted to *Nature* in January 1963 and rejected, and then it was rejected by the *Journal of Geophysical Research*. Morley suggested that a nearly unbroken record of the earth's magnetic field may exist in the permanent magnetization of the rocks on the ocean floors, and he suggested that determination of the rate of mantle convection might give the history of the reversals of the magnetic field. The paper by Vine and Matthews was accepted by *Nature* and published in September 1963.

In a letter to *Geology* in April 1974, N. D. Watkins wrote the following.

I recommend the reading of these articles to earth scientists, if only for the partial reproduction of L. W. Morley's paper on the interpretation of linear magnetic anomalies, polarity reversals, and his suggested 3- to 5-cm/year separation rate for mid-oceanic ridges. Submitted for publication twice in early 1963, and rejected on both occasions because (in one reviewer's opinion), "Such speculation makes interesting talk at cocktail parties, but it is not the sort of thing that ought to be published under serious scientific aegis," the manuscript certainly has substantial historical interest, ranking as probably the most significant paper in the earth sciences to ever be denied publication.

Morley's paper was published in 1964 with coauthor A. Larochelle but, by then, priority being as it is (see Chapter 12), the idea was established as the Vine-Matthews hypothesis.

Another manuscript rejected by the *Journal of Geophysical Research* in 1963 was J. T. Wilson's explanation for the origin of oceanic volcanic chains such as the Hawaiian Islands. This is now receiving great attention as the hot-spot hypothesis (Chapter 16), but the reviewer recommended rejection on the grounds that the paper contained no new data, had no mathematics in it, and disagreed with current ideas. Wilson was able to publish it immediately in the *Canadian Journal of Physics*, however, and priority was thus established despite the fact that few earth scientists read that journal.

By 1965, the two basic assumptions about magnetic anomalies and polarity reversals had become more widely accepted, and H. W. Menard of the Scripps Institution had identified a belt of ridges and troughs in the northeast Pacific, parallel to the linear anomalies. J. T. Wilson of Toronto University showed that the apparent absence of oceanic ridges elsewhere in the northeast Pacific is due to complications caused by the large fracture zones shown in Figure 3-12. Wilson also introduced the concept of transform faulting, proposing that these fault zones are a special kind of lateral fracture related to opening up of the crust along the ridge crests (Figure 5-3). This concept explained some problematic aspects of the sea-floor spreading model.

Despite this evidence, J. R. Heirtzler and his associates at Lamont Geological Observatory remained unconvinced. At a symposium in 1965, they wrote:

We are not sure of the ultimate origin of the ridge magnetic anomalies. . . . However we feel that the variation in amplitude of the axial anomalies as well as the completely different character of the flank anomalies argues against the Vine and Matthews hypothesis.

In a *Science* article in December 1966,

Vine reviewed new evidence from several magnetic surveys that linear magnetic anomalies can be correlated from one oceanic ridge profile to another, that these do parallel the ridge crests, and that the anomalies are remarkably symmetrical about the axis of the ridge in many regions. Vine considered this evidence to be virtual proof of sea-floor spreading.

Also in *Science,* in December 1966, W. C. Pitman and Heirtzler compared the magnetic anomaly patterns from the Pacific-Antarctic ridge with those from the Reykjanes ridge south of Iceland. They reported from the Pacific Ocean the existence of linear anomalies, symmetrically disposed about the ridge, that fitted perfectly with the polarity reversal time scale. They concluded:

We feel that these results strongly support the essential features of the Vine and Matthews hypothesis and of ocean-floor spreading as postulated by Dietz and Hess.

The definitive touch was the close match between Pitman's "magic profile" and N. D. Opdyke's results on magnetic reversals in sediment cores; these were measured in a room just across the corridor from Pitman's office (see Figures 9-7, 10-7b, and 10-9). With acceptance by the Lamont group, there followed about two years of almost unbearable excitement that spread to other geological departments. The revolution had begun.

A SYMPOSIUM,
TWO SCIENTIFIC MEETINGS,
AND A REVOLUTION

It was at a November 1966 Symposium on "The History of the Earth's Crust" at the Goddard Institute for Space Studies in New York that the various lines of evidence were brought together by and for a group of experts. Six of the 16 contributed papers were concerned with the earth's magnetism and its history as recorded in rocks. These papers convinced most of the symposium participants that a direct correlation existed between the polarity time scale measured in

millions of years, the widths of linear magnetic anomalies measured in kilometers, and the thicknesses of layers of alternately magnetized oceanic sediments measured in centimeters.

Another compelling item was the evidence presented by L. R. Sykes of Lamont. He reviewed the distribution of earthquakes on midocean ridges and the direction of motion associated with each break that caused an earthquake. He reported that the earthquake epicenters coincide precisely with either the ridge crests or the transverse fracture zones that offset them; they are absent from the fracture zones extending beyond the ridge crests. This evidence is shown in Figure 4-5. He also demonstrated that the relative movements for earthquakes along these lines is divergent along the ridge crest and is lateral, with neither divergence nor convergence, along the segments of transform faults. This is precisely as predicted by Wilson's concept of transform faults, as illustrated in Figure 5-3.

The evidence presented by P. M. Hurley and J. R. Rand was also persuasive. They demonstrated that portions of Africa and South America that come into contact on predrift reconstructions have matching age provinces, as illustrated in Figure 11-6.

In the United States, geologists and geophysicists gather at three large scientific meetings each year. They present their research results in a series of short, formal papers read from a podium and discuss the most recent developments informally in corridors between the lecture rooms, in social smokers, and in local bars. The same evidence reviewed at the Goddard Symposium was widely disseminated in the same month (November 1966) at the Annual Meeting of the Geological Society of America, and the effect was dramatically demonstrated at the April 1967 Annual Meeting of the American Geophysical Union.

If they intended to present a formal paper at the April Meeting, scientists were required to submit a short abstract by January, which would then be printed for distribution with the program at the Meeting. By

January 1967 nearly 70 abstracts on sea-floor spreading had been submitted to the organizing committee. Geologists and geophysicists had lost no time in reconsidering their own research data within the framework of sea-floor spreading, and many of them found that the pieces fit together more satisfactorily when they did so.

H. H. Hess, the geopoet, presented an invited lecture to a large audience at this meeting, recounting the history of the development of the concept of sea-floor spreading, the kind of retrospective review that one normally hears years after a series of events. But history moved fast in the mid-1960s, and here was a saga sung during the same months that the hypothesis was becoming known and accepted. Three years later, *Time* magazine recognized the hypothesis as "geofact." The bandwagon became a juggernaut.

H. W. Menard described the change of attitude in his 1971 book. He wrote that in 1966 and 1967, many members of the marine geology establishment reexamined the data they were familiar with, saying to themselves: "My observations are not compatible with sea-floor spreading, and I shall prepare a critical demonstration that this is so and thus demolish this nutty idea and we can all get back to work." Instead, the scientists found that their results were compatible with sea-floor spreading, and an elaborate network of confirmations appeared. This was the revolution: not new data, but new believers, a new conceptual model, a new paradigm.

The acceptance and application of sea-floor spreading to a wide variety of problems shed light on many, but it did not solve them all, by any means. Sir Edward Bullard, two months after the American Geophysical Union Meeting, summarized the situation in his Bakerian Lecture to the Royal Society of London:

The lecture on which this paper is based was given in June 1967; it was a well chosen time, the threefold story of the reversals of the field had just become clear and could be easily and elegantly set out. In the few months needed to write this paper there has been an avalanche of new results which has revealed many discrepancies and many matters needing elucidation; the usual chaos of the earth sciences is clearly about to be reestablished at a higher level of understanding.

PLATE TECTONICS

In the same year (1968) that J. R. Heirtzler and his Lamont associates applied their worldwide correlation of linear magnetic anomalies to sea-floor spreading, three of their colleagues at Lamont, B. Isacks, J. Oliver, and L. R. Sykes, published a significant review of the applications of earthquake studies to what they called "the new global tectonics." D. P. McKenzie and R. L. Parker (Cambridge University), W. J. Morgan (Princeton University), and X. LePichon (Lamont) published essentially concurrent papers in 1967 and 1968 dealing with the same topic from various viewpoints and, from this batch of papers, the theory of plate tectonics emerged in fairly well-defined form. We reviewed the theory and the earthquake evidence in Chapter 5.

PROCLAMATION AND CHALLENGE

J. Tuzo Wilson was an early champion of the revolution in earth sciences, and he wrote that the revolution was similar to, and as significant as, that which changed the approach to chemistry about 1800, that which occurred in biology about a century ago with Darwin's theory of evolution, and that which occurred in physics when classical views were replaced by modern views. The revolution shows a promise of advancing the study of the earth sciences from a stage of data gathering into the stage of formulation of a precise, comprehensive theory of global geology and use of the theory to make predictions.

The December 1968 issue of *Geotimes* printed (1) an exposition of the revolution by Wilson, (2) a letter from Russian geologist V. V. Beloussov maintaining that the concepts of continental drift and sea-floor spreading should serve only as working hypotheses along with other hypotheses, and (3) a reply from Wilson restating his contention that the revolution will unite formerly fragmented branches of the earth sciences into a new unified science of the dynamic earth.

Beloussov wrote a critique in 1970 entitled "Against the Hypothesis of Ocean-Floor Spreading." He found the simple models of plate tectonics, the divergent boundaries and the convergent boundaries illustrated in Chapter 5, to bear little or no relationship to the geology of the real earth. He argued that the rugged, mountainous relief of some ocean ridges could not be transformed into the level, deep ocean floor a few hundred kilometers away simply by horizontal movement. And he presented many other arguments of similar type.

Beloussov is especially concerned that our knowledge of the structure and development of the oceans is still much sketchier than geological data from the continents, yet the hypotheses of sea-floor spreading and plate tectonics lean heavily on data from the oceans. He believes that it is not appropriate to discard other hypotheses because of the current popularity of one model, and he presented a process of oceanization as an alternative working hypothesis. This is based on an earth model dominated by vertical movements within the earth rather than by horizontal movements over the surface of the earth.

He postulated that large areas of continental crust have been destroyed by the intrusion and extrusion of lava. Crustal blocks founder and sink beneath the lava, and the lava forms new ocean floor. This process has produced the ocean basins. The hypothesis is developed in detail and includes the historical perspective gained from structural studies of continental geology. It finds few followers among Western geologists at present.

Harold Jeffreys, the eminent geophysicist at Cambridge, has argued for years that the physical properties of the earth simply will not permit the earth to behave in the way required by the advocates of continental drift and, more recently, of plate tectonics. His views on the futility of paleomagnetic interpretations were stated in Chapter 12, where we also considered a few of the other problems and assumptions involved in paleomagnetic interpretations. One of his younger colleagues, P. Wesson, has been arguing the case against continental drift and generating long lists of items that are not explained satisfactorily by the mobile models.

Another challenge came from A. A. Meyerhoff (American Association of Petroleum Geologists) and C. Teichert (University of Kansas) in 1970 and 1971. Certain kinds of sedimentary rocks indicate the climate existing at their time of formation. Desert sandstones and coral reef limestones are familiar examples. Meyerhoff reviewed the global distribution of paleoclimate indicators in successive time periods back to 1,000 million years ago. He concluded that the climatic zones indicated by their distributions could only have existed if the planetary wind-current and ocean-current pattern had been essentially the same for 800 to 1,000 million years. If this conclusion is valid, Meyerhoff states that the only explanation is that the rotational axis of the earth, the continents, and the ocean basins have been in the same positions for the past 1,000 million years. This does not permit sea-floor spreading or plate tectonics to operate at all.

These and other related conclusions about rocks and fossils, based on "factual, observable data, in sharp contrast to speculations based on recent geophysical-oceanographic studies," convinced Meyerhoff and Teichert that, until advocates of the new global schemes find alternative explanations for the distribution of paleoclimate indicators, the concepts of sea-floor spreading and plate tectonics will have to be regarded as speculations supported by only a fraction of the known geological, paleontological, and paleoclimatological data. A bold position to take in front of a rolling juggernaut.

In his 1973 book, *A Revolution in the*

Earth Sciences, A. Hallam of Oxford University included a couple of pages (pp. 105-106) to "take brief note of some of the leading dissenters." He wrote: "It would be unfortunate if all such people were to be dismissed as diehards. The more intelligent, less bigoted ones may well be able to point out anomalies in current theory which require resolution, or to indicate fields where plate tectonics fails to supply a satisfactory explanation. Just such activities are part of the mopping up operations." He discussed Meyerhoff in the following words:

The American stratigrapher A. A. Meyerhoff falls in a different category from the others. He has somewhat heroically undertaken to challenge a large proportion of the evidence for continental drift, from the distribution of ancient evaporites and tillites to paleomagnetic data, in a series of lengthy, belligerent articles which, if nothing else, attest to an amazing capacity to cope with a vast literature. Meyerhoff must be credited with the only sustained and serious effort to tackle the relevant evidence, but so far he has not succeeded in raising serious doubts among the Earth sciences community at large.

Who has room for doubts once their minds have been made up? There have been few serious attempts to reply to the questions raised by Meyerhoff. Perhaps it is easier to ignore the questions.

In another 1973 historical account, *Continental Drift, the Evolution of a Concept,* U. B. Marvin of the Smithsonian Institution devoted pp. 182-190 to "Voices of Dissent," and she wrote:

In no cases can the dissenters be accused of simply trying to impede progress; they are asking very seriously whether the new view of the earth coincides more accurately with reality and will serve better than the old for solving geological problems.... New textbooks tend to omit all serious discussion of outmoded views. Indeed, although they would be totally helpless without the knowledge of the earth gathered over the last century, some geologists already regard as comic all ideas held before

1960....Meanwhile, the objections raised by the dissenters should, at the very least, inspire a critical attitude on the part of readers threading their way through the volumes of new articles published weekly in the scientific and news journals of the world.

A SECOND LOOK AT THE REAL MAGNETIC ANOMALIES

The Vine-Matthews hypothesis represented by Figures 10-4 and 10-5 is a keystone in the theory of plate tectonics. When Heirtzler and his Lamont colleagues in 1968 analyzed the magnetic anomalies on a global scale and extrapolated the time scale as in Figures 10-9 and 10-10, they noted that if the Vine-Matthews theory is basically in error, then the conclusions of their analysis do not apply. It has been pointed out by Tj. H. Van Andel, Beloussov, A. A. Meyerhoff, and H. A. Meyerhoff that the model picture of symmetrical, parallel, linear bands of anomalies displayed in so many published diagrams, such as Figures 10-4 and 10-5, differ significantly from the real magnetic anomalies, as illustrated in Figures 10-6 and 13-1.

The evidence from the oceanic magnetic anomalies was reviewed by Beloussov in 1970. He considered the spacing of anomalies, including some of those given in Figure 10-6, and concluded that the ratios of anomaly widths along most ridge segments are not strictly in accord with the concept of uniform sea-floor spreading. Three of the examples usually quoted "are the three whales that support the hypothesis." He also drew attention to a fact rarely mentioned by proponents of sea-floor spreading: for distances of thousands of kilometers along the ridges, including parts of the South Atlantic Ocean and in the Indian Ocean, the anomaly pattern does not show symmetry at all. Referring especially to the magnetic evidence, Beloussov concluded that:

This hypothesis is based on a hasty generalization of certain data whose sig-

Fig. 13-1 Some examples of magnetic anomaly profiles across oceanic ridges where the symmetry is not very good and where correlation of individual anomalies is less evident than in the ideal examples. Compare Figures 10-5c and 10-6. Examples selected by A. A. Meyerhoff and H. A. Meyerhoff, 1972, Amer. Assoc. Petroleum Geologists Bull. 56, 337-359.

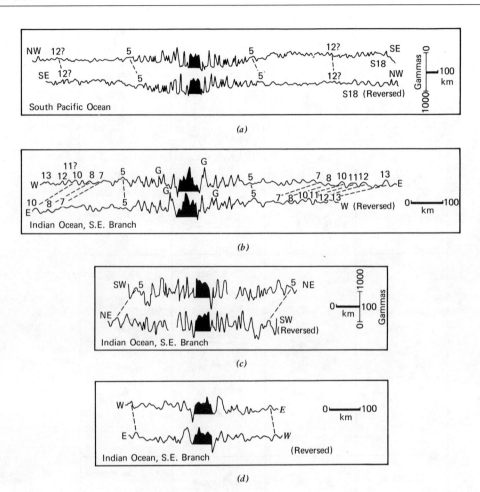

(a)

(b)

(c)

(d)

nificance has been monstrously overestimated. . . . It is evident that not a single aspect of the ocean-floor spreading hypothesis can stand up to criticism.

A more penetrating and critical review of the distribution and interpretation of the magnetic anomalies was published in 1972 by A. A. Meyerhoff and his father, H. A. Meyerhoff. They concluded that the model picture is fiction. They wrote:

The truth is that the magnetic anomalies form a series of irregular bands, parallel to subparallel–some not even subparallel . . .–associated in many places with segments of midocean ridges. A large anomaly is common at the center of the ridge but generally is discontinuous. The bands on either side of the central anomaly may be symmetrical, but more commonly are not. Locally, there are no linear anomalies, and those anomalies that are present show no obvious pattern. . . . In a few places, the anomaly bands are not associated with a midocean ridge, but cut across abyssal hills or turbidite plains. In numerous places, the anomalies strike directly into the continents.

Fig. 13-2 Examples of specific anomalies identified by number from different regions, illustrating the quality of characteristic magnetic signatures, or shapes of the anomalies. (Examples selected by A. A. Meyerhoff and H. A. Meyerhoff, 1972, see Figure 13-1.)

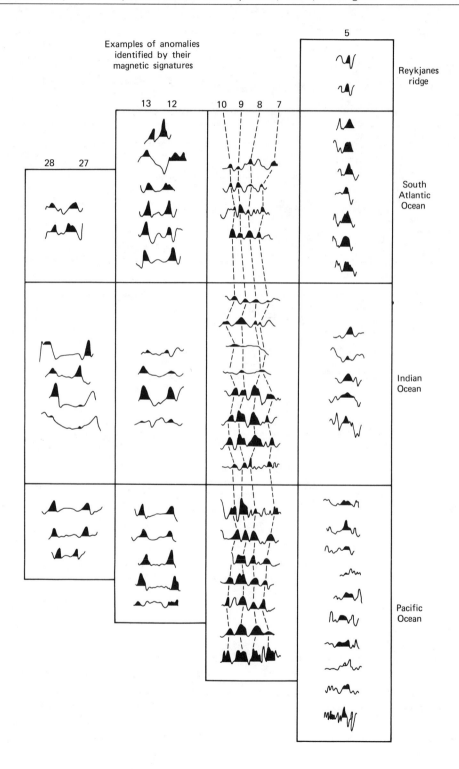

These anomalies do not sound like those described in Chapter 10 and illustrated in Figure 10-6. Is it possible that we have been following a plot for one of Tom Stoppard's plays? What are the real magnetic anomalies?

Meyerhoff and Meyerhoff noted that symmetry of magnetic anomalies had not been discovered before publication of the 1963 Vine-Matthews model (Figure 10-5), but that after publication of the Reykjanes Ridge and Pacific-Antarctic profiles in 1966 (Figure 10-6), symmetry was assumed to be the rule. Examples in Figure 13-1 show that the assumption is not always well founded. The farther the profile is away from the ridge, the greater is the asymmetry.

According to the model, the parallelism and linearity of the anomalies permit worldwide correlation of specific dated anomalies (Figure 1-3) by their magnetic signatures. Examples of numbered anomalies from three oceans are shown in Figures 10-6, 13-1, and 13-2. Despite some remarkably similar shapes, the correlation does not seem to be very well founded for all anomalies.

The magnetic anomalies plotted on the world map in Figure 10-10 are not continuous along all ocean ridges. Some of the gaps are due to incomplete mapping, but it appears that almost 20% of the ocean ridge system lacks magnetic anomalies that can be correlated with anything. Anomaly con-

Fig. 13-3 Polar view of the northern hemisphere showing the trends of magnetic anomaly patterns in the ocean basins around the old stable parts of the continents. (Map from A. A. Meyerhoff and H. A. Meyerhoff, 1972, see Figure 13-1.)

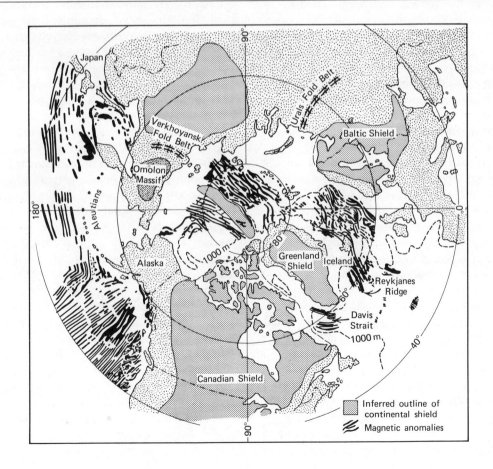

tinuity breaks down for 1500 km on the mid-Atlantic ridge south of Iceland, for 7000 km from south of Africa into the Indian Ocean, and for 3700 km across the Arctic Ocean north of Iceland (Figure 13-3). The gaps can be seen on Figures 13-5 and 13-7 and on Figure 10-10. It has been assumed that the absence of correlatable magnetic anomalies along the S.W. branch of the mid-Indian ridge means that the ridge is dead, without sea-floor spreading, but the earthquake activity along the ridge indicates that it is just as lively as the rest of the ocean ridge system. This is a contradiction for which the plate tectonics model appears to have no explanation.

Figure 13-3 is a polar projection looking down on the north pole with the black bands showing positive magnetic anomalies in several oceans. Note the classic example of Reykjanes Ridge just south of Iceland (Figure 10-6) and contrast this with the general lack of parallelism and symmetry north of Iceland, between Greenland and Norway. An enlarged map of this region shows that the anomalies are actually oblique to the ocean ridge as defined by the line of earthquake activity, which should correspond to the spreading plate boundary. In several places, there are parallel anomalies, apparently with no relationship to an existing ridge.

Clearly, the linear, symmetrically disposed magnetic anomalies are not as characteristic of the ocean ridges as they are for a few classic examples (Figure 10-6). Meyerhoff and Meyerhoff consider the magnetic tape recorder model and its extrapolation back through time as an example of assumptions built on assumptions, causing speculation to evolve into hypothesis, hypothesis to be transformed into theory, and this to become "established fact." They are especially critical of the claim that deep-sea drilling in the South Atlantic Ocean established the ages of the rocks believed to cause the magnetic anomalies, as we will see in the next chapter.

Even if the real magnetic anomalies do not conform everywhere to the ideal model, is this evidence against the "hypothesis" or "theory" or "fact" of sea-floor spreading? There has been little reaction so far in the scientific literature to the questions raised by the Meyerhoffs. Recall Hallam's comment: "an amazing capacity to cope with a vast literature . . . but so far he has not succeeded in raising serious doubts among the Earth sciences community at large." It is not easy to cope with these questions. Most scientists appear to remain happy with sea-floor spreading and plate tectonics, to assume that all problems will be explained, eventually, and to retain faith that it is the best model we have ever had for the earth. After all, what else is there?

ALTERNATIVE EXPLANATIONS

Figure 13-4 illustrates an alternative explanation for magnetic anomalies, which are approximately parallel to ridge axes and sometimes roughly symmetrical but not necessarily so. Beloussov suggested that the ocean ridges are young features built up of successive lava flows arranged like tile on a roof, with the older flows extending out from underneath the younger ones. The magnetic anomalies simply record the succession of lava flows (compare Figures 8-12 and 8-13). This kind of explanation is consistent with his concept of oceanization. It does not explain the regularity required according to the sea-floor spreading model, but we have seen that the regularity may not, in fact, be as regular as normally assumed.

Meyerhoff and Meyerhoff presented an alternative working hypothesis in 1972 that explains the ocean ridges as ancient features of the earth, inherited from 2,000 million years or more in the past. The associated magnetic anomaly bands have a complex origin and history, being caused by magnetic contrasts in lavas erupted intermittently along the ocean ridges through billions of years. They developed this idea from the global distribution of the ocean ridges and anomaly bands, as shown in Figures 13-3, 13-5, 13-6, and 13-7. These

Fig. 13-4 A structural interpretation of successive lava flows from the central rift valley region of an ocean ridge, with flows arranged something like tiles on a roof. If these flows are magnetized as shown, following polarity reversals, the effect is similar to that shown in Figure 10-4. This could produce a series of linear magnetic anomalies, approximately parallel to the rift valley. (Sketch based on proposals of Tj. H. Van Andel, developed by V. V. Beloussov in *Tectonophysics, 9,* 1970, 489-511.

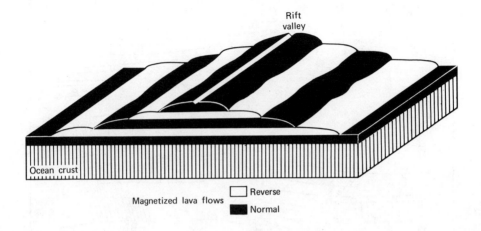

Fig. 13-5 Polar projection map of magnetic anomaly bands around North America. (From A. A. Meyerhoff and H. A. Meyerhoff, 1972, see Figure 13-1.)

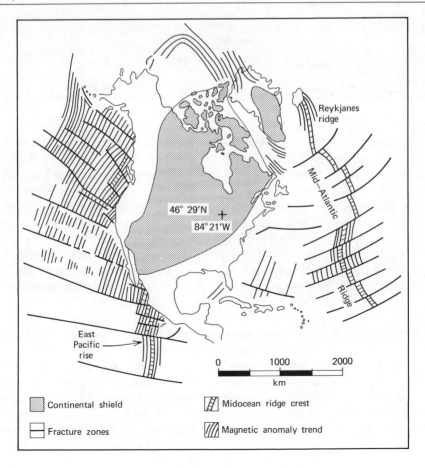

Fig. 13-6 Polar projection map of magnetic anomaly bands around South America. (From A. A. Meyerhoff and H. A. Meyerhoff, 1972, see Figure 13-1.)

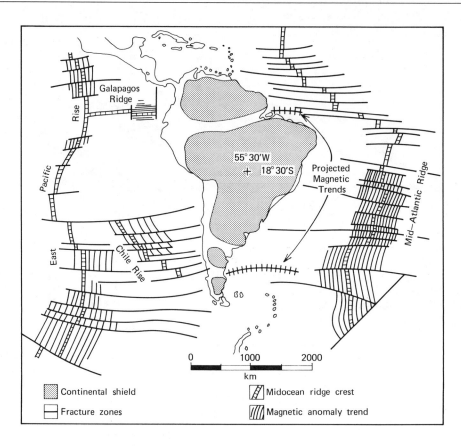

figures also show the distribution of the ancient shields or nuclei of the continents, which are made up of rocks older than about 2,000 million years. Parts of these are covered by layers of younger sedimentary rock, but the figures show the probable extent of the shields if the younger rocks were stripped off.

H. W. Menard, a fertile developer of many concepts in sea-floor spreading and plate tectonics, discussed the median position of the ocean ridge system in 1958 and then noted in 1965 that the ridges were more nearly median with respect to continental shields than to continental margins. This is the observation examined in more detail by the Meyerhoff team. They coupled it with the observation that the associated magnetic anomaly bands do not remain in the central regions of oceans, but they

commonly disappear at continental margins between the ancient shield areas. Meyerhoff and Meyerhoff listed 15 places where the anomaly bands "dive" beneath a cover of continental rocks younger than 2,000 million years. Readers can find them all on the figures for themselves.

The magnetic anomalies shown in Figure 13-3 exhibit many contradictions for sea-floor spreading and plate tectonics. There is a complex pattern of anomaly bands arranged in approximately concentric fashion around the continental shields bordering the Arctic Ocean. The trend of the anomaly bands in the sea between Greenland and Norway can be followed part way around the Greenland shield via Reykjanes Ridge and the Davis Strait, and around the Baltic shield with possible connection to the Urals fold belt in Asia. The trend of the anomaly

Fig. 13-7 Polar projection map of magnetic anomaly bands around Africa. (From A. A. Meyerhoff and H. A. Meyerhoff, 1972, see Figure 13-1.)

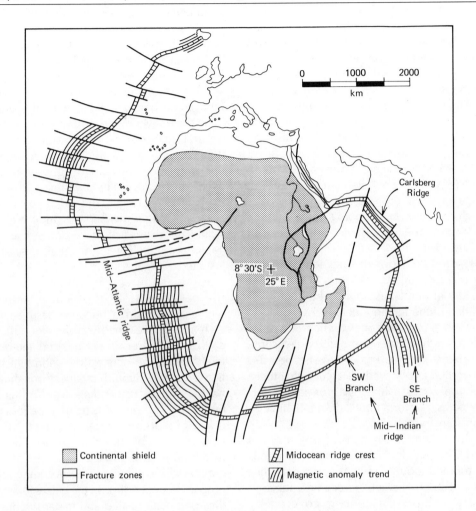

bands extending northward from Japan can be followed through the Verkhoyansk fold belt in Siberia and hence into the anomaly band in the Arctic Ocean; this separates the two Siberian continental shields. The trend of the adjacent set of anomaly bands in the Arctic Ocean can be extended to the east of the Omolon shield in Siberia to connect with the anomalies in the west Pacific Ocean. There is a third set of anomalies in the Arctic ocean that bends: one trend extending across Alaska to the anomaly bands in the northeast Pacific Ocean, and the other trend possibly connecting with the Davis Strait anomalies through the gap between the Canadian and Greenland shields.

The distribution of the anomaly bands around the Canadian shield in North America is shown schematically in Figure 13-5. Compare these trends of magnetic anomalies with the real anomaly pattern given in Figure 13-3. Meyerhoff and Meyerhoff suggested that the anomaly bands associated with the East Pacific Rise may divide into two branches, one extending northward on the western side of the continental shield and the other extending toward the northeast on the other side of the shield area. The latter anomaly band is not continuous because of the continental rocks, but they cited anomalies in the Gulf of Mexico and west Atlantic Ocean off

Florida that could represent a connection with the main Atlantic anomaly band. They contended that this is less speculative than the proposal that the North American plate has overridden the East Pacific Rise; this they considered a physical impossibility.

The symmetrical arrangement of the ocean ridges around South America, shown in Figure 13-6, was discussed by Menard. Magnetic anomaly bands associated with the Galapagos ridge and the Chile Rise can possibly be extended across the continent between pairs of ancient shields. Meyerhoff and Meyerhoff predicted that the eastern ends of such projected anomaly bands might be found by mapping in the Atlantic Ocean in the two places indicated on Figure 13-6.

Figure 13-7 shows the magnetic anomaly bands girdling Africa, which is formed mainly of one large shield. The Southwest branch of the Indian Ocean ridge is cut by numerous fracture zones, which could account for the destruction of the preexisting, ancient magnetic anomaly bands. Meyerhoff and Meyerhoff argued that these fracture zones, which are normally interpreted as transform faults, do not behave as transform faults at all.

Here, then, is a pattern of ocean ridges and associated magnetic anomaly bands distributed around ancient continental nuclei in approximately concentric fashion. Parts of these have been covered by younger rocks, giving the picture of magnetic anomalies ending abruptly against continents or continental shelves. It is not a perfect pattern, by any means. But is it any less perfect than the match claimed between the real magnetic anomalies (Figures 13-1 and 13-3) and the idealized models (Figure 10-5)?

Observed patterns are not really satisfying unless they have an explanation. Before plate tectonics was conceived, the dominant theory for global tectonics involved contraction of a slowly cooling earth. Meyerhoff and Meyerhoff proposed that the ocean ridges and associated anomaly bands are caused by compression resulting from earth contraction.

There is evidence that the asthenosphere may be thin or even absent beneath some continental shields, producing a very thick lithosphere, and certainly the lithosphere is thinnest beneath the oceans. This general relationship is shown in Figure 7-6, and Figure 13-8 illustrates the Meyerhoff scheme, schematically, with a much thicker lithosphere below the continental shields. If the outer layer of the earth is compressed, as in Figure 13-8b, we might expect that the pressure would be taken up by buckling of the thin oceanic lithosphere between the thick continental shields, producing an ocean ridge. The ridge would be fractured on its upper surface, and the result would be a series of more or less parallel fractures along the length of the ridge, as illustrated in Figure 13-8c.

These fractures are natural lines for the concentration of volcanic activity whenever the ridge becomes rejuvenated by renewed buckling because of continued earth contraction. Lavas erupted intermittently along the ocean ridges through billions of years would escape through the fracture zones. The ridges would thus be built up of a series of roughly parallel lava flows of different ages, different thicknesses, different directions of magnetization depending on the polarity of the magnetic field, and different magnetic properties, depending on the extent to which they had been heated and metamorphosed by subsequent lava eruptions. The general effect of the oriented stress field would be to produce a series of approximately parallel magnetic anomalies. Occasional symmetry can be explained logically in terms of structural geology, according to the Meyerhoffs.

THE GEOTECTONICS CREED IN HISTORICAL PERSPECTIVE

The history of continental drift has had its ups and downs, as illustrated in Figure 13-9, a diagram published in 1972 by Van Bemmelen to show how the voting for and against continental drift has varied during

Fig. 13-8 An interpretation of oceanic ridges involving compression rather than extension. Note the formation of fractures more or less parallel to the up-arched ocean ridge. (Based on model proposed by A. A. Meyerhoff, H. A. Meyerhoff, and R. S. Briggs, *Jour. Geology,* *80,* 1972, 663-692.

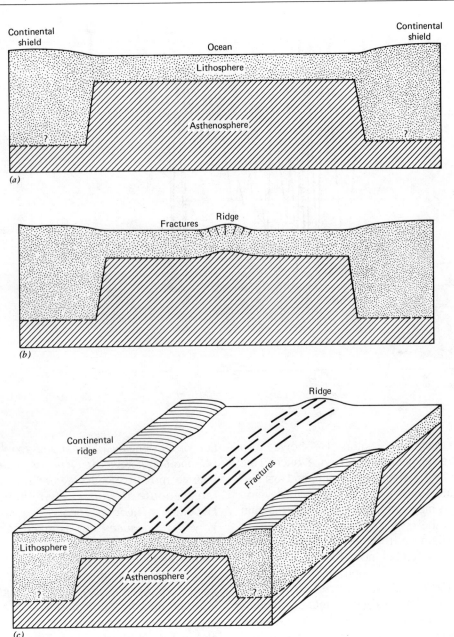

Fig. 13-9 The twentieth-century debate. Those for and against continental drift. (From R. W. Van Bemmelen, 1972, *Geodynamic Models,* Elsevier Publishers.)

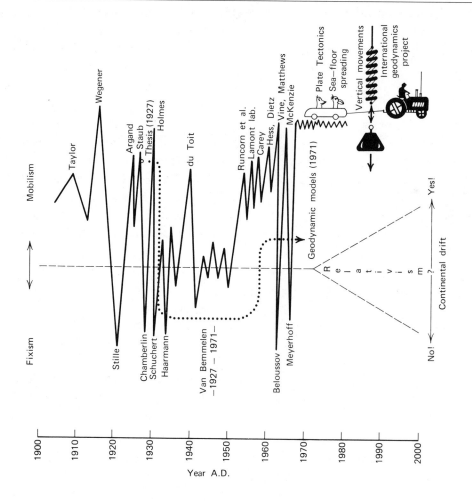

more than half a century. He showed also that he changed his mind twice, once in the early 1930s when he lost faith, and again in the late 1950s when he began to develop mobile models for the earth. The diagram shows the overwhelming present vote for mobilism, with even Beloussov and Meyerhoff left behind in the 1960s.

Geologists and geophysicists joke about working hypotheses and ruling theories, and it has been suggested that all students should be required to learn the New Geotectonics Creed before earning their degrees:

I believe in Plate Tectonics Almighty, Unifier of the Earth Sciences, and explana- *tion of all things geological and geophysical; and in our Xavier Le Pichon, revealer of relative motion, deduced from spreading rates about all ridges; Hypothesis of Hypothesis, Theory of Theory, Very Fact of Very Fact; deduced not assumed; Continents being of one unit with the Oceans, from which all plates spread; Which, when they encounter another plate and are subducted, go down in Benioff Zones, and are resorbed into the Aesthenosphere, and are made Mantle; and cause earthquake foci also under Island Arcs; They soften and can flow; and at the Ridges magma rises again according to Vine and Matthews; and ascends into the Crust, and maketh symmetrical magnetic anomalies; and the sea floor shall spread again, with continents, to make both mountains and faults, Whose evolution shall have no end.*

And I believe in Continental Drift, the Controller of the evolution of Life, Which proceedeth from Plate Tectonics and Sea-Floor Spreading; Which with Plate Tectonics and Sea-Floor Spreading together is worshipped and glorified; Which was spake of by Wegener; And I believe in one Seismic and Volcanistic pattern; I acknowledge one Cause for the deformation of rocks; And I look for the eruption of new Ridges and the subduction of the Plates to come.

This Creed was prepared by C. K. Scharnberger and was published in a letter to *Geotimes* by E. L. Kern in 1972, with apologies to the Council of Nicea.

Despite the jokes, there is little doubt that plate tectonics has become a ruling theory. Very few researchers these days can present a serious paper at a scientific meeting arguing against plate tectonics without receiving rude comments from a fair proportion of the audience.

This is not the first time in geological history that a theory has been acclaimed as virtually proven and has dominated geological thought for many years. There have been many great controversies in the earth sciences, of which the vigorous debate of this century (Figure 13-9) is just one. The lesson to be learned from these controversies is that a theory that appears in one generation to be unassailable may appear incredibly naive and wrong to the next generation. In these days of revolution and excitement, a brief outline of the fate of some earlier theories is a lesson worth repeating.

In 1775, at the age of 25, Abraham Werner was appointed lecturer at the Freiberg Mining Academy. In his lectures, he promulgated the Neptunian theory. He maintained that all geological formations and rocks of all types except for those actually observed to emerge from volcanoes as lavas had originated as successive deposits or precipitates from a primeval ocean. His stimulating lectures attracted many students, who then went forth and applied the Neptunist creed to the solution of geological problems in their various countries. Werner's thesis was challenged by the school of Plutonists and subsequently abandoned; this was one of the most celebrated and bitter controversies in science. James Hutton of Edinburgh became the leader of the Plutonists, who denied that a single primeval ocean was the source of rocks. He presented evidence eventually satisfying most geologists that many rocks had formed by the cooling and crystallization of hot material that had risen in fused condition from subterranean regions. Rocks formed in this way were accompanied by earthquakes and eruptions, in sharp contrast with the peaceful creative processes involved in the Neptunian conception of their origin and of the origin of the world.

Hutton's contributions are contained in his 1788 paper titled "Theory of the Earth" and in a book titled *Theory of the Earth with Proofs and Illustrations* published in 1795, two years before his death. The rival schools argued their cases, and Neptunism simply disappeared about 50 years after its sudden rise when Werner began to lecture in 1775. The controversy between Neptunism, considered by many to conform to the Book of Genesis, and Plutonism, considered by many to be atheistic, passed beyond the confines of scientific discussion, and it entered the contemporary literature. For example, it was incorporated in Goethe's *Faust* by dialogs between the Sphinxes and Seismos in Act II, and between Faust and Mephistopheles in Act IV.

Another controversy of long duration concerned the interpretation of fossils. Many imaginative explanations for these *figured stones* were proposed, and it was argued that they were not of organic origin because this was contrary to the book of Genesis. When fossils were eventually recognized as the remains of living things, champions of the theological cause maintained that fossils had been carried to their present positions above sea level by the Noachian deluge. Indeed, the existence of fossils in mountains was cited as scientific evidence that the Biblical deluge had occurred.

The science of paleontology developed from the study of fossils, and one of its early masters was Baron George Cuvier of Paris. He discovered that certain fossils were confined to specific rock formations; to explain this, he invoked a series of widespread catastrophes. Sudden floods and successive retreats of the water caused the disappearance of faunas characterizing certain formations. The doctrine of Catastrophism was dominant for nearly three centuries, with Cuvier's major contributions being published in 1811 and 1812. Hutton's *Theory of the Earth* not only led the Plutonists to victory over the Neptunists, but it also introduced the ideas that toppled Catastrophism. Hutton noted the continuity of geological processes and emphasized that in the geological record there is "no sign of a beginning—no prospect of an end." This idea that geological history should be explainable in terms of events occurring at present led to the controversy between Catastrophism and Uniformitarianism, as the competing principle was called by Sir Charles Lyell. The publication in 1835 of Lyell's book *Principles of Geology,* marked the end of Catastrophism, and Uniformitarianism became the new creed. However, it is now recognized that the uniform flow of geological history is punctuated by local, intermittent catastrophes such as the 1970 *huayco* in Chile, which was described in Chapter 4. The amount of erosion and transportation of rock material that was accomplished in a few minutes would take centuries by the steady process of stream action.

So here we are, after a long debate, with a new theory of plate tectonics. Let us remember that yesterday's theory is of historical interest only, and that today's theory is tomorrow's history. The Geotectonics Creed should not really be included among the requirements for a degree in earth sciences.

SUMMARY

The 1963 Vine and Matthews interpretation of linear magnetic anomalies in the oceanic regions was not immediately accepted. The revolution in earth sciences occurred during the winter of 1966 to 1967. At a symposium in November 1966, a group of experts reviewed available evidence on magnetic anomalies, magnetized rocks, earthquakes, and matching continental age provinces across oceans. The evidence was persuasive, and the word was passed around at the Annual Meeting of the Geological Society of America, attended by thousands, in the same month. Geologists and geophysicists reconsidered their own research data within the framework of sea-floor spreading and, at the Annual Meeting of the American Geophysical Union in April 1967, a very large number of papers was presented on this topic. In 1968 scientists from Lamont consolidated sea-floor spreading and plate tectonics with a worldwide correlation of linear magnetic anomalies and a review of the applications of earthquake studies to the new theory.

The proclamation of revolution was challenged by the Russian geologist Beloussov and by Meyerhoff and his associates, who argued that the simple models of plate tectonics bore little relationship to the complex geology of the real earth, that geological evidence for ancient climatic patterns required that the ocean basins had been in the same positions for 1,000 million years, and that the oceanic magnetic anomalies differed significantly from those required by the simple model. Meyerhoff and Meyerhoff concluded that the ocean ridges and associated magnetic anomaly bands were distributed around the ancient continental nuclei in approximately concentric fashion and suggested that this might be caused by compression resulting from earth contraction. This was the generally accepted global theory before plate tectonics.

SUGGESTED READINGS are at the end of the book.

chapter 14

Voyages of the Glomar Challenger, and Deep-Sea Sediments

There were two great developments in 1969 that promised to increase substantially our knowledge of the earth and the solar system, according to a quotation from P. H. Abelson cited in Chapter 1. These were exploration of the moon and the program of deep-sea drilling from the *Glomar Challenger*. The *Glomar Challenger* began the first of many cruise segments across the world's oceans on August 11, 1968. According to a pamphlet released from the National Science Foundation dated August 1970, the first two years of operation of the Deep-Sea Drilling Project

produced information of such significance as to mark it as one of the most successful scientific expeditions of all time. In this period approximately 195 holes were drilled at 125 sites in the Atlantic and Pacific Oceans. Sediment and rock cores were obtained from the earth's crust under water more than 20,000 feet deep. Several holes were drilled deeper than 3,200 feet into the ocean bottom. The drilling ship has used the longest drill string ever suspended from a floating platform—20,760 feet, almost four miles.

By the thirtieth cruise leg in August 1973, the score had increased to 450 holes at 300 drilling sites, and the total length of sediment and hard rock core recovered exceeded 25 km. Rates of deep-sea sedimentation vary considerably from place to place, but values typically range from 1 to 20 mm per 1000 years (Figure 9-7). The cores recovered thus represent much geological time, and their significance lies in the events recorded in the sediments during the time intervals covered by each core.

The sediments on the ocean basin floor consist of a mixture of sand, silt, and volcanic ash transported from the continents by rivers, the wind, or drifting icebergs, of minerals precipitated from the ocean, and of the shells and skeletons of microscopic animals and plants (plankton) that live in abundance near the ocean surface. The sedimentary layers provide a continuous historical record of geological and biological evidence of many kinds. The completeness of the oceanic sediment record contrasts with evidence from the continents, where the continuity of history usually has to be pieced together from many sources and locations. Within the sediment cores stacked in storage awaiting detailed examination, there are many undiscovered and unanticipated facts about the history of the continents around the ocean basins, about the earth's magnetism, about biological changes in the oceans, about the ocean waters, and about climatic changes.

The main aim of the Deep-Sea Drilling Project is to gather information about the age and processes of formation of the ocean basins. The new theories associated with the revolution in earth sciences led to specific predictions about the history of the earth's magnetic field (Chapter 9) and the origin of magnetic anomalies, the spreading of the sea floors and drifting of continents (Chapters 10 and 12), and the history of temperature changes in the oceans and atmosphere. These and other predictions could best be checked by direct sampling of the sediments from the deep ocean basins and continental margins and from the underlying rocks.

In this chapter, we will examine three results on different scales that were obtained from sedimentary cores and drilling the sea floor: first, the specific test for sea-floor spreading and the extrapolated polarity reversal time scale that was conducted in the South Atlantic Ocean; second, the remarkable history of the Mediterranean Ocean between 8 and 5 million years ago when it became isolated from the Atlantic Ocean, evaporated to a dry desert, and then filled again as the Atlantic Ocean cascaded over an enormous waterfall near Gibraltar; and third, we have the history of a specific volcanic eruption in the Aegean Sea that dispersed the Minoan culture and was probably responsible for the legend of the lost continent of Atlantis.

THE DEEP-SEA DRILLING PROJECT

The history of the Deep-Sea Drilling Project really begins with Project Mohole, which was an ambitious program designed to drill through the oceanic crust and the Mohorovicic discontinuity to obtain a sample of the earth's upper mantle. It is not true that the project was terminated because of fears that the ocean water would empty out through the hole drilled in the ocean floor, although many citizens did write letters of

protest to the government, insisting that this would happen. Geological mythology about a hollow earth apparently persists even today. The project was terminated amid political wrangling about procedures and estimated and actual costs. An original $5 million budget expanded to anticipated costs of more than $100 million. However, in 1961, some time before the project was abandoned by Congress, the Mohole group accomplished the first successful drilling of the sea floor in deep water, when the ship *CUSS I* drilled more than 180 m of sediments and lava in 3.7 km of water east of Guadalupe Island, Mexico. The new scientific slogan became "Not Mohole but more holes."

In May 1964, a deep-sea drilling program was formally established by four of the major American oceanographic institutions having strong interests and programs in the fields of marine geology and geophysics: the Institute of Marine Science, University of Miami; Lamont Geological Observatory of Columbia University; Scripps Institution of Oceanography, University of California; and Woods Hole Oceanographic Institution. These constituted the JOIDES (Joint Oceanographic Institutions Deep Earth Sampling) group, and they were later joined by the University of Washington.

In the summer of 1965, an experimental program off the coast of Florida by the drilling ship *Caldrill* confirmed the feasibility of drilling 300-m holes and recovering cores below water as deep as 900 m, and this stimulated plans for work in deeper water. Scripps was awarded $12.6 million by the National Science Foundation to operate the program, and they relied on advice from a number of advisory panels sponsored by JOIDES. Global Marine Exploration Company made excellent progress constructing and outfitting the *Glomar Challenger* in only 40 weeks; this is probably a record for a vessel as intricate as the *Challenger*. The project was right on schedule and operating within its budget, two remarkable accomplishments compared with many other government-sponsored projects. The operational and scientific success of the first

18-month series of cruises generated an additional $34.8 million in support from the National Science Foundation to cover operations between 1970 and 1973. Continued success assured funding after 1973, and drilling is scheduled to continue through August 1975.

It costs about $10 million per year to keep the *Glomar Challenger* at sea, and some scientists are concerned that the drilling is proceeding at a pace too fast for the research to keep up with it. The sedimentary cores have piled up in their repositories at Scripps Oceanographic Institution and the Lamont-Doherty Geological Observatory. The scientists taking part in the successive cruise legs examine the cores, and their preliminary dating and descriptions are published in a series of reports but, apart from this, many cores have hardly been studied at all. Yet, if a project of this magnitude were to stop for a period, many believe that it would be very difficult to raise funds for a second beginning, and thus there is considerable incentive for continuation, despite the lag in research. Now there are many more requests coming in for core samples to be used in varied research projects, and eventually the cores will be utilized. The cylinders of mud and oozes represent the greatest treasure trove yet yielded by Davy Jones' Locker; they accumulated through millions of years and it is too much to expect them to be unraveled, historically and scientifically, within a few months.

It has been proposed that for the period after 1975 the program should continue under a multinational charter as the International Program of Ocean Drilling (IPOD), with other nations contributing scientific direction and financial support. Many of the major geophysical questions have been answered by the preliminary results of the drilling program, and what may be needed now is a focus on specific detailed studies. The IPOD proposal would shift emphasis from general reconnaissance of the sea floor to study of the crust beneath the sedimentary layers and to the continental margins where mineral and petroleum resources are certainly present and require evaluation.

THE GLOMAR CHALLENGER

The *Glomar Challenger* is a unique ship with a striking profile that is clear even from the sketch in Figure 14-1. The ship is 120 m long with 10,000 tons displacement, and amidships there towers a 43-m drilling derrick, its top almost 61 m above the waterline. Most of the topside space forward of the derrick is occupied by an automatic pipe-racking device with more than 6.5 km of drill pipe stacked in 28-m lengths. During drilling, the drill pipe is suspended from the derrick and through an opening about 6 x 6.5 m extending through the bottom of the ship. At the tip of the drill string is a drilling bit, and above this is the core barrel, which captures and stores the cores of sediment as they are drilled. Once the drill bit touches the ocean floor, the entire drill string is rotated from the drilling deck. Drilling continues until the bit is worn out; then the string is retracted and the core is recovered.

What makes the drilling operation possible, in water too deep for anchors, is a dynamic positioning system that maintains the ship's position within a radius of about 100 m. A beacon that emits acoustic signals is dropped to the ocean bottom, and the sonar beams are received by four hydrophones beneath the ship's hull. A computer translates the pulses into directions and distances from the ship to a point directly above the hole and actuates some combination of the ship's main propellors and four side thrusters to move the vessel back to its station.

The layers of hard sediment encountered in the Pacific Ocean dulled the bits and frequently forced drilling to stop. At other sites, the first lava reached was interpreted as ocean crust, the basement beneath the sedimentary layer, but this could not be drilled deeply enough to make certain that it was not just a layer of lava with more sediments below it. In order to overcome this

Fig. 14-1 Sketch of the *Glomar Challenger* lowering its drill stem through the ocean toward the drill reentry funnel that has been secured in the ocean-floor sediments. (Based on a National Science Foundation report.)

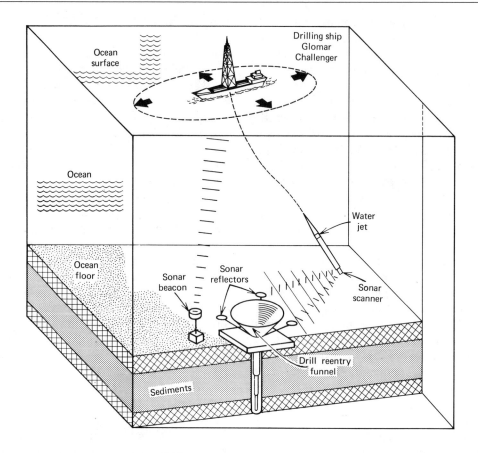

problem, a reentry funnel was designed. This ingenious device, illustrated in Figure 14-1, permits a drill core to be removed, the bit replaced, and the drill string to reenter the bore-hole on the ocean floor. The target is tiny: an invisible 12 cm diameter hole thousands of meters away, with both the drill stem and the vessel being constantly affected by ocean currents.

In June 1970 a new bit was successfully placed into a previously started hole for the first time. Before drilling of the first hole, the drill stem was inserted at the surface through the stem of a metal funnel 5 m in diameter, and the entire assemblage was lowered to the ocean bottom. Drilling began when the funnel's heavy base was secure on the ocean floor. This remained in position when the drill string was raised

and the bit replaced. The drill string was re-lowered with a sonar scanner on the bit assembly. This emitted sound signals that were echoed back from three reflectors spaced around the funnel brim. The information was relayed to the ship, and the stem was steered into the funnel by jets of water forced out of a hole on the side of the drill stem 18 m above the bit. This system is used selectively at sites where extensive drill wear is anticipated and, in particular, for penetration of the ocean crust beneath the sedimentary layers.

Despite the lag in detailed scientific research on many of the cores, the range and number of conclusions published are extensive and of such significance that probably even the most optimistic hopes of the program planners have been exceeded. In addition to the three specific items that we will examine in some detail, the results include the following.

The first discovery of oil and gas was made in deep-sea conditions in the Gulf of Mexico. This will influence approaches to the exploitation of marine energy sources.

Iron and manganese are enriched in a zone 5 to 10 m thick at the base of the sedimentary layer.

In many parts of the Atlantic Ocean evidence for strong bottom currents transporting sediments dispel the once-prevalent textbook view that the ocean depths display little erosion or current motion.

The glaciation of the northern continents began about 3 million years ago, more than 1 million years earlier than the date usually assumed.

Antarctica appears to have been glaciated for at least 20 million years, far longer than previous estimates of the most recent ice age, and it appears that the ice sheet rather suddenly receded to its present size about 5 million years ago. This change should have raised mean sea level by a significant amount.

The results obtained from the cores provide base lines permitting confident interpretation of data obtained from geophysical surveys over large areas.

There is evidence that Africa and Europe are approaching each other, with the compression causing parts of the eastern Mediterranean Ocean floor to rise in the initial stages of formation of a mountain chain.

The sediment cores include volcanic ash showing that dozens of major volcanic explosions have occurred from the Italian volcanic arc in the last 2 million years.

Several sites in the eastern Atlantic have a gap in the sedimentary sequence between 60 and 30 million years ago, which coincides with continental mountain building. Possibly both phenomena may relate to a change in the pattern of continental drift.

There is evidence that a narrow proto-Atlantic Ocean existed before the last episode of drift; perhaps North America and northwest Africa were never completely joined (see Figures 11-3 and 12-4a).

There is evidence from the cores that the Pacific Ocean plate has moved at varying speed and direction during the past 35 million years. In addition to the east-west motion away from the ocean ridge, the Pacific plate is moving northward. So is the floor of the Indian Ocean near Australia. The pattern of movements of plates appears to be more complex than previously suspected.

It has been established that first New Zealand and then Australia broke away from Antarctica between 60 and 80 million years ago (Figure 12-3c), and that the northward movement of India stopped for a period beginning about 60 million years ago (Figure 12-3c). The island of Madagascar has been a minicontinent for at least 100 million years (Figures 12-3b and 12-3c).

Theoretical treatments of plate tectonics do not incorporate complexities introduced by vertical movements, but there is now evidence for general subsidence and elevation in several parts of the ocean basins. In some oceanic regions there have been substantial vertical movements in short periods of geological time; one area sank 1.5 km in 5 million years, and another sank 1.8 km at a rate of 10 cm/year.

A submarine ridge in the Indian Ocean, 2000 km long and 1 km below sea level, was once a chain of islands with swamps and lagoons; more evidence for vertical movements.

The most significant finding of the whole project is generally considered to be the unequivocal establishment of the fact that the oceanic crust is geologically youthful, compared with most continental rocks, and that youthfulness is maintained by sea-floor spreading.

SEA-FLOOR SPREADING IN THE SOUTH ATLANTIC OCEAN

The provisional dating of the crust of the world's oceans (Figure 10-10) is based on the Vine-Matthews interpretation of magnetic anomalies and the assumption that spreading rates in the South Atlantic Ocean had remained constant for the past 80 million years (Figure 10-9). This is a long extrapolation, and the need for some independent test of the age of the ocean floor is obvious.

Early in the program, the *Glomar Challenger* drilled eight deep holes on a traverse across the mid-Atlantic ridge near 30°S, where the magnetic anomaly paterns are well defined (Figure 10-10), and recovered sediments that were assigned ages according to the microfossils contained within them. The ages of the deepest sediments from each drilling site corresponded very well with the ages of the ocean crust predicted by the magnetic anomaly time scale. In the "Initial Reports" published in 1970, A. E. Maxwell and seven coauthors wrote that the results "provided a critical test for both sea-floor spreading and magnetic stratigraphy."

In 1972, Meyerhoff and Meyerhoff took a critical look at the critical test, found the evidence flimsy, and wrote that because of the widespread, uncritical acceptance of the conclusion, "Speculation became 'fact' and science lost another battle." What is the evidence?

Figure 14-2a is a schematic representation of the sediments overlying the ocean crust, based on the more complete diagram of Figure 10-7b. Notice the difference between horizontal and vertical scales. The basalt lava forming the ocean crust is magnetized in blocks about 50 km across, and this is covered by blankets of magnetized sediments with the thicknesses of magnetized layers measured in meters. The obvious test would be to drill the basalt below the sediments, as indicated in Figure 14-2a, and to measure its age. This was not feasible before the reentry funnel was available (Figure 14-1), and there are, in addition, serious difficulties in obtaining reliable ages from these lavas even when they are recovered in sufficient quantity.

The deepest sediments, immediately overlying the basalt of the ocean crust, become younger toward the ocean ridge crest. The age of each block of magnetized crust is just a little older than the sediment

immediately above it; on the geological time scale, these ages are effectively identical. The test, therefore, was to recover sediment from just above the magnetized ocean crust and determine its age from its fossil content.

When Maxwell and coauthors plotted the ages of the deepest sediments recovered from each of the eight drill-holes against their distances from the ocean ridge crest, the relationship was clearly linear, and the slope of the line corresponded to a spreading rate of 2 cm/year. This spreading rate is identical with that estimated from magnetic anomalies (Figure 10-9), to the delight of the *Glomar Challenger* scientific team, and to the earth science fraternity in general. The fossil ages of the sediments are plotted against the magnetic age of the underlying crust in Figure 14-3 and compared with the 45° line which corresponds to equal values of both. In addition, the top scale shows the distance of each drilling site from the ridge axes.

The correspondence looks good, and all that is required is confirmation that the sediments dated do, in fact, have the necessary relationship to the magnetized crust that is being tested. Figures 14-2b to 14-2d show some of the problems involved.

Figure 14-2b illustrates an event that occurred after the deposition of the sediments shown in the diagram. Lava was intruded through the ocean crust and into the sediments but, instead of breaking through to the ocean floor, it spread out between layers of sediment, cooled, and solidified, forming a sheet of basalt between the sedimentary blankets. When a drill reaches this layer of basalt, it reacts just as it would if it reached the basalt of the magnetized crust; it grinds to a halt. Recovery of the sediment core with a little basalt as well could be interpreted as the same situation illustrated in Figure 14-2a, but the age of the sediment is younger than that deeper down in contact with the crust.

Figure 14-2c illustrates a lava flow occurring after the deposition of the sediments shown and spreading out over the ocean floor. In time, this becomes buried with ad-

Fig. 14-2 Schematic cross sections through the ocean floor on the flanks of an ocean ridge, showing magnetized lavas of the crust overlain by magnetized layers of sediment. Compare Figure 10-7b. The sediments are baked when hot lava comes into contact with them. (a) Conventional situation. (b) Possible situation if younger lava rises through the sequence established in part a and becomes trapped between layers of older sediment. (c) Situation if younger lava flows over older sediments on the ocean floor, and (d) is covered later by additional layers of younger sediments.

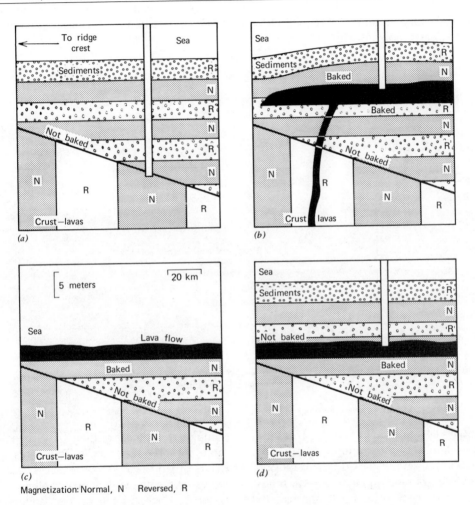

ditional sediments, as shown in Figure 14-3d. The response to drilling is the same as for Figure 14-2b, as shown in the diagram.

The descriptions of the sediments and lavas recovered were studied by Meyerhoff and Meyerhoff, and they concluded that the samples from five and possibly six of the eight sites clearly show that the bottom of the sedimentary sequence had not been reached. Their conclusion is based on the fact that when hot lava comes into contact with sediment, it bakes it, and this hardened metamorphic rock is easily distinguished from the normal sediment.

Figure 14-2 shows the regions where sediments would become baked by young lava, intruded into or through the sediments. Note that sediment deposited onto the magnetized ocean crust, or the surface of

Fig. 14-3 Some of the results from deep-sea drilling in the South Atlantic Ocean. The age of the deepest sediments recovered at various places is plotted against the distance of each drilling site from the ridge crest. The magnetic anomalies give a time scale, Figure 10-9*b*, which provides an estimate of the age of the crustal lavas below the sediments. The straight line through the points shows good correlation between the age of the deepest sediments and the estimated age of the lavas below, thus fitting the picture in Figure 14-2*a*. But the fact that many of the deep sediments recovered are baked and associated with lavas (basalts) suggests that the situation corresponded to Figures 14-2*b* and 14-2*d*. (Drilling results from A. E. Maxwell and others, *Science, 168,* 1970, 1047-1059, with interpretations as recorded by A. A. Meyerhoff and H. A. Meyerhoff, 1972, see Figure 13-1.)

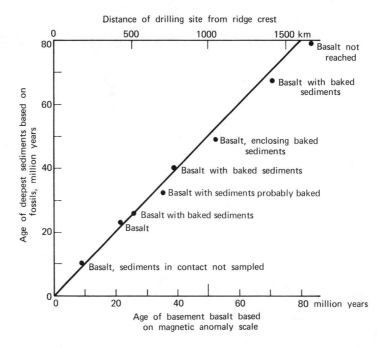

a cooled lava flow, is not baked. This introduces another uncertainty, because the situation shown in Figure 14-2*d* is indistinguishable from that in Figure 14-2*a*, unless the drill is able to penetrate through the lava flow and demonstrate that there are more sediments below.

Figure 14-3 lists the distribution of baked sediments among the eight drill sites. The baked sediments are surely not the deepest sediments in contact with the magnetized ocean crust. The results do not fit the theoretical model of Figure 14-2*a* but do fit the models of Figures 14-2*b* or 14-2*d*. Perhaps the sills or flows of lava are close to the bottom of the sediments, near the ocean crust, but the critical test remains to be completed. Deeper drilling may reveal a more complex sequence of sediments and lavas above the magnetized crust. Nevertheless, the very good correlation between the ages of the deepest sediments reached by drilling and the magnetic anomaly time scale is generally accepted as satisfactory confirmation for sea-floor spreading in the South Atlantic.

THE DEATH VALLEY OF THE MEDITERRANEAN

The sun beats down on the Mediterranean Sea comforting the pleasure seekers on the beaches and evaporating more than 4000 km³ of water each year. Less than 500 km³ of this is replaced by rain and fresh water from the rivers. The Mediterranean is almost landlocked, being connected to the world's oceans by the narrow Straits of Gibraltar. It is from the Atlantic Ocean that about 3500 km³ of water flows each year to maintain the Mediterranean Sea at a constant level, ensuring that the people on the beaches do not have to walk too far to reach the water's edge.

If the Straits of Gibraltar were closed, the Mediterranean Sea would evaporate completely in about 1000 years, leaving a great valley covered in salt flats. The *Glomar Challenger* has produced evidence that this actually happened between about 8 and 5 million years ago.

The Mediterranean Sea is relatively small, but the existence of regions with depths of more than 3 km show that it is a genuine ocean basin (Figure 14-4a). It is the remnant of the vast Tethys Ocean that once existed between Laurasia and Gondwanaland (Figures 12-3 and 12-4). Northward movement of Africa and India gradually narrowed the ocean, but it remained in continuous connection with the bodies of water that were to become the Atlantic and Indian Oceans through many millions of years. Figure 12-3c shows the connection 65 million years ago. The compression caused by the converging plates was responsible for the folding and uplift of the deep ocean sediments into the Alpine mountain chains. Previous continental and marine geological and geophysical surveys suggested that the key to many problems might be found by deep drilling into the Mediterranean floor.

Fig. 14-4 *(a)* Map of the Mediterranean region, showing three deep ocean basins with depths greater than 3 km. *(b)* Sketch of the great basins and valleys that would be produced in the Mediterranean region if all of the water were removed.

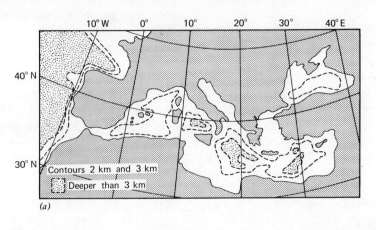

(a)

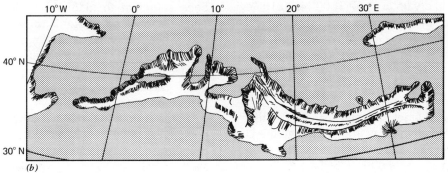

(b)

On the thirteenth leg of the Deep-Sea Drilling Project, the *Glomar Challenger* started from Lisbon on August 13, 1970, for a Mediterranean cruise, with an international group of scientists headed by cochiefs W. B. F. Ryan of Lamont-Doherty Geological Observatory and K. J. Hsü from Zurich. The ship drilled many deep holes below deep water and probed the sediments to depths of more than 300 m. The sediments recovered tell a fascinating story of the history of the Mediterranean Sea for the past 20 million years. The evidence is incomplete and some geologists dispute the interpretation, but the story is worth telling.

The oldest sediments from the deepest parts of the cores are deep-sea oozes containing fossils that show that the Mediterranean was still an open seaway from the Atlantic to the Indian Ocean during the period 20 to 15 million years ago. Soon after this period, the connection with the Indian Ocean was closed; the fossils in the Mediterranean sediments evolved differently from those in the Indian Ocean, but retained kinship with those in the Atlantic Ocean.

The fossils in the younger sediments changed from normal deep-sea types to hardy types that could withstand changing salinities. There was evidence that the waters of the Mediterranean were stagnating, and the western end of the sea probably closed between 10 and 8 million years ago as a result of converging plates and the slow uplift of mountains.

Once the Mediterranean became a landlocked sea, evaporation losses must have reduced its level rapidly. We saw that with present rainfall and inflow from rivers it would dry up in about 1000 years. The evidence from the cores, however, shows that the Mediterranean did not dry up quite so quickly, 8 million years ago. What was the source of the surplus water? The answer comes from the geology of Eastern Europe, which was covered at that time by a vast body of brackish water, Lac Mer. This extended all the way from Vienna to the Ural Mountains and the Aral Sea. The present Caspian Sea and Black Sea are its last remnants. The water from Lac Mer drained into the Mediterranean basin in sufficient quantities to slow down the complete evaporation to dryness. Possibly there were also intermittent influxes of water from the Atlantic Ocean until the barrier was stabilized.

As ocean water evaporates, the concentration of dissolved solids increases and the solids are then precipitated out in the form of minerals, which form layers on the sea floor. There is a very precise correlation of different minerals, according to the amount of water evaporated. Calcium carbonate precipitates first in the form of limestone. When 70% of the water has evaporated, the mineral gypsum is precipitated, and only when 90% of the original seawater has evaporated does the brine become strong enough to precipitate rock salt. Gypsum, salt, and other minerals precipitated as sedimentary layers in this way are described as evaporites. The sequences and types of evaporites found at various places and at various depths in the Mediterranean floor are consistent with the conclusion that, of the original water trapped in the ocean basin, more than 90% evaporated away, yielding layers of rock salt in the deepest basins, where the last pools of water persisted, as indicated in Figure 14-4, until they, too, disappeared.

The story is complicated by the fact that there is sediment within some of the evaporite deposits enclosing fossils of creatures that can exist only in fresh or brackish water. The bottom of the Mediterranean basin therefore must have been partly covered by a series of large, brackish lakes for some time between 8 and 7 million years ago. The supply of water from Lac Mer must have been sufficient to maintain these lakes. They disappeared about 7 million years ago, when the converging plates elevated the Carpathian Mountains sufficiently to change the drainage pattern, and the waters of Lac Mer then escaped to the north. The main supply of water for the Mediterranean was thus cut off, and the lakes evaporated completely. Rain and water from the surrounding rivers collected intermittently in shallow, salty lakes on the Mediterranean floor, evaporated rapidly, and added more salt deposits to those left by the vanished ocean.

Figure 14-4*b* gives some idea of how the Mediterranean region might have looked without water. The continents and continental shelves would have been high plateaus around the several deep basins of the former ocean floor, and the steep continental slope falling down from the edge of the shelf would have been one of the most striking features of the scenery. The islands and formerly submerged submarine volcanoes would have risen from the salt flats of the sunken valleys and basins as great peaks or isolated plateaus. It was a desolate, desert region, with evidence from minerals in some of the salt deposits indicating that temperatures reached 150°F. All life was exterminated. The sunken desert, 3 km and more below sea level, has been compared with Death Valley in California, but it was much larger and much deeper, and probably much more inhospitable.

Above the evaporite deposits in the drill cores is a layer of sediment indicating that all parts of the Mediterranean studied were flooded simultaneously 5.5 million years ago. This was followed by deep-sea oozes confirming that the Mediterranean Ocean once again had connection with an ocean large enough to maintain its water supply. The explanation of this great deluge is that erosion of the rock barrier damming the Straits of Gibraltar at first admitted a trickle of water from the Atlantic Ocean, which soon grew to a cascade, and before long to an enormous waterfall.

The rate of inflow of water during the flooding has been estimated by considering the types of fossils present in the sediments above the evaporites. If the rate of inflow was too slow, the evaporation would have produced water that was too salty for their survival. In order to support the life forms represented by the fossils preserved in the sediments, the inflow of seawater would have had to exceed the evaporation loss of water by a factor of ten. The amount of water required to maintain this condition is equivalent to 100 times the flow over the

Victoria Falls, and 1000 times the flow over the Niagara Falls. At this rate, it would have taken about 100 years to fill the Mediterranean basin with seawater again and to restore the former relationship with the Atlantic Ocean, causing the deposition of normal deep-sea oozes on the ocean floor.

Imagine the immense volumes of water cascading over the Gibraltar Falls down to the desert thousands of meters below. It must have been quite a spectacle, had there been human eyes to view it. But this will not serve as an explanation for Noah's flood, because the first true man and woman did not walk the earth until about 1.3 million years ago. It is conceivable, however, that a tribe of the first ape-man/ woman to walk erect, *Australopithecus Africanus,* could have wandered to and looked over the edge of the continental shelf to see the water plunging down to the depths.

The volume of water required to fill the Mediterranean Sea is considerable. It would have lowered mean sea level throughout the world by 10 m. Similarly, when the Mediterranean Sea had evaporated several million years previously, the level of the other oceans would have been raised by 10 meters.

When sea level changes, this has significant effects on the rivers draining the continents and flowing into the oceans. Sea level represents the base level down to which the rivers flow, and they cut their channels down to this level and not much deeper. Rivers cannot flow uphill. If sea level rises, as it did when the Mediterranean Sea evaporated, then the mouths of the world's rivers would have been flooded, and their channels would have become partly filled with sediments to bring the river flow up to the level of the ocean. When the Mediterranean was filled and the world's oceans were lowered again, the rivers would have cut their channels deeper to maintain their level with respect to the sea.

Evaporation of the Mediterranean Sea had a drastic effect on the rivers flowing into the basin. The base level for the rivers was progressively lowered and their cutting power was thus greatly increased. For 2 or 3 million years, the rivers carved great canyons in Europe and Africa as they sought to establish a new base level 3 or 4 km below the height of the continents.

It has been discovered by drilling that rivers such as the Rhone and the Nile might have matched the Grand Canyon of Arizona in splendor. Figure 14-5 is a schematic cross section showing the general pattern discovered for the major rivers. The deep canyon that relates to the Mediterranean desert period is shown, cut into the hard rocks of the continental mass. These canyons may extend down to 500 or 1000 m

Fig. 14-5 Deep drilling through the continents near several rivers flowing into the Mediterranean Sea has revealed structures similar to that illustrated in Figure 14-5*b*. An interpretation is illustrated in part *a*. If the Mediterranean Sea was dried up 6 million years ago, the rivers would have cut great canyons into the continents and continental shelves as they flowed down to the bottom of the deep basins sketched in Figure 14-4*b*. Subsequent breaching of the Gibraltar dam with flooding from the Atlantic Ocean would have filled the Mediterranean again, as well as the deep canyons. The canyons would therefore become filled with sediments deposited from ocean water until the former level was more or less restored, and then the rivers would build up their flood plains of freshwater sediments.

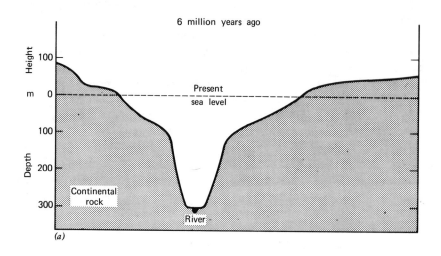

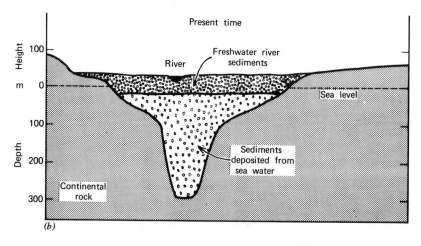

below sea level, depending on distance from the sea. When the Mediterranean Sea filled again, 5.5 million years ago, these canyons were flooded, becoming inlets of the sea, and they were partly filled with marine sediments. Then the rivers built up layers of freshwater sedimentary deposits as they regenerated flood plains corresponding to the restored level of the Mediterranean Sea.

There is another story here about how the current orthodoxy in science controls what scientific ideas are published. Deeply buried river channels like those illustrated in Figure 14-5 had been discovered by oil geologists drilling in Libya. Frank T. Barr and his colleagues of the Oasis Oil Company concluded that when these channels were cut, the Mediterranean Ocean must have been thousands of meters below its present level, and that the incised channels were filled with marine sediments because they had later been flooded by a rising sea. Their report was not accepted by scientific journals because editors considered the idea too outrageous to be respectable.

It also happens that published scientific papers may languish without attention if they differ too much from the current orthodoxy. G. Ruggieri of the University of Palermo studied the fossils of the Mediterranean Ocean and sought an explanation for the marked change in the species of fossils preserved in sediments older and younger than about 5.5 million years. Many others had considered the same problem, and Sir Charles Lyell, as long ago as 1833, had noted that the change had a special significance. Ruggieri, in 1955, suggested that the change was caused by evaporation of the Mediterranean Ocean with the destruction of one set of fauna, followed later by the influx of a different fauna when the Mediterranean basin was flooded again. Here we have the essence of the history that was pieced together 15 years later from a series of drill holes recovered from the floor of the Mediterranean Sea.

We noted above that the evidence for desiccation and flooding is fragmentary and incomplete, and not all geologists are prepared to accept the interpretation of Ryan, Hsü, and some of their colleagues. Some geologists advocate a process of salt deposition in deep water, without the requirement for evaporation of a whole ocean. Other geologists prefer models involving shallow water deposition of the evaporites at a time when the ocean floor was much higher. There is evidence elsewhere for considerable vertical movements of ocean floor in periods of a few million years. These alternatives have been considered by Ryan and Hsü, but they concluded that the evidence is best explained by desiccation and flooding.

THE LOST CONTINENT OF ATLANTIS

The legend of Atlantis has been one of the most popular stories of all time. Innumerable books and articles have discussed the story, and many experts have presented evidence that locates the lost continent "definitely and indisputably," or so they claim. Atlantis has been located in south, north, and west Africa, the Azores, the Canary Islands, Ceylon, Spitzbergen, the Baltic Sea, and 4000 m high in the Andes. Colonel Fawcett went into the Amazon jungle seeking Atlantis and was never seen again. Other expeditions went in search of Colonel Fawcett. Many insist on placing Atlantis in the Atlantic Ocean, as a bridge between early Egyptian and American Indian civilizations. An early version of continental drift had America separated from Europe by the sinking of Atlantis. Velikovsky argued that a comet from Jupiter sped past the earth in 1600 or 1500 B.C., swamping Atlantis in the same tide that parted the Red Sea for the children of Israel. Flying Saucers may have been involved with Atlantis.

The legend came to Plato from the Egyptians, and he wrote in "Timaeus":

There occurred violent earthquakes and floods, and in a single day and night of misfortune, all your warlike men in a body sank into the earth, and the island of Atlantis in like manner disappeared in the depths of the sea. For which reason the sea in those

*parts is impassable and impenetrable be-
cause there is a shoal of mud. . . .*

There is now convincing evidence that
the original Atlantis may have been the volcanic island of Santorin, now consisting of
five islands including Thera, about 125 km
north of Crete. The evidence has come from
excavations by Marinatos on the island and
from exploration of the deep-sea sediments
in this part of the Mediterranean, not by the
Glomar Challenger but by earlier oceanographic expeditions that recovered sedimentary sections from piston-cores.

In the 1969 edition of W. A. McNeill's
History of Western Civilization, it is reported that a civilization arose on the island
of Crete in the third millenium B.C. The
Cretan (or Minoan) civilization carried on
trade with the Egyptians, but was independent in such things as artistic style, religious cult, and method of writing. About
1400 B.C., the Cretan civilization was destroyed, probably by invaders from mainland Greece.

It is now widely believed that the civilization was destroyed when Santorin exploded
in a tremendous volcanic eruption that occurred about 1500 B.C. There are many
small lithosphere fragments squeezed between the African and European plates, and
Crete lies along a convergent boundary
south of the small Aegean plate. The lithosphere sinks in a zone extending northward
beneath Crete and under the Cyclades Islands. Santorin is the only active volcano
associated with this sinking plate.

Small eruptions have occurred throughout the history of Santorin, most recently in
1928, 1938, and 1950, as the volcano attempts to rebuild in its former conical glory.
Its location near a plate boundary was indicated by a 1956 earthquake that badly
damaged the main town on the islands, and
killed dozens of people.

Study of deep-sea cores from the eastern
Mediterranean has made it possible to date
two specific events (see Chapter 9). Volcanic ash thrown out from eruptions of Santorin has been found at two levels in the
sedimentary cores, and these have been

dated at about 23,000 B.C. and 1500 B.C.
The second eruption has been compared
with the more familiar eruption of Krakatoa,
in 1883, which was carefully observed and
well documented. Krakatoa is similarly associated with the Indonesian convergent
plate boundary. The Krakatoa volcanic explosion demolished the cone, leaving an
open space, or caldera, about 4 km across.
About 80 km³ of rock disappeared, wind-
blown ash fell 5000 km away, and tidal
waves rolling ashore in Java and Sumatra
killed more than 30,000 people. The remnants of the volcano were covered in ash
to a depth of 60 m. The ash covering Santorin is still up to 45 m thick, even after almost 3500 years of erosion, and the surface
area of the caldera, the area of the volcano
lost in the eruption, is four times that of
Krakatoa.

The evidence uncovered by the ar-
chaeological work of Marinatos as he digs
carefully through the thick layers of volcanic ash is quite remarkable. On one limb
of the island there is a whole city buried by
the great volcanic eruption of 1500 to 1470
B.C. Apparently the city was partly shattered by an earthquake, and organized
rescue teams cleared the streets, rebuilt
walls, and made possible the escape of the
inhabitants with their personal valuables.
The rescue teams possibly came from
Crete. Then a volcanic eruption buried the
city in ash and pumice to a depth of about
5 m. Some years later, the volcano began to
boom, the way Stromboli booms today, and
columns of steam, dust, and pumice were
puffed toward the sky at frequent intervals.
Lava rolling down the volcano slopes and
glowing at night could probably be seen as
far away as Crete. Finally, there came the
indescribable explosion as the volcano
burst, throwing out tons of ash that blew to
the southeast, according to the deep-sea
sedimentary cores, and sending tidal waves
to ravage neighboring coasts.

Marinatos believes that the Minoan city
on Santorin was "Atlantis," and he
suggested that this was the mother city of
the Cretan Minoan civilization. He was the
first to suggest that the widespread desola-

tion of Crete was due to Santorin's eruption. Tidal waves smashed the boats and coastal settlements, while ash, dust, and poisonous gases spread across Crete, spoiling the farmland and sending starving survivors to the less hospitable western end of the island and then to mainland Greece and other islands as refugees. Whatever the details may have been, there seems to be little doubt that the explosion of Santorin caused a shift of power from Minoan Crete to Greece. The Minoan spirit that was carried abroad by the refugees has been described as the leaven of later Greek civilization.

We could conclude, therefore, that plate tectonics, which caused the explosion of one small volcano, was responsible for the loss of Atlantis, the dispersal of Minoan culture, and for planting the seeds of Western civilization in the developing communities of ancient Greece. The course of history might have been quite different if the cultured, rich civilization that had turned the Aegean Sea into a peaceful Minoan lake had been spared the disruption caused by a belch from the earth's interior.

SUMMARY

The Deep-Sea Drilling Project aims to gather information about the age and processes of formation of the ocean basins by direct sampling of the sediments beneath the oceans. Long sediment cores are recovered in deep water by the remarkable drilling ship, *Glomar Challenger.*

In 1970, results of deep-sea drilling at sites along a line transverse to the South Atlantic ocean ridge were described as a "critical test for both sea-floor spreading and magnetic stratigraphy"; the fossil ages of the deepest sediments conformed to the age of the underlying lava according to the magnetic anomalies. The evidence from the rocks recovered, however, indicates that the main oceanic crust was not penetrated; the evidence is persuasive, but the definitive test has yet to be made.

Deep drilling beneath the Mediterranean Sea recovered layers of salt. The sediments in the cores are interpreted as follows. The Mediterranean Sea was an open seaway from the Atlantic to the Indian Ocean 20 to 15 million years ago. Then the connection with the Indian Ocean was closed. About 8 million years ago, the Atlantic opening was closed by convergence of Africa and Europe and, within 1000 years, the water evaporated completely, leaving a great valley, 3 km deep, covered in salt flats. The rock barrier damming the Straits of Gibraltar was breached 5.5 million years ago, and the basin filled with water from the Atlantic in about 100 years.

Study of deep-sea cores from the eastern Mediterranean reveals the wide areal extent of volcanic ash thrown out from eruptions of Santorin at about 23,000 B.C. and 1500 B.C. It is now widely believed that the explosive eruption of Santorin in 1500 B.C. was responsible for the destruction of the Minoan civilization, which had been centered on the island of Crete, just 125 km to the south. The claim that Santorin is the legendary lost continent of Atlantis is more compelling than previous guesses. Plate tectonics destroyed Atlantis, and spread the seeds of western civilization.

SUGGESTED READINGS are at the end of the book.

chapter **15**

The Diversity and Extinction of Species

When Darwin discussed *The Origin of Species* in 1859, and *The Descent of Man* in 1871, he convinced most naturalists that living creatures were not created in fixed and immutable forms, but that they had changed through the generations by natural processes operating over great periods of time. Those that changed into forms better adapted to their environment survived, and the others declined and became extinct. This concept of evolution revolutionized biology. Darwin's evidence and arguments did not include the possible effects of continental drift and of polarity reversals of the earth's magnetic field because these phenomena were neither known nor conceived at the time. Now, however, their effects on the evolution, diversity, and extinction of species are the topics of active discussion by paleontologists.

Sedimentary rocks contain a record of life through the ages. The record consists of the fossilized remains of creatures that died in environments where their shells or skeletons became buried by younger sediments and of vegetation similarly preserved on land or beneath the sea. The organisms living in a given region during a given period of time are represented by their fossils in a particular layer of sediment. The record is very incomplete because only a fraction of the living organisms leave remains that are preserved and because the sequence of sedimentary rocks is incomplete. Many sedimentary rocks have been destroyed by weathering, transportation, and redeposition elsewhere, as outlined in Chapter 6.

We saw in Chapter 12 that paleontological evidence was important in the development of the continental drift debate. The debate continues today. Some cite evidence as proof that continental drift has occurred, and others cite evidence to support arguments for the presence of ocean basins in their present positions for the past 600 million years. We will not review that evidence and the associated arguments. Instead, we will examine the theme that plate tectonics provides a whole new framework for the study of past life forms; the drifting continents may have regulated the ebb and flow of life. We will also examine one of the ideas arising from evidence that played an important role in the revolution, the suggestion that reversals of the earth's magnetic field (Chapters 9 and 13) may have exerted a selective force on the extinction of species.

THE EQUILIBRIUM MODEL OF BIOGEOGRAPHY

Biogeography is a branch of biology that deals with the geographical distribution of animals and plants. The paleogeographic distribution of fossils in sediments of particular ages and of variable ages is thus concerned with biogeographical models. Operation of the equilibrium model is illustrated in Figure 15-1.

The model starts with an isolated area initially devoid of life. Consider, for example, a new volcanic island rising from the ocean basin floor. There are many processes that can transport seeds and animals to such an island and, almost as soon as the eruption ceases and the lava has cooled down, the process of colonization by plants and animal species begins. The number of species on the island increases with time.

Through millions of years, the diversity increases as additional species arrive and because the existing species evolve into new species as the fauna becomes adapted to the environment. The rate of extinction of the older species is less than the rate of increase of new species. In a stable environment, the animal and plant populations become specialized, each species occupying its own niche in the ecological scene. This continues until the area is effectively saturated, with all available niches occupied. The fauna on the island then remains in a steady state of equilibrium and the rate of production of new species remains equal to the rate of extinction of old species. Minor changes in the environment, or habitat, produce minor fluctuations in the number of species around the equilibrium value, as indicated in Figure 15-1.

The total number of species that can be supported in a given geographic region is a function of the habitable area. This has been determined by counting the number of species of animals or plants on islands of various sizes, the islands having relatively uniform climate and similar environments in other respects. Figure 15-2 illustrates the orderly type of relationship that exists. Each point gives the total number of species on a particular island with the area as plotted.

Fig. 15-1 Development of an equilibrium species diversity model.

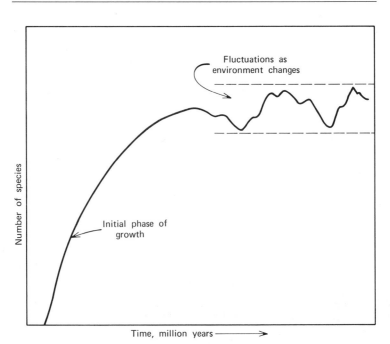

The graph through the points shows that if the area of an island is increased by a factor of ten, the number of species at equilibrium increases by a factor of about two. Biologists express results of this kind in a simple mathematical formula, which is refined for various environments by theoretical and empirical studies. For a given area, the number of species can be predicted according to the equilibrium model.

FACTORS INFLUENCING SPECIES DIVERSITY

According to the equilibirum model of biogeography, the habitable area of a faunal province is the main factor limiting the number of species. Plate tectonics causes great changes in the areas of faunal provinces. Large continents are disrupted into smaller continents that become isolated from each other by deep ocean basins; these are effective barriers between the organisms on the dispersed continents.

Smaller continents collide, forming super-continents and combining faunal provinces.

Consider two continents of equal area, each populated by a large number of species, S. Assume that these two continents have been isolated from each other for millions of years and that there is no faunal overlap. Between them, these two continents support $2S$ species. If the continents collide, we have a new province with double the area, but this cannot support $2S$ species. In fact, the number of species that can be supported on the doubled area is not much greater than S, as shown by the shape of the curve in Figure 15-2 for large areas. Therefore the rate of extinction exceeds the rate of production of species until the new equilibrium diversity is reached.

Similarly, consider the subdivision of a continent, populated by Z species, into two separate continents of equal area that become isolated from each other by intervening ocean basins. As shown in Figure 15-2, each smaller continent cannot support as

Fig. 15.2 Area-species curve based on the number of species living at equilibrium on islands of different sizes.

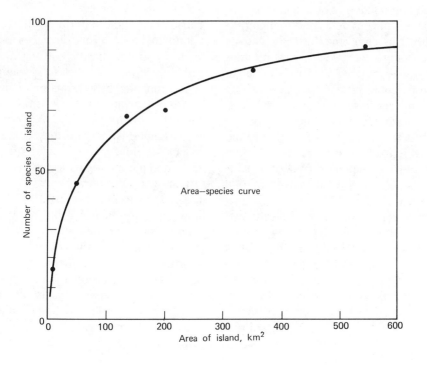

Area—species curve

many as Z species, but the equilibrium number for each is not far below Z when we consider continents with large areas. If the species on each continent evolve in isolation from each other, the total number of species that can be accommodated on the pair of continents is significantly more than those on the original continent, and species diversification occurs.

In the context of the equilibrium model, plate tectonics acts as a bellows to expand or contract the effective area of continents, causing the diversification or extinction of species. It operates by creating or by destroying ocean basin between continents and thus creating or destroying faunal barriers.

There are many factors influencing the environment and modifying the theoretical equilibrium value for species diversity based on the habitable area.

As a continent drifts across the globe, the environment changes. Figure 15-1 shows, for one region, fluctuations in species diversity after the initial phase of increase. More favorable environmental conditions cause increased diversity, and less favorable conditions decrease diversity because the rate of species extinction then exceeds the rate of production of new species.

If fragmented continents drift to the north or south, the changing latitude causes climatic changes that provide opportunities for new species to develop. This is a change in environment leading to species diversity. Stability in food resources can cause high diversity of specialized organisms, whereas fluctuation in abundance and type of food resources is likely to produce fewer more flexible species. Migration of continents is likely to be associated with changes in food resources.

Changes in sea level cause large changes in the environment by exchanging a habitable area of dry land with the equivalent area of submarine continental shelf. Figure 3-11c shows that the continental shelf is simply a portion of the continental masses that is covered by seawater overflowing from the ocean basins. The geological record demonstrates clearly that the extent of the continental shelf has increased and decreased periodically through geological time. If the height of sea level increases, the ocean overlaps further onto the continents, causing a marine transgression. If sea level is lowered, the water retreats across the continents causing a marine regression. Widespread shelf seas develop during periods of major transgression, and this provides a larger area for the divergence of marine species than the narrow shelf seas existing during periods of major regression. The relationship corresponds to the graph for island areas in Figure 15-2.

Two factors that can cause worldwide increase or decrease in sea level involve the history of ice sheets and of small ocean basins. Sea level rises if the ice sheets melt (Chapter 6). If the Mediterranean Sea were to become separated from the Atlantic Ocean, the water would evaporate and be precipitated in the other oceans, raising world sea level by 10 m (Chapter 14).

Plate tectonics provides another mechanism for transgressions and regressions through changes in volume of the submerged ocean ridges. It has been argued that during periods of rapid sea-floor spreading, the ocean ridge expands because more hot material wells up from within the earth. This displaces seawater from the ocean basin, causing a marine transgression. A decrease in spreading rate is accompanied by contraction of the ridge, return of seawater to the ocean basins, and hence by marine regression. When continents collide, spreading may cease at the associated ridges, with resultant ridge collapse and marine regression.

During a pulse of rapid sea-floor spreading that occurred between 110 and 85 million years ago, it has been estimated that marine transgressions almost doubled the area of continental shelves. When the rate of sea-floor spreading decreased to normal, sea level fell, causing marine regressions and, by 60 or 50 million years ago, the area of continental shelf had been reduced almost to its former value.

THE DIVERSITY AND EXTINCTION OF SPECIES RELATED TO PLATE TECTONICS

Many paleontologists have studied the diversification of species through geological time by counting the numbers of species found in sedimentary rocks from successive age intervals. The results for individual species or for families of related species may be plotted in the form of a histogram (constructed in the same way as Figure 3-11a), or the histogram can be represented as a curve (Figure 3-11b).

Figure 15-3a shows the type of diversity pattern obtained for the past 600 million years. The results of various investigators differ in details, but there is general agreement about the following features: (1) a rapid rise in the number of species between about 500 and 470 million years ago; (2) high diversity for a period of 200 million years, with a maximum near 370 million years ago; (3) a rapid decrease in the number of species between about 280 and 225 million years ago; and (4) a rapid increase in diversity to levels higher than the previous maximum, up to geologically recent times, about 2 million years ago. This could be interpreted as a general trend for an increase in diversity from 600 million years ago to recent times, with a major interruption caused by a period of extinctions that decreased the number of species to a minimum near 225 million years ago.

This method of plotting diversity data averages the diversification through fairly long intervals of time, and many short-term changes are not shown very clearly. The arrows indicate periods in which many species became extinct, and other kinds of graphs are required to show the details. In addition, there are several sources of error that could influence the diversity data. One of these concerns the survival rate of the sedimentary rocks. We noted above that both the fossil record and the sedimentary rock record are incomplete.

If all of the sedimentary rocks deposited within a specific time interval were pre-

served for careful study by teams of paleontologists, they would find the maximum number of species leaving fossil records for that time interval. Their results would give the maximum diversity that could be determined for the interval. If half of these sedimentary rocks were not available for study, either because they were destroyed and dispersed by erosion or because they remained buried by younger sediments, the paleontologists would find fewer fossils and almost certainly fewer species. The apparent diversity would therefore be lower. We may conclude that the greater the survival rate for sediments formed within a specific time interval, the greater the recorded diversity of fossils.

Many geologists have compiled estimates of the volumes of sedimentary rocks of various ages. Figure 15-3b shows the type of results obtained, expressed as the volume of sediments per year of formation. Note the correspondence between parts of Figures 15-3a and 15-3b. The survival rate of sediments with ages between 250 and 200 million years ago is only about 50% of that for sediments preserved in comparable time intervals before and after this period. This suggests that the decrease in diversity during the same period, which is normally attributed to an increase in the extinction rate compared with the production of new species, could be caused simply by the relative paucity of sediments of these ages.

This and other considerations led D. M. Raup of Rochester University to state, in 1972, that "In the meantime, it would be prudent to attach considerable uncertainty to the traditional view of Phanerozoic diversity." The traditional view is Figure 15-3a. Raup emphasized the uncertainties involved in both sets of data illustrated in Figure 15-3. Knowing that these uncertainties exist, we can now indulge in speculative interpretations of the diversity data within the framework of plate tectonics.

In 1972 J. W. Valentine and E. M. Moores of the University of California considered the factors involved in plate tectonics that could influence species diversity. They

Fig. 15-3 *(a)* An estimate of the diversity of species during the past 500 million years, with times shown that were characterized by many extinctions. (Various sources, especially J. W. Valentine and E. M. Moores, *Jour. Geology, 80,* 1972, 167-184.) *(b)* Volume of original world-wide sediments produced each year that have actually survived to the present time (avoiding erosion, Figures 6-5 and 6-6). A generalized estimate of relative changes. (From various sources, especially as summarized by D. M. Raup, in *Science, 177,* September 22, 1972, 1065-1071.)

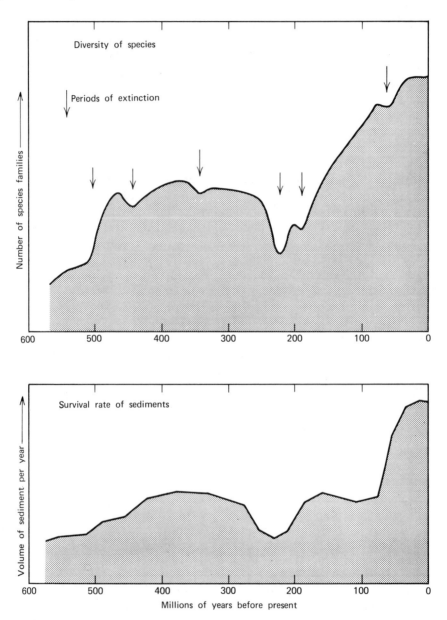

concluded that increasing diversity would be expected as continents break apart and that the coalescence of continents should cause a decrease in diversity. We stated that conclusion above in connection with Figures 15-1 and 15-2. Valentine and Moores then compared a species diversity graph (Figure 15-3a) with their best estimates of the distribution of continents through time and concluded that the diversity data were satisfactorily explained by the history of continental drift. They examined diversity data specifically for marine animals living on or near the floor of the continental shelves, but it is known that intervals of diversification and extinction have tended to occur simultaneously in marine and land species of animals and plants.

The history of continental drift from 510 million years ago to the present is outlined in Figures 12-4 and 12-3. Figure 12-4d shows three continental masses 510 million years ago, during the first period of rapid diversity increase shown in Figure 15-3a. Valentine and Moores related this increase to the breakup of a supercontinent some millions of years earlier. Continued drifting of these continental masses, as shown for 380 and 340 million years ago in Figure 12-4, maintained environments favorable for the high diversities shown for this period. By 225 million years ago, the continents had reassembled into the supercontinent of Pangaea, as shown in Figure 12-4a, and this was accompanied by major marine regression. The reassembly and marine regression are associated with a marked decrease in diversity, as anticipated. Pangaea began to break up 200 million years ago. The fragmentation and dispersal of the several continents illustrated in Figure 12-3 produced a large area of shallow marine environments on continental shelves, several ocean basins, great lateral spread of the continents, as well as movement toward high latitudes with associated climatic changes. This produced environments favoring high diversity, and this is in accord with the diversity data.

REVERSALS OF THE EARTH'S MAGNETIC FIELD AND FAUNAL EXTINCTIONS

The earth's magnetic field protects much of the earth's surface from cosmic rays, the solar wind, and other charged particles and radiation in space. These become diverted and trapped in the magnetosphere, a huge doughnut-shaped radiation belt surrounding the earth, extending from an altitude of about 1000 km above the surface to a distance of more than 5000 km. Charged particles spiral around the lines of magnetic force surrounding the earth (Figure 8-5), and as these are discharged into the atmosphere in the polar regions, where the lines of force direct them toward the earth's surface, they cause the spectacular auroras.

It is now established that the earth's magnetic field has reversed polarity frequently through geological time (Figures 9-3 and 9-4), and the intensity of the field is significantly reduced as it reverses polarity. According to Figure 9-9, the magnetic field intensity is lower than normal through an interval of about 20,000 years during each reversal.

There has been much speculation about what happens during the interval when the earth's magnetic field is of very low intensity. Charged particles from the solar wind and an increased percentage of cosmic rays would reach the surface, and the particles normally trapped in the radiation belt would fall to the surface. In the upper atmosphere, cosmic rays might increase the production of radioactive elements, and the solar wind might convert more oxygen into ozone, which could lead to increased absorption of the sun's radiation and changes in climate down below.

Starting in 1963, before the occurrence of magnetic polarity reversals was generally accepted, there has been considerable discussion about whether or not these effects could cause extinctions and transformations of species. The increased radiation might produce permanent changes in the genes of animals. The mutations, inevitably harmful, would build up in generations through thousands and millions of years making the animals less fit for survival until they eventually became extinct.

These claims have been disputed on the grounds that the earth is adequately protected from charged particles and radiation by the atmosphere, even in the total absence of a magnetic field. Most investigators have concluded that the increased radiation doses received at the earth's surface would be insufficient to produce significant effects in the evolution of species.

It is conceivable that changes in the upper atmosphere such as increased production of ozone could cause changes in climate, and climatic changes can contribute to species diversity. Few physicists and paleontologists, however, consider that polarity reversals have a causal relationship with climatic changes that have any bearing on the extinction of species.

Despite the general denial of magnetic reversals as a factor in evolution, the idea continues to have some champions. Evidence was presented between 1964 and 1967 that several species of marine microorganisms represented by fossils in deep-sea cores either became extinct or made their first appearance in the geological record at a level very close to the most recent reversal of the earth's polarity, 0.7 million years ago (Figure 9-3). It was not certain, however, that the extinction of any species was simultaneous all over the world.

J. D. Hays of Lamont-Doherty Geological Observatory presented results in 1971 from a study of 28 deep-sea piston cores from locations throughout the world. The sediments in the cores were dated by their polarity reversals and the reversal time scale (Figure 9-3). Hays studied the distribution of certain microfossils that were widely distributed through the cores. He found that eight species were abundantly represented in the lower parts of cores but were absent in the upper parts. Disappearance of a fossil from the core corresponds to extinction of the species.

The eight species became extinct during the past 2.5 million years, and each species became extinct at the same time throughout its geographical range. Six of the species disappeared immediately following magnetic reversals recorded in the sediments. Hays concluded that this degree of correlation is too high to be the result of chance, and that the evidence is strongly suggestive that magnetic reversals, either directly or indirectly, had exerted a selective force on the extinction of the microorganisms that populate the upper ocean levels. In addition to the possible causes of radiation and climatic change, he suggested that biomagnetic effects merited some consideration. There is now an extensive literature on the biological effects of magnetic fields, although the mechanism by which magnetic fields affect organisms remains speculative.

Encouraged by the correlation between magnetic polarity reversals and the extinction of marine microorganisms within the past 2.5 million years, Hays moved on to bigger things and explored the possibility that changes in the frequency of magnetic reversals might be related to the extinction of dinosaurs 65 million years ago.

EXTINCTION OF THE DINOSAURS AND OTHER BEASTS

Dinosaurs dominated the earth for about 135 million years. Contrast that with less than 2 million years for the existence of true man. The dinosaurs adapted successfully to the changing environment until 65 million years ago. Then, throughout the world at about the same time, they all died out. Their extinction did not occur in a day or a year, but through several millions of years. Nevertheless, they became extinct quite abruptly in terms of the geological time scale. Whatever happened also caused extinction of the flying and swimming reptiles and of many species of primitive mammals.

Many hypotheses have been proposed to explain the demise of the dinosaurs, but none has received general acceptance.

Most of them have been farfetched ideas notably lacking in evidence. They include some kind of plague, sterility, racial senility, destruction of eggs by small, hungry mammals, and passage of the earth through a zone of increased cosmic radiation as it rotates around the center of the galaxy. Another suggestion is that a star exploded close enough to affect the earth. A nearby supernova would blast the earth's surface with radiation and cause great turbulence in the atmosphere. The atmospheric turbulence might have caused an abrupt drop in temperature, sufficient to exterminate the dinosaurs and other organisms.

This catastrophic extinction of fauna has been repeated on a less dramatic scale many times. Various studies indicate that mass extinctions have occurred near 500, 440, 340, 225, 195, and 65 million years ago, as shown in Figure 15-3a. A wave of animal extinctions began about 20,000 years ago, and it has been estimated that as many as 20% of the earth's animal species are on the way to extinction. Familiar beasts on the recent casualty list include mammoths and sabre-toothed tigers.

The extinctions occurring 225 and 65 million years ago were the most profound. There are explanations for these in the context of plate tectonics. The 225 million year extinction period is shown clearly in Figure 15-3a, and this we related to formation of the supercontinent, Pangaea. Recombination of dispersed continents cannot be invoked to explain the extinctions of 65 million years ago (Figure 12-3), but this does correspond to the period when water was draining off the continents in a major marine regression following a pulse of rapid sea-floor spreading.

It was noted by Hays that the mass extinctions of 225 and 65 million years ago coincide with changes in polarity intervals. Figure 9-4 shows that at these times, long intervals of normal polarity were followed by mixed polarity intervals. Hays suggested that during the millions of years of constant polarity, many of the species that evolved were potentially susceptible to the effects of

polarity reversals and that these became extinct when reversal activity resumed. The reversing magnetic field may have exerted a selective influence on evolution.

What are the chances that the dinosaurs became extinct because of their vulnerability to magnetic polarity reversals, with associated minor increases of radiation and perhaps climatic changes? Few paleontologists would bet on them. It seems more likely that the extinction of dinosaurs may be explained by a biological model involving populations and environments at a time when the environment changed drastically through major marine regression. The regression was caused by subsidence of large areas of the ocean floor following a period of unusually rapid sea-floor spreading. But that is another story.

SUMMARY

Plate tectonics provides a new framework for the study of past life forms. When continents collide, the rate of extinction exceeds the rate of production of species. When a continent is subdivided into two separate continents, the total number of species that can be supported increases. During periods of rapid sea-floor spreading, the ocean ridge expands, displacing seawater onto the continental shelves. Transgression of the sea provides a larger area for the divergence of marine species. Subsequent regression of the sea reduces the area of flooded continental shelf and causes extinction of marine species.

A graph for the diversification of species through time shows that increased and decreased diversity can be correlated with the dispersal and combination of continents. The drifting continents appear to have regulated the ebb and flow of life.

Mass extinctions of species have occurred several times, the most recent one 65 million years ago, including the demise of the dinosaurs. This coincides with a change from a long period of normal magnetic polarity to a mixed polarity interval. It has been proposed that the species that evolved during constant polarity became potentially susceptible to the effects of polarity reversals. Most investigators have concluded that the increased radiation dose from space reaching the surface during a polarity reversal would be insufficient to affect the evolution of species. But it has been discovered that several species of marine microorganisms represented by fossils in deep-sea sediment cores became extinct immediately following magnetic reversals. The earth's magnetic field may exert a selective force on the extinction of some species, but few geologists would list the dinosaurs among them.

SUGGESTED READINGS are at the end of the book.

chapter **16**

Causes of Plate Tectonics

In 1960, an avowed belief in continental drift by a young applicant for a faculty appointment in an American university would probably have been sufficient to ensure that the candidate was not appointed. Similarly, in 1975, a forthright announcement by a similar candidate that he or she did not believe the evidence for continental drift would probably have the same effect.

It is a moot question as to whether it is more serious to express faith in a heretical hypothesis or to express doubt about interpretation of evidence cited for a popular hypothesis; but it is probably true that whether the hypothesis is heretical or popular often is more significant than whether we are dealing with faith or with evaluation of evidence.

Continental drift did not become respectable during the first half of the century mainly because no one could work out a satisfactory mechanism. Authoritative geophysicists asked: "What's a nice continent like this doing, drifting over an earth with these properties?" They said it could not be done. If this opinion was believed, then what was the point of trying to evaluate the evidence presented by drifters?

The early debate stagnated in circumstances where earth scientists believed or disbelieved in drift without evidence that was considered to be definitive. Several lines of evidence have been claimed as proof for plate tectonics, however, with its rider of continental drift, and we have examined some of these in preceding chapters. The theory is now popular, despite the fact that we still do not have full understanding of what drives the plates.

A mechanism involving convection in the solid mantle was proposed in 1928 by A. Holmes of Edinburgh University, and a comprehensive scheme of mantle convection to explain mountain building was presented in 1939 by D. T. Griggs of the University of California. Because continental drift was not popular, however, few earth scientists considered convection in the solid mantle to be a feasible mechanism. Now that the theory is popular, it is generally accepted that the driving forces involve mantle convection, as outlined in Chapter 2 and Figure 10-5.

THE MANTLE DRAGS THE PLATES

The oldest ideas are illustrated in Figure 16-1. Large-scale convection currents rise in the mantle, producing an ocean ridge with a central rift valley and, as this slowly moving material flows laterally at the top of the asthenosphere, it drags the rigid lithosphere shell along with it. Imagine a piece of wood floating on a bowl of viscous molasses. If the molasses is stirred so that it flows, it carries the wood with it. Where the convective motion turns downward, back into the mantle, the lithosphere plate is dragged down with it, as shown in Figure 16-1. This kind of process has been observed in lava lakes in Hawaii, where the solidified surface crust of the cooling lava is moved around by the viscous but still mobile lava below. Many features simulating plates and plate boundaries are produced by the large slabs of lava crust as they are broken up and forced into collision.

Mantle convection of this kind may drive the plates, but what drives the mantle convection? The mantle rock can rise upward under the influence of gravity only if its density is less than that of the surrounding rock. The most likely cause of such a situation is localized heating at depth. If the rock in a certain region is heated, it expands, and its density decreases. This would make little difference in the lithosphere, where

the rock is cool and rigid but, beneath the lithosphere, the temperatures are high enough that rock can be slowly deformed. It flows in plastic fashion like red-hot steel. It can be shown that quite a small expansion and decrease in density is sufficient to make the rock buoyant. The process is illustrated schematically in Figure 16-2a. The rock rises, very slowly, at a rate of perhaps 1 cm per year. As it rises, the denser rock around it flows inward beneath it.

It is just as difficult to imagine these motions occurring at slow rates as it is to conceive the vastness of geological time and of astronomical distances. A rate of 1 cm per year is less than 1 mm per month, about 3/100 mm per day, and about 1/1000 mm per hour. The end of the hour hand on an average household clock moves at a rate of 5 cm per hour. However patiently you sit watching the hour hand of a clock, you cannot see it move. Yet this is 50,000 times faster than the motions illustrated in Figure 16-2. Obviously, the mantle is effectively motionless within our normal time framework. But a rate of 1 cm per year is the same as 1 m per century, and the distance covered in 1 million years is 10 km.

Fig. 16-1 Mantle convection drags the lithosphere plates across the asthenosphere.

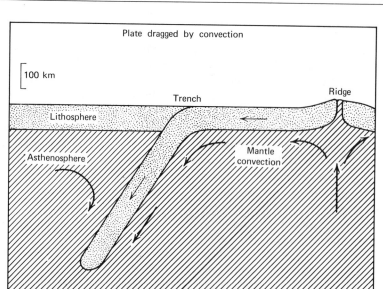

Fig. 16-2 Schematic cross sections through portion of the earth's interior, where temperatures and pressures are very high; solids become capable of slow flow. If a portion of the interior is less dense, or more dense than the surrounding material, buoyancy forces will cause that portion to rise or to sink, respectively. *(a)* Floater. *(b)* Sinker.

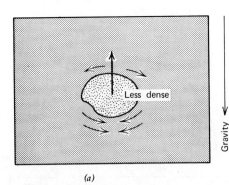 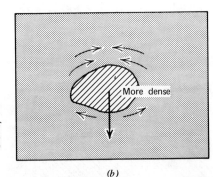

(a)　　　　　　　　　　　　　*(b)*

Given the geological time scale, small changes in density, and the action of gravity, then masses of hot, solid material will move through significant distances in the earth's interior.

In Figure 16-1, the mantle cools as it flows horizontally beneath the rigid lithosphere plate. As it cools, its density increases until it eventually becomes heavy enough to sink down through the surrounding mantle, forming the downward limb of the convection cell. The cell must be completed somewhere at depth, with approximately horizontal flow transferring material from the region of downflow to the region of upflow.

The scale of these large convection cells is uncertain. Figure 16-3 illustrates some of the models that have been proposed. One extreme has the cells extending through the whole thickness of the mantle. The other has most of the motion limited to the asthenosphere layer. Figure 16-3*d* has everything.

A major unanswered question is the source and distribution of the heat that produces the density differences required to make the convection cells move. The mantle rock contains a small proportion of radioactive elements that generate heat as they decay, but why there should be any increased concentrations in certain regions

to cause localized heating remains speculative.

The only plausible explanation for the existence of the earth's magnetic field is the occurrence of electric currents resulting from the convective motions in the molten portion of the earth's metallic core. Interaction between the molten outer core and the lower mantle probably is accompanied by transfer of energy from the core into the mantle, and this energy could be translated into some kind of motion, which could initiate a convection cell.

There are many unsolved problems and, at times, geophysicists are inclined to seek easier solutions, invoking mechanisms such as that illustrated in Figure 16-4. If mantle convection is the work of the Devil, then we can leave the explanations to philosophers and theologians.

THE PLATES DRAG THE MANTLE

Figures 16-1, 16-3, and 16-4 show movement in the mantle being followed by the lithosphere plates. It has been pointed out that if the plates move for some other reason this, in turn, will cause flow in the asthenosphere. The lithosphere and the asthenosphere are said to be coupled together; one cannot move without the other.

Fig. 16-3 Mantle convection drags the lithosphere plates. *(a, b, c)* illustrate the scales of internal convection cells that have been proposed. *(d)* J. C. Holden's version of many kinds of convection in the atmosphere, mantle, and core. The funnel at the top represents a hot-spot, see Figure 16-8. (Reproduced by kind permission of J. C. Holden, from Chapter 11.2 by R. S. Dietz and J. C. Holden in *Implications of Continental Drift to the Earth Sciences,* Vol. 2, 1973. Edited by D. H. Tarling and S. K. Runcorn, Academic Press.)

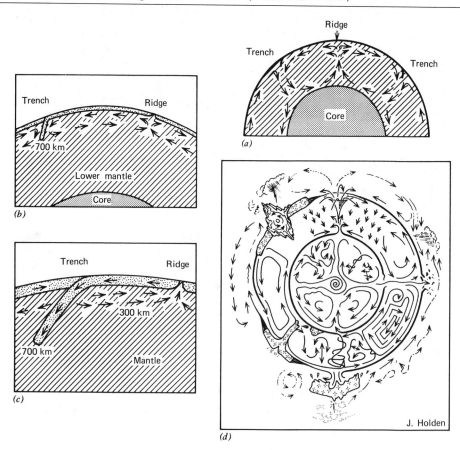

Try moving a piece of wood over a bowl of molasses; the molasses will be dragged along because of its high viscosity.

There are several factors that could contribute to movement of the plates, quite apart from mantle flow. Figure 16-5 shows a plate with a horizontal upper surface, the sea floor, but with its lower surface at an angle to the horizontal; the lithosphere thickens with distance from the ridge. Given this situation, the lithosphere plate has a tendency to slide downward and across the asthenosphere under the influence of gravity.

The forces involved are similar to those acting on a block of wood sliding down an inclined plane, as illustrated in the inset. The scales are quite different, and the inclined slope beneath the lithosphere plate is very small. It has been calculated that if the bottom of the lithosphere has a slope of only 1/3000, the lithosphere could slide laterally at a rate of 4 cm per year. This produces flow in the asthenosphere, as illustrated in Figure 16-5.

Lateral movement on the gently inclined boundary between lithosphere and asthenosphere also involves downward movement. Downward movement gives a decrease in potential energy, and calculations show that the amount of potential energy lost per year is approximately equal

Fig. 16-4 One possible source of mantle convection currents, according to J. C. Holden. (Reproduced by kind permission of J. C. Holden, from R. S. Dietz in *Sea Frontiers, 13,* 66-82, 1967.)

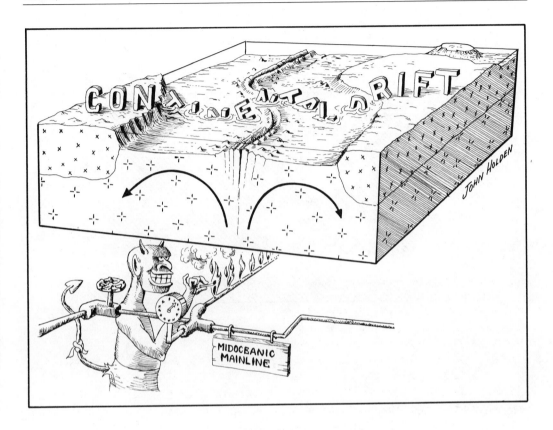

Fig. 16-5 Lithosphere plates slide down a gentle slope and cause the mantle to flow.

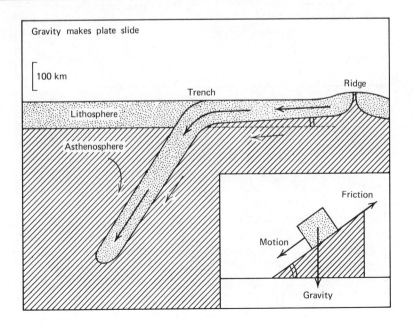

Fig. 16-6 The cold, dense lithosphere slab within the earth sinks (see Figure 16-2) dragging the plate behind it.

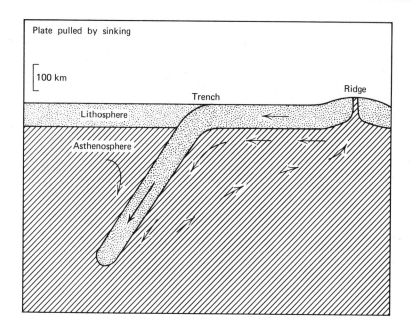

Fig. 16-7 Lava rising through fissures beneath the rift valley at the crest of an ocean ridge exerts enough pressure to push the lithosphere plates apart.

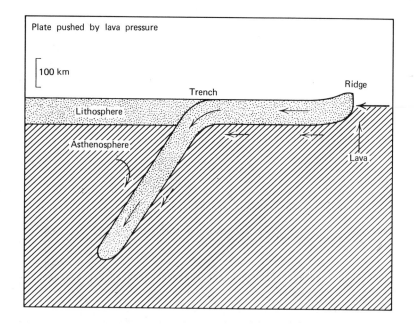

to the rate of energy released from earthquakes at plate margins.

The presence of cool lithosphere slab in the hot mantle to depths of 700 km is an unstable situation, and it must be accompanied by gravity-effected movements corresponding to that illustrated in Figure 16-2b. Figure 16-6 shows a form of convective motion in which the lithosphere slab is the active part of the convection cell. The lithosphere slab extending into the interior is cooler than the surrounding mantle, and it would take millions of years to warm up because rock is such a poor conductor of heat. Calculations show that the center of a sinking lithosphere slab at a depth of 400 km may be as much as 1000°C cooler than the surrounding mantle. Because it is cooler, it is more dense, and the force of gravity makes it sink, pulling behind it the sheet of lithosphere.

This has been described in terms of a "bathtub analogy," with a towel laid carefully on the surface of the water. If one edge is pushed beneath the water, this will begin to sink. The more towel that goes under water, the heavier it gets, the faster it sinks, and the faster the part remaining on the surface is pulled along. If the lithosphere moves in this way it generates movements in the asthenosphere, too. Figure 16-6 shows a return flow from the sinking slab through the asthenosphere at a shallow level. If the plate is dragged away from the ocean ridge by the sinking slab, there must be compensation by upward flow from the interior from beneath the ridge.

The uprise of material from beneath the ocean ridge crest, with partial melting and eruption of lavas, as illustrated in Figures 16-1 and 16-7, can also contribute to the motion of the plates. If compressed lava fills a deep fissure beneath the ocean ridge, its pressure could force the plates apart. Some of the lava is erupted at the surface, and some solidifies on the walls of the fissure, generating new lithosphere. Once initiated, the fissure never dissipates, and more lava rises from the asthenosphere. Movement of the pushed plate causes movement in the underlying asthenosphere.

The motions of lithosphere plates could be caused mainly by a combination of the two processes shown in Figures 16-7 and 16-6. The rise of hot, partially melted, less dense material from the asthenosphere pushes a plate at one end, and the sinking of cool, dense lithosphere pulls a plate from the other end. The lithosphere and the asthenosphere cannot move independently of each other because they are coupled together. Probably all of the mechanisms illustrated in Figures 16-1, 16-5, 16-6, and 16-7 are involved, and the relative contributions of each will not be known until a detailed three-dimensional mathematical and physical analysis has been completed on the problem. Apparently, a complete solution is beyond the capability of the present generation of computers, and we have inadequate knowledge of the properties of the rock materials at the high pressures and temperatures within the earth.

There is another problem. The two-dimensional cross sections illustrated in this chapter look reasonable, but fitting these schemes into three dimensions and taking into account the areas and distributions of plates and offset plate boundaries (Figure 3-10) is quite another matter. Fitting the model in Figure 16-1 to the mid-Atlantic ridge, for example, would be difficult because the ridge is displaced by so many offsets.

THERMAL PLUMES CREATE HOT-SPOTS AND DRIVE THE PLATES

Another type of model for the plate driving mechanism involves thermal plumes, only a few hundred kilometers in diameter, rather than the more conventional convection cells. This concept was developed by W. J. Morgan of Princeton in 1971, following an explanation proposed in 1963 by J. T. Wilson of Toronto to account for the existence of volcanic island chains that have no connection with plate boundaries. We noted in Chapter 13 that Wilson's article was first rejected by one journal

Fig. 16-8 Diagrammatic representation of the hot-spot hypothesis, with thermal plume of solid mantle rock rising and spreading out laterally in the asthenosphere. *(a)* Cross section. *(b)* Map.

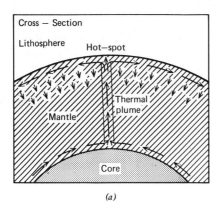

(a)

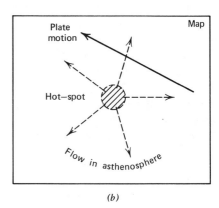

(b)

because it disagreed with ideas then fashionable.

Figure 16-8 illustrates the main features of the model. All upward movement of mantle material is restricted to about 20 thermal plumes, rising upward from the core-mantle boundary, and possibly initiated by transfer of energy from the fluid outer core. Where these vertical plumes reach the lithosphere, the flow becomes horizontal, spreading radially in all directions. It creates a hot-spot with volcanic activity and may cause upward doming of the lithosphere. The return flow to balance the upward flow concentrated in the plumes is accomplished by very slow downward movement of the whole of the rest of the mantle. The direction of movement of each plate is determined by the combined effects of forces exerted on the lithosphere plates by the radial flows, and the forces generated by the interaction of plates along their boundaries.

The relationship of hot-spots to volcanic island chains is illustrated in Figure 16-9. The most recent volcanic activity occurs at one end of the chain, and the ages of lavas from the dormant or extinct volcanoes in the line appear to become progressively older with distance from the active end of the chain. The direction of the island chains

correspond closely with the directions of movement of the plates on which they sit. Contrast this with the geometry of volcanic activity along ocean ridges and volcanic island arcs (Figures 5-4 and 6-8).

The Hawaiian Islands represent just one example of a linear volcanic island chain in the ocean basins. These islands are thousands of kilometers away from the nearest plate boundary. Wilson suggested that the island chains showed the direction of plate movement, and subsequent work on earthquake studies and magnetic anomalies supports his prediction.

Wilson proposed that the source of the lava for such volcanoes must be deep in the mantle, below the asthenosphere, as shown in Figure 16-9a. The source remained fixed in position, so that as the lithosphere plate moved, the volcano V_1 was cut off from its source of lava, and it became extinct. A new volcano, V_2, was formed in time interval T_2, as shown in Figure 16-9b. In this way, a line of volcanoes growing older from V_4 to V_1 was generated, with the line following the direction of movement of the plate, as shown in Figure 16-9c.

The 20 hot-spots identified by Morgan include the active ends of volcanic island chains, but most of the hot-spots are near ocean ridge crests, and there is a hot-spot

Fig. 16-9 Diagrammatic representation of how a fixed hot-spot can produce a line of volcanoes increasing in age from one end to the other. The Hawaiian volcanic island chain is an example. Older volcanoes V_1, V_2, and V_3 are carried away from their lava source and become extinct.

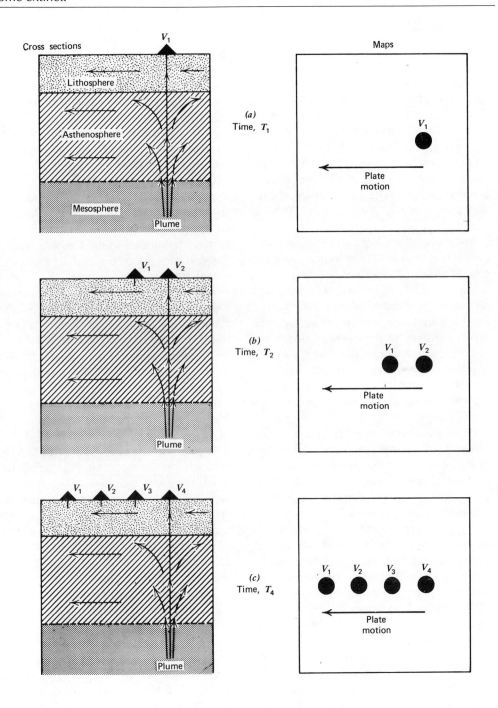

near each place where ocean ridges meet in junctions. Examples are Iceland, Tristan da Cunha, the Azores Islands, and the Galapagos Islands. The chemical characteristics of the lavas erupted from the proposed hot-spots differ in some respects from the lavas of ocean ridges and island arcs, which could be attributed to their derivation from deeper source material.

One of the aims of Leg 33 of the Deep-Sea Drilling Project was to study linear volcanic island chains in the South Pacific Ocean. In November and December of 1973, the *Glomar Challenger* drilled through sedimentary rocks and into the volcanic rocks of the Line Islands. This chain of volcanoes straddles the equator, between Tahiti and the Hawaiian Islands, and approximately parallels the Hawaiian island chain.

Results obtained from three drilling sites over a length of 1270 km indicate that the Line Island chain does not fit into a simple hot-spot model. If the line of islands was produced by successive eruptions from a fixed hot-spot beneath the moving Pacific plate, the ages of the volcanic rocks beneath the ocean floor should increase from one end of the chain to the other, as indicated in Figure 16-9. In fact, the geological histories of all three sites appear to be nearly identical.

The preliminary report of Leg 33, published in *Geotimes* in March 1974, included reference to some of the problems involved in deep drilling. The picture of the ship with its fancy electronic equipment makes the whole business look easy (Figure 14-1). But several of the experiments had to be terminated or abandoned for reasons such as these: "the ship had moved too far away from the beacon"; "drilling difficulties culminated in bending the bumper subs of the bottom-hole assembly"; "a fragment of a pump was washed into the drill pipe and prevented retrieval of the core barrel"; and "a bolt accidentally fell down the pipe and prevented recovery of the core barrel." The nuts and bolts aspects of science are just as important as the generation of great ideas.

THERMAL PLUMES OR GRAVITATIONAL ANCHORS?

An accurate physical model for thermal plumes on a global scale is not even remotely possible with the present state of knowledge, but the idea has generated much discussion among geophysicists and geologists. Many earth scientists have adopted the plume hypothesis, and they are using it as a basis for all kinds of schemes and interpretations. Various aspects of the volcanic activity, and the effect of doming, which produces a "bumpy" asthenosphere over which the lithosphere plates must ride, have been described and debated.

Other earth scientists express doubt that a plume could retain its coherence as it rises through 2800 km of mantle. They consider it more likely that such a plume would expand too much to exist as any kind of focused hot-spot by the time it reached the asthenosphere.

H. R. Shaw and E. D. Jackson of the U. S. Geological Survey have proposed an alternative model in which the thermal focusing effect that produces a hot-spot is localized in the asthenosphere by a mechanism involving flow, friction, and melting. The complex process is illustrated schematically in Figure 16-10, which is similar to Figure 16-9c, except that some directions of movement have been changed.

Figure 16-10a shows a cross section through the earth at time T_4, some time after volcano V_3 had been severed from its supply of lava by movement of the lithosphere plate. Figure 16-10b is a map showing the volcanoes and plate movement, the direction of flow of the asthenosphere below, and the region where melting produces lava for eruption.

The source of lava for V_3 was the white region in the asthenosphere. When the lava was erupted, the residual rock in the white region became denser than the surrounding material in the asthenosphere and the mantle below in the mesosphere. Therefore material from the white region sank downward in the manner illustrated in Figure 16-2b. This modified the direction of

Fig. 16-10 Diagrammatic representation of the gravitational anchor hypothesis, showing how sinking of more dense rock left over after melting and eruption of lava might localize shear stresses caused by the flow of rock in the asthenosphere. The localization of stress could cause further melting in that region as new rock flows in. Older volcanoes V_1, V_2, and V_3 have drifted away from this source of lava and become extinct. (a) Cross section. (b) Map.

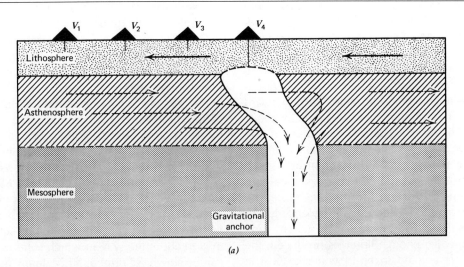

(a)

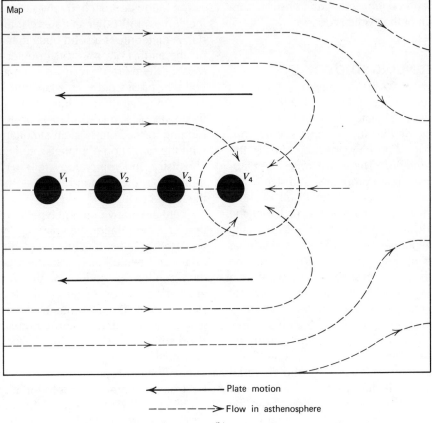

Plate motion

Flow in asthenosphere

(b)

flow in the asthenosphere, as illustrated in Figure 16-10. Imagine a flowing river with an enormous drain-hole opened up below it. Most of the water flows on undisturbed but, in the region of the drain-hole, flow-lines sweep toward the hole and down into it. Now slow everything down to the incredibly slow rates that we have considered for mantle movements.

The material sinking into the static mesosphere acts as an anchor, like a drain-pipe maintaining the region of downward flow in position more or less directly above it. The white region, because of the downward movement, has a more complex flow pattern than the surrounding asthenosphere, and this causes localized heating by friction. In time, therefore, the white region becomes hot enough to cause another episode of partial melting and, when enough liquid is produced, the lava is erupted to build volcano V_4. The white material again becomes more dense, and continued sinking ensures continuation of the cyclic process.

A CONTINUING STUDY

A prerequisite for understanding the underlying causes of plate tectonics is an understanding of the structure, composition, mineralogy, and physical properties of the earth's interior. There are four main approaches to these problems, and significant advances have been made in all of them during recent years. Yet the fact that the diametrically opposed hypotheses of rising thermal plumes and sinking gravitational anchors can be presented to explain the same phenomenon suggests that we have a long way to go before we can work out what is really happening in the mantle.

The first approach involves the study of earthquake waves emanating from an earthquake focus. We saw in Chapter 7 that the waves are affected by the material through which they pass, and analysis of the waves provides information about the structure and properties of the interior.

The second approach is study of lavas and the fragments of rock that they sometimes carry up from the deep source regions. These and other rock types that may have been derived from the mantle provide information about the chemistry and mineralogy of the upper part of the mantle.

The third approach is a laboratory study. High-pressure experiments using minerals and rocks of appropriate composition are now capable of reproducing the conditions of temperature at various depths within the earth. These experiments provide direct information about the conditions of melting, the formation of lava, and the depths at which various reactions and density changes occur in mantle materials. Measurement of the physical properties of minerals and rocks at selected pressures and temperatures has provided valuable information about the way in which solid rocks can be slowly deformed, and actually flow, under mantle conditions.

The fourth approach involves the study of meteorites and other extraterrestrial material. A meteorite is a solid body that arrived on the earth from an orbit in outer space. Most of them are derived from the asteroid belt that passes between Mars and Jupiter. They are generally considered to represent fragments of disrupted planets or of bodies forming in the solar system simultaneously with the formation of planets; study of their chemistry and mineralogy tells us something about the size and thermal history of the former planetary bodies.

There are many different types of meteorites, and the relationships among them all are used to infer the history of planetary origin and evolution. These inferences are then applied to models for the formation and differentiation of the earth. There are more than 1500 well-authenticated meteorites, many of which comprise numerous individual pieces of an original fragment that entered the earth's atmosphere. Despite this large number of specimens, the evidence is far from complete. Models for planetary evolution still rank as speculation, and inferences about the earth's interior based on extraterrestrial rocks and chunks of nickel-iron alloy must be treated with caution.

Much more extraterrestrial material is needed to fill in the many gaps in the story of the origin and evolution of the planets, and the Apollo missions to the moon provided the first samples other than meteorites. The first manned landing on the moon was achieved in 1969, the year after publication of the formal theory of plate tectonics. In 1969, earth scientists found themselves looking downward into the earth's interior in attempts to explain the origin of continents on which our civilization developed and upward to the moon in the hope that this might improve our knowledge of the earth's interior.

SUMMARY

The driving forces for plate tectonics involve some kind of mantle convection. The dynamics of mantle convection is a major unsolved problem in plate tectonics. If rock deep within the earth is heated by some process, it expands and its density decreases. This is sufficient to make red-hot rock buoyant, and it rises very slowly, flowing in plastic fashion in much the same way that red-hot steel is slowly deformed by a blacksmith's hammer. The hour hand of a clock moves thousands of times faster than the solid mantle material. When it reaches the lithosphere, the mantle flow is diverted to a horizontal direction, and it drags the overlying lithosphere plate with it. Where mantle flows converge, a downward limb of a convection cell is produced. It is not known whether the convection cells extend through the whole thickness of the mantle or whether they are restricted to the asthenosphere layer.

There are other factors that could cause the lithosphere to move over the asthenosphere, including a push at one end, where lava rises between two lithosphere plates at a divergent boundary, and a pull at the other end, where cool, dense lithosphere is sinking into the hot mantle at a convergent plate boundary. The lithosphere and asthenosphere cannot move independently of each other because they are coupled together.

According to one hypothesis, all upward movement of mantle material is restricted to about 20 thermal plumes rising from the core-mantle boundary, with the return flow accomplished by very slow downward movement of the rest of the mantle. Oceanic volcanoes and volcanic island chains are cited as evidence for hot-spots where thermal plumes reach the lithosphere. An alternative explanation proposed for the volcanoes involves a hot-spot in the asthenosphere caused by sinking "gravitational anchors," plumes of rock sinking down through the mantle because the eruption of lava increased the density of mantle rocks left behind.

SUGGESTED READINGS are at the end of the book.

chapter 17
Exploration of the Moon

Exploration of the moon in six Apollo space missions during 3.5 years, between July 1969 and December 1972, provided a harvest of 380 kg of rock and results from numerous scientific experiments conducted on the moon's surface and from spacecraft orbiting around the moon. In Chapter 1 we examined the overlap between this exploration program and the revolution in earth sciences. In Chapter 16 we noted the hope that the space program might provide a better understanding of the earth's interior and, therefore, a better understanding of how plate tectonics works.

Never before has such a major scientific problem been tackled with such intensity by so many scientists in such a short period using so little material. The way in which hypotheses were formulated, developed, and abandoned or modified through the six Lunar Science Meetings between 1970 and 1975 would make a fascinating study in the philosophy of science and the psychology of scientists.

There are still several schools of thought about some of the major problems, and the ways in which these have developed have much to do with the influence of strong personalities. In contrast, the theory of plate tectonics is now challenged by only few

scientists, but strong personalities were similarly involved during its development through continental drift and sea-floor spreading, as outlined in Chapters 2, 11, 12, 13, and 14.

When all of the data and evidence had been reviewed at the Sixth Lunar Science Conference of 1975, it was clear that the earlier models for the origin of the solar system, the earth, and the moon had been grossly oversimplified. Summaries of the main sessions at this Conference were published in Geotimes, May 1975. Although there is no consensus about the origin of the moon, many facts have been established, many problems have been resolved, and many details about the early history of the moon have been worked out. This information can be related to the history of the earth during the first billion years of its existence.

BEFORE THE APOLLO MISSIONS

On May 25, 1961, President John F. Kennedy made a commitment to land men on the moon before the end of 1970. The Apollo missions were possible only because of the very careful preparation in earlier programs during the 1960s, following this commitment.

The major features of the near side of the moon's surface were well known before the space program commenced because they had been studied for centuries through earth-based telescopes. Galileo used his newly invented telescope to examine the moon and, in 1610, he described the level, dark areas that resembled seas among the lighter-colored mountainous regions, as well as the ring-shaped features, or craters. Figure 17-1 is a sketch of the near side of the moon, showing the distribution of the dark maria (singular: mare) and the light highlands.

U.S. Air Force photographs of the moon have been used to construct a detailed topographic map with contour interval of 300 m. Members of the U.S. Geological Survey used this as a base for a geological map of the moon, using standard photogeological techniques. The most important geological work was that of E. M. Shoemaker and R. J. Hackman in 1962. They determined the relative ages of the major and minor physiographic features objectively and unambiguously. The mare basins are younger than the heavily cratered highlands that they cut, and the dark material filling the mare basins and flooding some of the neighboring regions is still younger. Younger features include many craters, ridges, scarps, and peculiar valleys called rilles. The youngest craters are those with rays extending from them; rays are interpreted as rock material ejected from a crater during its explosive formation.

Fig. 17-1 Sketch map of the near side of the moon showing the major mare. Also shown are the positions on the moon's surface of landings by spacecraft of the Surveyor, Apollo, and Russian Luna programs.

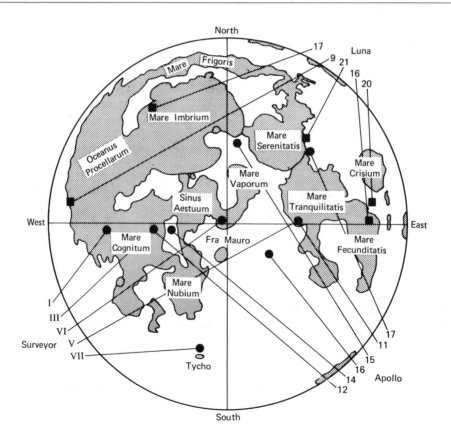

Photography of the moon from space began with the launching of the spacecraft Luna 3 by the Russians on October 4, 1959. Luna 3 transmitted enough photographs of the far side of the moon to permit construction of a map. The most notable feature discovered was the dearth of large maria compared with the near side.

The U.S. Ranger program was designed to obtain photographs of the lunar surface. In 1964 and 1965 Rangers 7, 8, and 9 obtained more than 17,000 close-up pictures, with features as small as 30 cm becoming visible for the first time. Resolution was 1000 times better than that obtainable with earth-based telescopes. This was followed by the Orbiter program, with the first launching on August 10, 1966. The five Orbiter spacecraft obtained complete photographic coverage of the moon and considerable geophysical data. Measurements of the moon's force of gravity revealed surprising and unexpected concentrations of mass beneath about 13 mare basins. These were called mascons.

A Russian spacecraft, Luna 10, orbited the moon on April 3, 1966, with a gamma-ray spectrometer. This is an instrument capable of providing chemical data about the rocks below. The results were reported as consistent with the composition of basaltic rock, which is a common lava on earth.

Data from automated spacecraft making soft landings on the moon were added to the results from orbiting spacecraft. The positions of selected landing sites are shown in Figure 17-1. The first soft landing was made by the Russian spacecraft Luna 9 on February 2, 1966.

Some distinguished scientists feared that the mare basins might be filled with loose dust, and this gave NASA engineers some headaches in the early stages of the lunar program. It was argued that thick layers of fine dust would not support the weight of a spacecraft. The suggestion that a spacecraft landing delicately on the surface of the moon could keep on going down through the cushion of dust until it was swallowed up was sufficient to spawn a whole series of research projects designed to test the

hypothesis. Scientists succeeded in building "fairy castles" of electrically charged dust particles in conditions simulating the lunar environment. Perhaps it was possible that the lunar surface could be covered with a thick layer of dust composed mostly of the holes between particles! When Luna 9 landed on a surface that was stable, despite the presence of dust, NASA engineers were relieved of one worry. Luna 9 transmitted close-up photographs of the lunar surface.

A series of soft landings, the Surveyor program, was designed by NASA to transmit engineering and scientific data from the lunar surface as a basis for planning landing sites for the Apollo manned flight program. Surveyor I landed on the moon on June 2, 1966, and Surveyor VII landed on January 10, 1968.

One of the early high points in the space program for earth scientists was the success of the alpha-scattering experiment on Surveyor V, which landed on Mare Tranquillitatis on September 11, 1967 and transmitted the first direct chemical analysis of the lunar rocks back to earth. A small cube with edges about 15 cm long was lowered to the surface, and this emitted alpha particles. The box enclosed detectors for alpha particles back-scattered by the surface and for protons generated in the surface material by the alpha particles. The spectra of these particles transmitted back to earth provided quantitative data on all elements present in a thin surface layer about 10 cm in diameter, except for hydrogen, helium, and lithium. This was science fiction turned to reality.

The director of the experiment, A. L. Turkevich of the University of Chicago, reported that the surface analysis corresponded more closely to basalt than to any other of the half dozen different rock types that had been proposed as candidates for mare material. Similar results were obtained from Surveyor VI, which landed in Sinus Medii, and this convinced most earth scientists that the first results could be extrapolated with confidence to the maria as a whole. These were composed of lava, similar to the basalt common on earth, which confirmed the conclusions from the experiment on Luna 10. This suggested that the moon had been hot at some stage in its history.

Surveyor VII landed near the ray crater Tycho and sampled the rugged highlands. The chemical analysis differed in significant respects from the maria analyses and from terrestrial basalts. It corresponded more closely to a rock called anorthosite, which contains a high proportion of the mineral feldspar. Fragments of this kind of rock were subsequently found among the Apollo 11 samples. The chemical differences between mare and highland material provided the first direct evidence for chemical differentiation of the moon.

A striking feature of the Surveyor V analysis was the high content of titanium. It was much higher than in terrestrial basalts, and some of Turkevich's geological colleagues suggested that he recheck his numbers very carefully, since geologists would not believe such high titanium values. Turkevich maintained that his results were reliable. The rocks later returned from Mare Tranquillitatis by Apollo 11 and analyzed in many laboratories yielded the same high titanium values recorded by Turkevich's instrument and returned to him from a distance of almost 400,000 km, by radio transmissions.

THE APOLLO PROGRAM

The first year of the Apollo program was marked by one incredible success after another. The Apollo 8 mission from December 21-27, 1968, was the first manned orbit of the moon, a reconnaissance to provide information for landing approaches. In the Apollo 10 mission from May 18-26, 1969, the manned lunar excursion module (LM) was separated from the command module and flown to within 15 km of the moon's surface.

On July 20, 1969, the LM landed on Mare Tranquillitatis during the Apollo 11

mission. Neil Armstrong and Edwin Aldrin then spent three hours walking on the moon and performing specific scientific tasks, including the collection of 22 kg of rocks and lunar soil.

The landing site for the Apollo 12 mission, which began on November 16, 1969, was a location on Oceanus Procellarum, within 200 m of the Surveyor III spacecraft. Astronauts Al Bean and Charles Conrad spent eight hours on the moon's surface setting up experiments and collecting 34.3 kg of rock samples.

The exhilaration created by these early successes was dampened by near-disaster for Apollo 13. After blast-off on April 11, 1970, Apollo 13 was headed for a landing on the Fra Mauro highland region near the Apollo 12 site when a pressurized oxygen tank burst and damaged the spacecraft. The crippled spacecraft orbited the moon without landing and fortunately made a safe return to earth on April 17. This mishap caused a reevaluation of many aspects of the Apollo program, and Russian achievements in 1970 with automated devices intensified a developing debate in the United States about the relative merits of manned and unmanned exploration.

Official enquiries were conducted to determine the cause of the accident and to prevent a recurrence. Public interest in the lunar program was waning, and criticism of the high cost of the program was increasing. The budget for NASA was cut by Congress to its lowest level in years, and this forced a review of the whole program. The engineering, technological, and theatrical goal of landing a man on the moon had been achieved, and the value of the scientific research programs had to be reconsidered.

Originally, 10 lunar landings were scheduled, but the budget limitations forced NASA to cancel three of the seven scheduled Apollo missions remaining after Apollo 13. The scientific repercussions resulting from this cut were severe, and most scientists were dismayed by the reduced prospects. In an editorial in *Science* on September 18, 1970, P. Cloud wrote:

People are asking why man should return to the moon again and again. . . . The moon is the only other planet we can hope to study in sufficient detail for close comparison with our own. We have only just begun that study. It is as if we were trying to understand North America by examining Plymouth Rock. . . . Many billions of dollars were spent to get within reach of these goals. Only a small fraction of the investment already made would see the job to a fruitful conclusion. To stop short for reasons within our control would, in retrospect, be seen as one of history's most irresponsible follies.

On the other hand, some scientists at the December 1969 meeting of the American Association for the Advancement of Science in Boston felt that the U.S. space program had already become a luxury too expensive for continuation in its present form. The question of the relationship between science and society had been brought into prominence by contemporary social and international problems, and this question was a major topic for discussion at the meetings. Some argued that automated spacecraft should replace manned missions in future exploration of space beyond the moon; others felt that the main thrust of the space effort should be directed toward earth applications projects. Discussion continued. The program resumed.

Apollo 11 and 12 had landed on the maria. When the Apollo program was resumed on January 31, 1971, Apollo 14 blasted off to a safe landing in the Fra Mauro Hills. Alan Shephard and Edgar Mitchell trundled their instruments around in a two-wheeled cart.

Six months later, on July 26, Apollo 15 rose from its pad in Florida carrying a lunar roving vehicle, something like a dune buggy. This enabled astronauts David Scott and James Irwin to drive for more than 25 km in the region of the Apennine Mountains, which rise to a height of nearly 5000 m. They drove the Rover close to the Hadley Rille, a spectacular canyon about 300 m deep.

The flight of Apollo 16 began on April 16, 1972; it landed John Young and Charles Duke on the Descartes Highlands with the lunar Rover, not too far from Apollo 11 site. There, John Young had the misfortune to trip over a cable that disconnected a $1,200,000 experiment designed to measure the flow of heat from the moon's interior.

Apollo 17, the final flight, took off on December 7, 1972. Eugene Cernan and Harrison Schmitt drove the Rover across more than 35 km of the Taurus-Littrow valley floor in the highlands. They used their geological hammers and other tools to good effect, collecting more rock samples than any other team, about 115 kg in all. Misapplication of his geological hammer, however, caused Eugene Cernan to break off one of the rear fenders of the moon car. Eventually, they improvised a paper fender to protect them from the flying dust.

During the planning and operation of the Apollo program, there had been many disagreements between the engineers who managed the program and the scientists who were primarily concerned with the science content of the successive missions. The selection of experiments to go with each mission involved not only scientific decisions, but decisions about payload and how the precious hours on the lunar surface would be divided among setting up experiments, actual exploration, and the ordinary "household chores" involved in establishing and maintaining a temporary space base.

Not until NASA was confident that military pilots had perfected the landing techniques was a scientist permitted to participate directly in lunar exploration. All crew members had intensive training in field geology before their missions, but Harrison Schmitt was the first geologist to win a coveted place on an Apollo mission, Apollo 17, the last mission. He was also a jet pilot.

At the Fourth Lunar Science Conference in Houston in March 1973, scientists were concerned about the future of the research program. Many rocks remained to be studied, and streams of data were still being transmitted back to earth from the five instrument stations left on the moon. Recent budget cuts suggested that lunar science might be in for harder times. They were delighted to hear reassuring words from G. M. Low in his opening speech:

We at NASA have a firm commitment, first to preserve and protect the resources we already have at hand, and second to set aside substantial funding to support the scientific effort of lunar analysis.

THE RUSSIAN PROGRAM

While the American space effort concentrated on landing the first man on the moon, the Russian Luna series of spaceflights achieved an impressive list of "firsts": the first to hit the moon, the first to orbit the moon, and the first to make a soft landing on the moon.

In 1970 there were two more remarkable achievements. Luna 16 was the first spacecraft to land on another body and return to earth automatically. It landed on Mare Fecunditatis on September 20, picked up samples of lunar soil to a depth of 35 cm, sealed them in a container, and blasted off on a journey back to earth, leaving the landing stage behind to send back additional measurements from the surface.

In November 1970 Luna 17 made a soft landing in Mare Imbrium, and a remote-controlled, robot machine rolled down a ramp and moved forward under the guidance of operators on earth. Lunokhod I (the "moonwalker") was powered by solar cells and driven by eight independent spoked wheels. The ground around the moonwalker was monitored by TV cameras at front, back, and sides. It carried an X-ray spectrometer to analyze soil scooped up by the machine and special mirrors to reflect laser beams sent from earth, similar to reflectors left by Apollo missions. Lunokhod I survived the extreme cold of the two-week lunar night and continued to explore the moon for more than 10 months.

These automated achievements were repeated. Luna 20 landed in the highlands

south of Mare Crisium in February 1972 and brought back about 50 g of lunar soil. Lunokhod 2 rolled down the gangplank of Luna 21 in January 1973 and began reconnoitering the area in a mountainous region at the edge of Mare Serenitatis. It continued to operate until May. The Russians have designed similar automated stations, Planetokhods and Marsokhods, for future exploration of Venus, Mercury, and Mars.

THE EFFECT OF APOLLO 11 ON PRE-APOLLO CONTROVERSIES

The origin of the moon and its surface features have been debated for centuries. While the moon remained an inaccessible globe in the sky, speculation had few limits. Relatively few scientists were seriously involved in the debates, and controversies raged without hope of resolution. When the lunar exploration program began, however, many more scientists and engineers were recruited, and others were consulted. They were asked to evaluate topics such as the prospects for landing a spacecraft safely, the prospects for establishing manned lunar bases, and the prospects for finding and recovering essential materials on the moon—oxygen for breathing, water for drinking, and construction materials for building base huts. Many fertile minds wove diverse explanations to account for all observations, expectations, and hopes. Ancient controversies intensified, and new ones developed.

When a lunar landing became imminent, so that the rival hypotheses and speculations could be tested, one might think that the debaters would have become somewhat more cautious in their claims. In fact, the effect seemed to be just the opposite. Protagonists of hypotheses polished their arguments and presented them once again to their audience of colleagues. If they were plugged in to a public relations circuit, they also presented them to their press agents. Many of these hypotheses were rapidly relegated to the status of exploded legends as the lunar exploration program developed and, by the time the Apollo 11 results were reported at the First Lunar Science Conference, some of the former controversies had been effectively resolved.

One controversy of fundamental significance for the origin and history of the moon was whether the moon is hot or cold. Since about 1950, the views of Nobel Prize-winner H. C. Urey had dominated cosmological thought. He argued, convincingly and with authority, that planetary origin is a cold process. G. P. Kuiper and R. B. Baldwin, on the other hand, maintained that the moon's interior must have been hot and at least partially molten.

Many lunar experts believed that the moon had always been a cold, dead planet. They saw no evidence for the existence of lava flows on the moon. These scientists followed a tradition that began long before Urey placed the model on a basis of sound chemical and physical principles. Ancient civilizations revered the cold Moon Goddess as a contrast to the hot Sun God. The moon looks cold, as confirmed by numerous references in the literature of successive civilizations.

Other experts interpreted features observed on photographs in terms of volcanoes and lava flows, which requires that temperatures of about 1200°C were attained at least locally on the moon at some time in its history. The Apollo 11 rocks proved that red hot lavas had flowed over the mare basins. The controversy then shifted, and the questions became how much of the moon was hot, what caused the high temperatures, at what stage of the moon's history, and for how long.

Another controversy related to the hot- or cold-moon debate concerned interpretation of the dark material within the mare basins. There were four hypotheses, each cogently argued. Some had interpreted this as lava erupted from the moon's hot interior. Others had concluded that the energy released from the impacts of large meteoritic bodies was sufficient to melt the rock at the surface, which then flowed like lava across the basins. A third interpretation was that the basins and craters produced by impacting meteorites were refilled with fine dust derived from the original shattered rock that was blasted out by the explosion. According to a fourth hypothesis, the maria really did represent ancient seas that existed as transient features in the early history of the moon, before the water was lost to space by evaporation. The dark mare material then consisted of sediments deposited beneath the vanished oceans. The rocks from Apollo 11 proved that the maria were filled with lavas, but there is also a cover of soil, the regolith, consisting largely of fragments of lavas.

Because of the intense cratering of the moon's surface, many scientists anticipated that it would be littered with meteorites and meteorite fragments, accumulated through billions of years. Chemical analyses of the Apollo 11 soils indicated the presence of no more than 2% of admixed meteoritic material. The rest was derived from lavas and other lunar rocks.

Meteorites of several varieties have been discovered on the earth's surface. It had been proposed that some of these represented lunar splash—fragments of rock blasted from the lunar surface by impacts and thrown into range of earth's gravity. The chemistry and mineralogy of the Apollo 11 rocks did not favor this hypothesis.

An important question for long-range lunar exploration plans is whether or not water is available on the moon in any form. Research grants had been awarded to scientists who were prepared to think about the possible history of water during the evolution of the moon, where and how it might be preserved at the present time in near-surface rocks, and how it might be made available to teams of astronauts.

Speculation was rampant. The dark lowlands that had been named "seas," or "maria," by Galileo were interpreted again as dried-up ocean basins, filled with layers of sediments. Some of the rilles on the moon are deep canyons following sinuous paths, and they look very much like rivers; it was claimed that these had been carved by flowing water at an earlier stage of the moon's history. If water had once flowed on the moon, much of it would have evaporated into space, but perhaps some of it had seeped into the ground and become frozen by the intense cold of the lunar night. It could persist in ice caves shaded from the heat of the lunar day. Perhaps the whole surface of the moon was underlain by vast deposits of ice.

Even if water had never flowed across the lunar landscape, it was proposed that water might be leaking from the moon's interior in certain fractured regions. As the water approached the surface, it would freeze, expand, and slowly develop the low mounds and ridges of rock that were observed on the high-resolution photographs. Water was needed on the moon, and some scientists were determined to find it!

The Apollo 11 rocks proved to be drier than any rocks known on earth. The maria are covered by bone-dry lavas, not sediments. The rilles were not carved by flowing water; they are probably collapsed tunnels through which streams of lava once flowed. There is no evidence for the existence of ice beneath a layer of rock or soil and no evidence for water on the moon.

Before the flight of Apollo 11 there were three main hypotheses for the origin of the moon, with a variant of one of them making a fourth. The results presented at the First Lunar Science Conference were derived from the most intensive study of a small quantity of material in the history of science. Replicate experiments conducted independently by many groups of investigators had determined the mineralogy and chemistry of the rocks down to the finest detail, and the general agreement of the data was remarkably good. Despite this, there were many different interpretations of the data. There were still four hypotheses for the origin of the moon. There were serious objections to all four, and this made it easier for protagonists of each one to find among the new data presented some information that would support their preferred hypothesis.

The four hypotheses are: (1) capture hypothesis (formation of the moon

elsewhere in the solar system followed by capture when it passed too close to the earth); (2) double planet hypothesis (the earth and moon were formed simultaneously from condensation of the solar nebula); (3) fission hypothesis (the moon is a portion of the earth that was broken off and ejected into orbit); and (4) precipitation hypothesis (a variant of the fission hypothesis). A massive primitive atmosphere of hot silicates present in the early stages of formation of the earth formed a ring of planetesimal bodies circling the earth, which then coalesced to form the moon.

The uncertainty was demonstrated by publication of three papers in the September 1970 issue of the *Transactions of the American Geophysical Union,* each arguing the case for a different hypothesis. S. F. Singer of Washington, D. C., favored the capture hypothesis; J. A. O'Keefe of the Goddard Space Flight Center, NASA, preferred the fission hypothesis; and A. G. W. Cameron of Yeshiva University, New York, supported the precipitation hypothesis as it had been developed by A. E. Ringwood of Canberra University.

Singer presented new calculations and discussions of the consequences of capture, but the capture hypothesis had forceful opponents. In a November 1970 issue of the *Journal of Geophysical Research,* A. E. Ringwood contended that chemical data from Apollo 11 rocks could not be explained by the capture hypothesis, and he wrote:

A hypothesis that merely sweeps these problems under the table is hardly worthy of serious consideration.

O'Keefe attempted to:

show that the Apollo 11 data support the idea that the moon was formed by the breakup of the earth, and that they suggest that after the breakup, the moon went through a heating episode that boiled away most of its mass. I will go on to discuss the possibility that the planets of the solar system might have been formed in a similar way, that is by breakdown of Jupiter-sized objects rather than by buildup from smaller ones.

This shows how conclusions from study of the lunar rocks are applied to problems of the origin of the solar system as a whole, but the state of uncertainty was shown by the conclusion of B. Mason and W. G. Melson of the Smithsonian Institution, based on the same chemical data. In their book, *The Lunar Rocks,* they wrote:

The results of the Apollo mission are the strongest argument against the fission hypothesis.

Cameron wrote that although the precipitation hypothesis at first seems to be the most implausible of the four,

the conditions that seem to me most probable in the formation of the solar system lead naturally to this picture.

Ringwood argued for the precipitation hypothesis, and he also discussed the probable disappointment of the world at large in the January 1970 Lunar Conference. Instead of yielding the secrets of the universe, the Conference presented an impression to nonscientists of considerable confusion. Despite the agreement in basic data, most interpretive aspects were submerged in controversy. Ringwood wrote:

In retrospect, it is difficult to see how this situation could have been avoided. With over 130 scientific teams working on specialized problems and ignorant of the findings of their colleagues (and rivals), it is not surprising that broad interpretations that were frequently based on investigations with narrow scientific perspectives should have been so often in conflict. Now that the detailed scientific results on Apollo 11 rocks have been published, there is no longer any excuse for this degree of confusion.

Ringwood then reviewed critically all available data and hypotheses, including the new results from the Apollo 12 rock collection, and fitted it into his 1966 model for the origin and evolution of the moon, beginning with the precipitation hypothesis.

The high hopes of many optimistic scientists and a news-hungry public were not fully realized by the First Lunar Conference. Nevertheless, several problems that had been controversial for many years were settled unequivocably, and some very significant facts were discovered that provided boundary conditions for the development of hypotheses and pointed the way for further investigations. This was the beginning of a revolution in planetology, to accompany the revolution in earth sciences.

THE HISTORY OF THE MOON

The rocks from Apollo 11 were soon followed by additional samples from the Apollo 12, 14, 15, 16, and 17 missions—too soon for many investigators, who found that they could barely complete their planned studies on one batch of rocks before the next batch arrived, and that they could barely keep up with the flood of results presented at the annual Lunar Conferences. Each conference was followed by publication of three large volumes of research papers. The total number of printed pages was 2490 for the first conference and 3290 for the fourth conference.

After the Sixth Lunar Conference in April 1975, although there was still no consensus about the origin of the moon, the range of possibilities for the internal structure of the moon had been more closely defined, and the history of the moon's surface was fairly well established. All four classic theories for the moon's origin are still with us, but each one has been changed considerably from its original form.

The history can be considered in four stages, following a scheme outlined by P. D. Lowman of the Goddard Space Flight Center in a March 1972 issue of the *Journal of Geology.* He synthesized the available data and used three approaches to provide a consistent history of the evolution of the moon's surface that is still generally accepted, although there are many variations in detail. The relative ages of the lunar landforms and associated rock types were known by 1962. Age dating by radiometric methods of the rock samples gave absolute ages to the relative lunar time scale, and the detailed study of the rocks provided information about the conditions and processes of formation.

In Stage I, the moon was formed by unknown means about 4.6 billion years ago (see Figure 1-3). It may have been formed hot or cold, but it must have been heated rapidly to temperatures of at least 1200°C, becoming partly melted. This permitted the formation of the light crust, preserved in the lunar highlands, which is composed of crystals that floated up in the molten material. Other heavier crystals sank within the moon, and possibly a liquid metal core was formed. This process caused the first major chemical differentiation of the moon.

Stage II was a period when the lunar crust was bombarded by large and small bodies orbiting around the earth with the moon or following trajectories in the solar system that brought them into collision with the moon and earth. The impacts of the largest bodies, 100 to 200 km in diameter, formed the mare basins and greatly modified the surface of the lunar crust. Smaller bodies impacted to produce smaller craters. The rocks became covered by a layer of soil, the regolith, consisting of fragments from the rocks below, fragments from rock blasted by impacts elsewhere on the moon, and fragments of the projectiles themselves. Many of the fragments were partly or completely fused by the impacts, and rapid cooling of the molten rock produced glasses. Glass fragments and spherical glass beads are widely distributed in the lunar regolith. The intense bombardment ceased about 4 billion years ago, although lesser impacts by meteorites have continued to the present time.

Stage III was a second period of chemical differentiation. The mare basins produced by the early, major impacts were filled with basaltic lava in successive eruptions occurring during an interval of about 600 million years, between 3.7 and 3.1 billion years ago. The mineralogy and chemistry of these lunar lavas are complex, and their interpretation is disputed. They show significant chemical differences compared with terrestrial basalts. They were formed under very dry and reducing conditions. The rocks are greatly depleted in volatile elements with low boiling points such as alkalis; they are enriched in titanium and a number of other refractory elements with high boiling points.

For the past 3 billion years, in Stage IV, the moon has been relatively peaceful. After its dynamic beginning, it ceased to be an active body. There may have been minor volcanism during the past 3 billion years, but no definite proof of this has been found. Sporadic impact by asteroid belt meteoroids and comet nuclei modified the landscape by degradation, turning the soil over in a process termed "gardening" by lunar geologists.

The instrument stations left on the moon continue to transmit data to earth. Moonquakes occur, but they are much less frequent than earthquakes, and they have smaller magnitudes. Their focal centers are about 800 to 1000 km deep, about half of the radius of the moon (Figure 17-2).

Measurement of the velocities of moonquake waves has provided a sketchy picture of the moon's internal structure, which is compared with that of the earth in Figure 17-2. The outer layer is composed of rubble or extensively fractured rock, which becomes completely consolidated by a depth of about 20 km. There is a boundary between crust and mantle at a depth of about 60 km, and other kinds of data suggest that this crustal layer varies in thickness from 40 to 60 km on the near side of the moon to more than 150 km on the far side. The mantle is rigid. It extends down to a depth of about 1000 km, where the moonquakes originate. The observation that shear waves cannot pass efficiently through the region below the mantle suggests that the core, about 700 km in radius, is partly molten. This requires internal temperatures of about 1500°C.

The moon presently has no detectable magnetic field, but all of the rocks returned

Fig. 17-2 The moon's internal structure compared with that of the earth (see Chapter 7). This is an interpretation of the moon's interior based on scanty data, and it will undoubtedly be modified in the future.

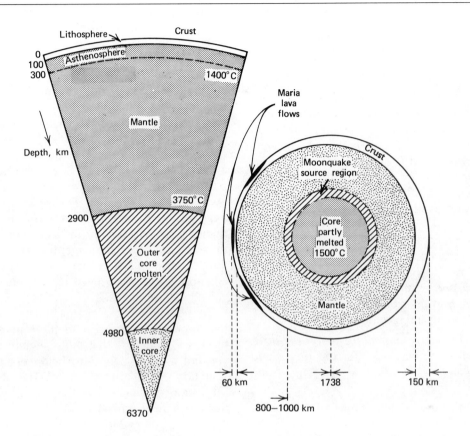

from the moon are magnetized, very weakly but stably. This proves that when the rocks cooled they did so in a magnetic field (Chapter 8). The simplest interpretation is that the moon's core is metallic, composed largely of iron, possibly with dissolved sulfur and other elements, and that between at least 4 and 3 billion years ago the extent of melting in the core was sufficient for convective motion to generate a dynamo. The dynamo and the magnetic field switched off when the proportion of molten metal decreased through cooling.

The earth is still an active planet, with an atmosphere and hydrosphere. The processes involved in plate tectonics have destroyed most of the surface features formed during the early history of the earth. The ocean floors which cover almost 70% of

the earth's surface are mostly younger than 200 million years, and most of the continental surface was formed during the past 2.5 billion years. The oldest rocks known on earth (in Greenland) formed 3.8 billion years ago.

The history of the moon, inferred for the period from the origin of the solar system to 3.1 billion years ago when significant lava eruption appears to have ceased, thus complements the known history of the earth, which is covered for the most part with rocks very much younger than 3 billion years. Geologists are now speculating about the early history of the earth, using the events inferred for the history of the moon as a guide. For example, did the earth once have a crust similar to that forming the lunar highlands? Did the earth's surface share the intense meteoritic bombardment experienced by the moon until 4 billion years ago?

THE ORIGIN OF THE MOON, AND THE MATTER OF PHILOSOPHY

There is no concensus about the origin of the moon, nor is there conclusive evidence against any of the four hypotheses extant before the Apollo 11 landing, although each one faces difficult problems. The ways in which various groups of investigators have used the lunar results illustrates differences in scientific philosophy. Several groups used the available, admittedly scanty data from the Apollo 11 mission to erect comprehensive schemes for the origin and evolution of the moon. The alternative philosophy shared by many investigators is expressed by the statement of Mason and Melson in their book covering the Apollo 11 landing and previous research.

It is clearly premature to propound a comprehensive theory for the origin and evolution of the Moon at this stage in the space program.

Premature or not, enthusiastic lunar investigators did their best to absorb the enormous amount of new data presented and published each year as more and more lunar samples were returned to earth and distributed and as data from orbiting spacecraft and the lunar scientific stations increased in quantity. Each new item was either grafted onto the existing hypotheses or onto modified hypotheses. Controversies, often bitter, developed at the lunar conferences. The data more and more demonstrated the complexity of the evolution of the moon and solar system, but they also placed more specific constraints on the processes involved.

Some scientists found the time or emotional demands of the Apollo program to be excessive, and many of the original research teams experienced attrition. Responding to the pressure of other commitments, associate investigators tended to reduce their lunar efforts until they left al-

together, resuming the full-time research programs that had occupied them before September 1969. Those who were hooked on lunar studies, however, apparently received an exhilarating rush with each Lunar Science Conference.

It is generally agreed that the planets formed by the local condensation of gas and aggregation of particles within the solar nebula about 4.6 billion years ago. The solar nebula consisted of a large, central concentration of matter at high temperature, which became the sun, surrounded by a thin, disc-shaped nebula of gas and dust particles, with temperature decreasing away from the center. The process of aggregation of particles to form small bodies was followed by the accretion of these to form larger and larger bodies, until they reached planetary size. The material composing the moon must have been involved in this process.

Consideration of the chemical data from the lunar rocks has forced most investigators to conclude that the material was very hot at some stage, either before or after accretion into a discrete moon. The only satisfactory explanation for the chemistry is that the more volatile elements of the moon were boiled off and lost at an early stage in the moon's history.

At the Fifth Lunar Science Conference in 1974, it appeared that some version of the capture hypothesis was preferred by an increasing number of scientists. J. V. Smith of the University of Chicago presented a model for the catastrophic capture of the moon, and J. Wood of the Smithsonian Observatory independently developed a similar theme. Smith proposed that the precursor of the moon was a chemically differentiated planetary body formed at approximately the same distance from the sun as the earth-sun distance, but in a different part of the nebula. The moon approached the earth and was broken up by the powerful gravitational forces, but it was not completely disintegrated. The problem with most capture models, without this constraint on sun-moon distance, is that it is difficult to capture a passing planetary body without disintegrating it completely. The captured fragments of the broken lunar body were hot. They encircled the earth in orbit and, as a result of collisions and differential gravitational forces, the composition of the debris changed compared with that of the original body. The more volatile elements were boiled away with concomitant enrichment in the more refractory elements. The orbiting debris gradually accreted into larger bodies as a natural consequence of successive collisions, and a new moon took shape from the material with changed composition. The intense bombardment of the lunar surface in Stage II of its history could correspond to the end of the accretion process.

The problem of the origin and early evolution of the moon is a difficult one, and clearly it has not been solved. Despite the obvious uncertainties about the whole business, many scientists become zealous guardians of the validity of their hypotheses. Others adopt a more lighthearted attitude, as shown by Figure 17-3, a Christmas card distributed in 1970 by Orson Anderson, now at the University of California.

Continued studies of the earth, moon, and planets should lead eventually to a general theory of planetology that will give us a better idea of the chemistry of the earth's interior and of how the earth works. The earth is an active planet, with a mobile interior and a shifting outer shell that changes the architecture of ocean basins, continents, and mountain ranges in cycles of a few hundred millions of years. The lunar program has established, in contrast, the fact that the moon, after a lively beginning, has been a dead, rigid body for about 3 billion years.

THE FUTURE

Approximately three quarters of the Apollo rock samples stored at Houston have not even been opened yet, and the instruments left on the lunar surface are still operating. Lunar research will therefore continue. But much scientific effort has been diverted from the moon to programs for the unmanned exploration of Mars, Venus, Mercury, Jupiter, and other planets, with valuable results already achieved.

The United States' first manned orbital space station, Skylab, was launched in May 1973; it was then occupied by three astronauts who docked their Apollo ferry ship with Skylab a day later. The third and last team of astronauts to occupy Skylab landed in the Pacific Ocean in February 1974, having established that astronauts can live and work in space long enough to make round trips to distant planets. NASA has plans all ready for a giant spacecraft if the U. S. Congress should appropriate funds for a manned trip to Mars in 1986.

In 1974, however, NASA was hardpressed for funds. In addition to the unmanned missions to distant planets, there was concentrated effort on development of a reusable space shuttle that could transport people and materials back and forth between the earth and an orbital space station. The future of manned exploration remains uncertain.

We can close this chapter on the moon with a vision of the future presented in a 1970 book, *Where the Winds Sleep,* by N. P. Ruzic. He wrote the book as if he were based on the moon in 2045 and looking back over the history of man's colonization of the moon. He emphasized that this is not science fiction, but extrapolated science based on current NASA programs.

The author ranged widely but systematically through aspects of science, industry, and sociology. He discussed the use of lunar bases as meteorological stations surveying the earth's weather patterns, as astronomical observatories surveying the stars without the obscuring veil of the earth's atmosphere and as stepping stones to planetary exploration. He described the growth of mining and industry on the moon. Lunar factories made use of the hard vacuum, sterile environment, extreme cold, and low gravity. They manufactured ultrapure metals, semiconductor materials, thin-film materials, certain pharmaceutical products, optical glass products, and many other

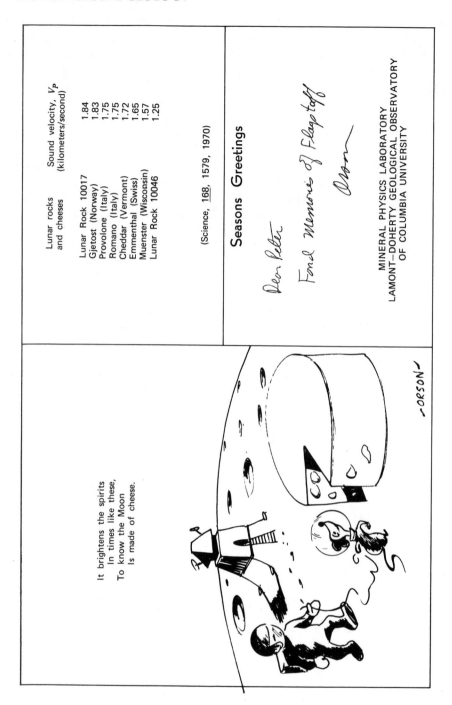

Fig. 17-3 Christmas card prepared by Dr. Orson Anderson after studying the rocks from Apollo 11. He measured the sound velocity through two lunar rock samples, which correspond to the P-wave velocity of earthquake waves *(Vp)*, and compared these with similar experimental measurements for various kinds of cheeses. The results, published in *Science*, showed that the velocities in the cheeses lie between the limits obtained for two kinds of lunar rocks.

items. The development of cryogenics led to the use of superconductors, which formed the basis of lunar transportation systems utilizing a superconductive levitating effect, and of large, economical cyclotrons. The moon was farmed to feed the 50,000 lunarians living in its cities and the tourists who came to see the wonders of the moon.

In his epilog, Ruzic wrote:

More important, if this book can influence, in even the smallest ways, Soviet and American leaders to move toward a single space program, then the new economic efficiency and automatic sharing of scientific findings could escalate to other spheres of economics, such as step-by-step diminution of funds for war. At least it is worth a try. Without such co-operation, the nuclear arsenals of nations cannot long continue unused.

We will see what develops after the successful space rendezvous between American astronauts and Russian cosmonauts of July 1975.

SUMMARY

Topographic and geologic maps of the moon's surface were constructed from photographs. Chemical analyses of lunar rocks were measured by unmanned Russian and U.S. spacecraft between 1966 and 1968. Rock samples were returned from six manned Apollo lunar missions between 1969 and 1972, and the lunar explorers left scientific instrument stations operating at each station. The Russian Luna program sent an unmanned spacecraft to the moon to collect rocks and return to earth automatically and two remote-controlled robot machines, Lunokhods, which explored the lunar surface for many months.

The lunar rocks were intensely studied by many investigators in many countries, and the results of these studies and of the data transmitted from the automatic instrument stations were presented at annual Lunar Science Conferences, starting in 1970. The mare basins are filled with dark lavas, similar to the basalts of earth but with distinctive chemical characteristics. The light highland rocks are quite different. Before Apollo 11, the dominant theory held that the moon had always been cold. The discovery of lavas shows that at least part of the moon had reached temperatures of about 1200°C. There is virtually no water on the moon.

Evidence from the five Apollo missions produced no conclusive evidence against any of the four hypotheses for the origin of the moon. Some version of the capture hypothesis seems to be preferred. Once formed, the moon was heated rapidly and the light-colored crust was produced. Intense bombardment of the crust by large and small bodies orbiting the earth produced the mare basins and many craters between about 4.5 and 4 billion years ago. Between 3.7 and 3.1 billion years ago the basins were filled by erupted lava flows. Since then, the moon has been a rigid and dead body. In contrast, the earth remains active, 70% of its surface is younger than 200 million years, and probably 99% of the continental surface is younger than 3 billion years. The history of the moon thus complements the known history of the earth.

SUGGESTED READINGS are at the end of the book.

chapter **18**
Environmental Geology and Earthquake Prediction

Plate tectonics and the revolution in earth sciences during the late 1960s caused geologists to take a new look at their familiar subject, and the space exploration program of the 1970s placed the early history of the earth in the wider context of the solar system. During the same period, as people have become more conscious of our polluted terrestrial environment, some aspects of Ruzic's vision of life on the moon begin to appear quite attractive.

Our industrial civilization and our demand for fuels that energize the industrial processes, our transportation systems, and our homes have combined to disrupt the normal geological cycles through rocks, water, and the atmosphere. Earth scientists seek and extract chemicals from the ground, and the users, industrial and individual, debouche the remains into water or air.

The questions raised about whether research funds should be spent for space exploration or for improvement of the human environment, both social and physical, were mentioned in Chapter 17. We read a statement by W. R. Dickinson in Chapter 1 that one of the key ideas of human civilization may be the realization that mankind must learn to live with his physical envi-

ronment. This whole topic, with rather vaguely defined boundaries, is generally termed "environmental geology."

Environmental geology includes the discovery and acquisition of natural resources such as water, minerals, rocks, metals, and fossil fuels, the planned utilization of land in rural and urban communities, and the avoidance or control of natural hazards such as floods, landslides, and earthquakes.

Plate tectonics has direct applications to several aspects of environmental geology. As we will see in the following pages, the theory may provide prospecting guides for resources such as mineral and metal deposits, fossil fuels, and geothermal energy. Natural hazards exist for the many cities in earthquake belts, and plate tectonic theory may help people to avoid the hazard by providing more information and understanding and by guiding us to prediction and perhaps even to control of some earthquakes.

NATURAL RESOURCES

The progress of man is paralleled by his progressive mastery of the immediate environment and by his recognition of the earth materials around him as useful resources. Successive cultures are named according to the most sophisticated material that was used for tools, as in the Stone Ages, the Bronze Age, and the Iron Age. In our Atomic Age, we use an enormous range of materials for the construction and maintenance of a bewildering array of technical apparatus. We accept this apparatus and its products as a part of everyday life, but the time is approaching when we will be forced to realize that they all come from exploitation of the earth and that the earth resources are not inexhaustible.

Several books in recent years have reviewed the theme that we live on spaceship earth with finite resources, and that this fact, combined with the growth of world population, will lead to a catastrophic collapse of world society within the next century. Probably the best-known volume promulgating this doomsday thesis is The Limits to Growth, by D. H. Meadows and coauthors. This draws attention to the fact that world population and our consumption of natural resources are increasing at exponential rates, and that this will bring us to natural limits sooner than most of us realize. If unpleasant consequences are to be avoided, detailed plans on a global scale should be instituted in the near future.

With exponential growth of anything, be it population or consumption of resources, a limit does come rapidly. If something grows at an exponential rate, this means that it increases by a constant fraction each year; it may double, treble, or quadruple each year. The upper line in the graph of Figure 18-2 shows exponential growth. The effect of a constant rate compared with an exponential rate can be illustrated by considering withdrawals from a bank account. Suppose that you have $189 in the bank, and you spend a constant $3 per day; the money will last for 63 days. Now suppose that you spend the money exponentially, multiplying your withdrawal by a constant factor each day, say by times two. Then, on succeeding days, you withdraw $3, $6, $12, and so forth, and on the sixth day you withdraw $96 and empty the account. If

your savings were greater than you thought, and your initial bank account was doubled to $378, the account would be emptied on the seventh day, when you withdraw $189. Doubling your reserves keeps you solvent for only one more day.

The Limits to Growth is based on a computer simulation of the world. The variables in the model include food, resources, industrial output, death rate, and birthrate. The variables follow actual historical values from 1900 to 1970, and the model assumes that there are no major changes in the physical, economic, or social relationships that have historically governed the development of the world system. With this oversimplification of the reaction of a real world to the predicted effects, the computer predicted that the present exponential world population growth and exponential increase in consumption of natural resources inevitably leads to collapse and catastrophe. Pollution increases as natural resources diminish, and the population eventually outstrips food and industrial supplies. The world's population then dwindles rapidly because of the shortage of food and health services, with a return to the bad old days of famine and plague.

Concentrated deposits of minerals, metals, and fossil fuels are generated in cycles with durations of tens or hundreds of millions of years. Most of those existing now will have been discovered and exploited within the foreseeable future. There is nothing we can do about replenishing these deposits. We are faced with the facts that natural resources are finite in amount and that the easy pickings are disappearing fast.

The explosive growth of world population and concomitant growth of technology and industry have already placed severe strains on the availability of some natural resources. We have lived through the brief period of abundance when we could simply pluck from the crust the materials that we need. But the natural resources of the earth, although finite, will take an extraordinarily long time to become exhausted. In future years, what we now discard as waste will be routinely recycled. New materials will be discovered as substitutes for many in current use. Many elements will eventually be extracted from the ordinary rocks of the earth's crust, where they are present but in much lower concentrations than their present sources. This will be expensive, but feasible. And the moon is a large chunk of chemical elements awaiting exploitation when we need it.

The computer prediction will not come to pass because the world system will *have* to change before the stage of catastrophic collapse is reached. Already, we experience local pollution at intolerable levels, famine in large areas of the world, and oil shortages or distribution problems. The intermittent adjustments in our social and economic systems forced by successive crises such as these will hopefully lead to eventual attainment of a reasonably harmonious balance between the geological cycles and a world society with stabilized population.

Our future activities must become integrated into geological cycles in such a way that our requirements are diverted from the cycles, used efficiently, and then replaced with minimum disruption and contamination. We cannot do this for minerals and metals because the cycles are much too slow. But we can integrate with water and carbon in their fast cycles.

The main problems facing this goal are not scientific and technological, but economic and political. A realistic global approach to a suitably managed world system of cycles in the earth sciences may well be delayed by personal greed and the democratic right of personal freedom. Unfortunately, short-term delays in global planning at present can have very large effects on exponential curves tens of years in the future.

AIR AND WATER

Air and water are the most basic of natural resources because our lives depend on them; so also do the lives of the animals and vegetables that constitute our food

supplies. We reviewed the water cycle in Chapter 6, tracing the circulation of water through the atmosphere. The total quantity of water in the hydrosphere is fixed, but by evaporation and then precipitation it falls, again and again, as fresh water on land.

Most of us take water for granted. It flows from a tap or flushes away wastes at our command. Engineers have made available to us a small portion of the enormous volumes of water that are circulated among ocean, atmosphere, and land. They have diverted one of the great cycles of nature. After use, we return the water either from individual homes or from industrial plants, but this water contains additional solid and dissolved ingredients; it is polluted. Similarly, we discharge effluents into the atmosphere, changing the composition of the air.

Although there are local problems of inadequate water supplies in the United States, the official position of the U. S. Geological Survey is that the present supply is adequate. This conclusion applies on a global scale as well, but the water is not always available where it is needed, and there are extensive regions where water supplies are inadequate to sustain large populations.

We have no control over regional climatic changes, and any slow change toward more arid conditions can place severe strains on the ability of society to adjust its needs to the diminishing supply. This was demonstrated by the effects of the drought of 1932 to 1937 in Oklahoma and neighboring states. The combination of poor agricultural practices and the shortage of water turned the region into a vast "dust bowl," and many areas were simply abandoned. The effect of climatic changes has been demonstrated in more tragic fashion by the drought of recent years that has devastated parts of Africa and India.

There is evidence that a major change in world weather patterns began about 1930 was temporarily halted in the 1950s and has continued since then. This is causing more rain to fall on the oceans of the southern hemisphere and less on the densely populated land masses of the northern hemisphere. Weather patterns carried across Africa on westerly winds are shifting toward the south. In some places, the edge of the Sahara desert has moved 150 km southward in six years. Similarly, the summer monsoons that bring rain to India are also shifting toward the south, and regions in the north of the old monsoon belt are now arid and becoming desert. The monsoon failed for an unprecedented series of four years between 1970 and 1973.

The worst-hit part of Africa is a 6500 km band extending east-west from Ethiopia to the Sahel countries (Chad, Niger, Upper Volta, Mali, Mauritania, and Senegal). After six years of light rainfall between 1968 and 1973, grazing lands withered away, wells and rivers dried up, thousands of cattle, sheep, goats, and camels died, and nearly one third of the 51 million people in the drought-stricken belt were threatened with starvation. In 1973 estimated deaths by starvation and rampant diseases in shantytown refugee camps was nearly a quarter of a million. In addition to drought, India has a problem of overpopulation. Half of its 600 million people live at a bare subsistence level. Refugees from the dried-up villages across the Deccan Plateau of central India have streamed into the cities looking for work, or alms when they find no work. The back streets of Bombay have become home for thousands of refugees.

Water under control is a natural resource, a boon to society. But a shortage of water causing drought through long enough periods can lead to misery, famine, and even to termination of societies. At the other extreme, an excess of water out of control is a disaster. Air and water from atmosphere and hydrosphere regularly combine to produce storms over which we have no control, despite attempts to modify them by various experiments. The power of nature manifest in a single hurricane or tornado is indeed awesome when measured in terms of property damage and destruction of life.

There is nothing we can do now to control global climatic patterns or the disastrous overabundance of air and water delivered locally in severe storms. But we can use better methods to control and preserve the water once it is delivered. Many water problems arise directly from neglect by governments that have the power to make decisions for flood control and water regulation. In 1950 a research team warned New York City that it would need additional water by 1970, and the team made specific recommendations, most of which were ignored. By 1965, after four years of low rainfall, the reservoirs supplying water to New York had fallen to one third of their capacity, and Manhattan Island was running short of drinkable water. New York City was declared a disaster area.

Problems like this have led to an increased awareness of the water cycle, its influence on the needs of modern civilization, and the ways in which communities modify the cycle. Starting in 1965, the research talents and experience of scientists from more than 70 countries were combined in the International Hydrological Decade, a program sponsored by UNESCO. Let us hope that decision makers in government and industry pay attention to the conclusions reached.

ROCKS AND MINERALS

In Chapter 6, we outlined the geologic cycle and the processes involved in the formation of igneous, sedimentary, and metamorphic rocks. Many of these rocks are natural resources used in the construction industry, ceramics industry, chemical industry, and as fertilizer for agriculture. The cliffs of concrete rising from city streets and the concrete highways leading away from the urban skyscrapers are made from various kinds of rock. The rocks are quarried, crushed, and treated. These source materials mixed in appropriate proportions constitute cement powder which, when mixed with water, becomes reconstituted into a synthetic rock. But it has the advantage that it can be shaped as desired before it dries out and hardens.

Mineral resources are normally special rocks in which one or more element has become concentrated. If the concentration is high enough that it is economical to mine the rock and extract the required element, the rock is called an ore deposit. Iron ores and aluminum ores are relatively abundant, but the known reserves of some of the other industrial metals are so low that, even at present rates of consumption, the ore deposits will be mined out before the end of the century. These include copper, lead, and tin. With exponential growth in their utilization, following the industrialization of underdeveloped countries and world population growth, shortages will develop even sooner.

Shortages will drive prices up, which will make it economically feasible to mine lower-grade ore deposits (i.e., rocks with lower concentrations of the required elements). If no substitutes are found for these elements, we may reach the stage where selected mountain ranges are mined and crushed and separated into different elements by technologies not yet developed. The technology must also be developed for the recycling of materials that have been transferred from the geological cycle into the industrial system. The rock cycle that regenerates ore deposits on a time scale of tens or hundreds of millions of years is too slow for our needs, so ways must be found to organize our own cycles.

At present, however, it is still far cheaper to discover hidden ore deposits, the remainder of the easy pickings concentrated by the rock cycle. As long as these remain, they will exert a growing influence on international politics. The wealth and power of nations depend to a large extent on the production of mineral resources. Discovery is the first step, and this is followed by political jockeying for control of the deposit.

The theory of plate tectonics is being used as a guide to promising locations for geophysical exploration of hidden ore deposits. We know from present-day activity that the igneous processes occurring at ocean ridges and at volcanic arcs lead to the formation of various ore deposits. For example, most of the world's copper is mined from deposits associated with volcanic island arcs or continental arcs that are situated above a sinking lithosphere slab. These geological environments are illustrated in Figure 6-8. Rocks of ancient ocean ridges and volcanic arcs have become incorporated into continents through successive collisions and separations of the continents, and these are potential sites for ore deposits formed earlier in their history. One aim of modern prospectors, therefore, is to identify ridge and arc rocks in the ancient geological record.

FOSSIL FUELS

For centuries, the energy requirements of society were supplied by domestic animals, human slaves, firewood, and wind and water wheels. New energy sources were developed with the discovery and utilization of fossil fuels, essential ingredients of the Industrial Revolution, which changed economic and social patterns in the nineteenth century.

The fossil fuels include coal, oil, natural gas, oil shale, and tar sands. These are produced in the carbon cycle. Carbon is not an abundant element in the earth, but it plays a most important role as the essential element of organic compounds, which constitute living organisms. Carbon circulates through the atmosphere, ocean, rocks, and the biosphere. The biosphere controls many important chemical reactions.

Carbon dioxide in the atmosphere was derived originally from the earth's interior, and volcanic eruptions bring additional carbon dioxide to the surface. In the presence of sunlight plants, which contain a substance called chlorophyll, are able to convert carbon dioxide and water into organic molecules. This complex chemical process of photosynthesis converts solar radiation into chemical energy with the release of oxygen. The reverse process, involving oxidation (reaction with oxygen), converts organic molecules into carbon dioxide and water and releases energy.

Most of the carbon in plants and animals becomes oxidized and returned to the atmosphere or ocean in a fast cycle, in geological terms, but some organic carbon compounds become trapped in the slow rock cycle. If the conditions for preservation are favorable, they may form concentrated deposits. If the subsequent geological history is suitable, the deposits are preserved, and it is these that we seek and mine for energy.

Coal is produced by the preservation of decaying vegetation in fresh-water swamps. Leaves, twigs, branches, and whole trees fall into the swamp, where the waters protect them from oxidation by the atmosphere, and the concentrated organic material becomes peat. Burial of the peat by later sediments, with subsequent mild metamorphism, converts it into coal.

Organic matter from plants and animals is present in small amounts in most sediments forming today. In some geological environments, such as sedimentary basins on continental shelves, the proportion of organic debris is higher than elsewhere. If the organic material is buried before it becomes oxidized, it will undergo chemical reactions as the marine mud is converted into a rock, and it becomes transformed into liquid and gaseous hydrocarbon compounds. These light fluids are squeezed out of their host rocks, and they migrate upward toward the surface, tending to follow paths through the more porous rocks, such as sandstones. Under favorable circumstances the fluids become trapped in one of several kinds of geological structures, and continued accumulation of the oil and gas through millions of years generates a natural resource that we call an oil field. All we have to do is to find these buried fields, drill a hole through the trap, and allow the fluids to continue their escape to the surface through the oil well at a controlled rate that is convenient.

Oil shale and tar sands are sedimentary rocks with high proportions of organic material that can be treated to yield hydrocarbons. In oil shales, the organic material did not experience the chemical reactions required to convert it to oil and gas. Tar sands, on the other hand, probably represent former oil deposits in traps that have had a complex geological history.

Figure 18-1 illustrates schematically the annual rate of consumption of energy from each of the fossil fuels based on known val-

Fig. 18-1 Sketch illustrating the relative amounts of annual energy consumption throughout the world, from fossil fuels and hydropower. The dashed lines are estimates of future consumption rates. Different authorities produce somewhat different estimates, but this general pattern appears to be widely accepted.

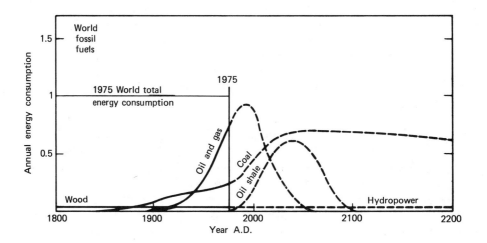

ues in the past and estimated future values. The energy generated from water, hydropower, is shown for comparison; this remains low throughout the time interval considered. As recently as 1800, most of our energy came from burning wood but, during the nineteenth century, this was replaced by coal. Oil and gas were used only in relatively small amounts until about 1925, but then their rate of consumption increased dramatically. By 1975, approximately 40% of the world's energy came from oil, 20% from natural gas, and 40% from coal.

There is only a finite amount of oil and gas trapped in the earth's crust, and this is being used up at an increasing rate. M. King Hubbert of the U.S. Geological Survey has considered the oil fields already known, estimated the amount of oil as yet undiscovered, and analyzed the increasing annual rate of world consumption. The results are shown in Figure 18-1. World production will probably peak near the year 2000, and all sources will be exhausted somewhere between 2050 and 2075. Figure 18-1 includes an estimate of the amount of energy that may be derived from oil shale and tar sands to supplement oil and gas in the future. This may last until 2100.

Coal is available in large quantities. Figure 18-1 shows an estimate of doubled coal production by 2000 and approximately tripled coal production through the year 2200. At this rate of use, there are coal reserves to last for several centuries.

There is an urgent need to discover new reserves of oil and gas for immediate exploitation. The theory of plate tectonics is being used as a guide to the location of possible deposits. One obvious approach is to examine reconstructions of the supercontinent Pangaea, as in Figures 11-3 and 12-4a. The presence of oil fields on continental shelves of one continent suggests that similar fields may exist on the continental shelf of its matched neighbor.

The sites of ancient river deltas are promising locations for oil fields, but it is difficult to locate deltas older than 200 million years. Deltas must have developed around Pangaea, for example, but many of these have become incorporated into mountain ranges by the subsequent movement of plates. Unraveling plate tectonics and reconstructing the ancient positions of continents and mountain ranges may permit the reconstruction of ancient drainage patterns and the locations of ancient deltas.

Another idea is that when a continent overrides an ocean-ridge system, which is one interpretation for the termination of the East Pacific Rise in the Gulf of California (Figures 3-7 and 3-12), then the additional heat flow from below (Figures 16-1 and 16-3) is just enough to make the hydrocarbon compounds dispersed in the sedimentary rocks migrate upward. This could lead to the development of oil fields. Continental shelves close to ocean ridge systems are accordingly promising regions for prospecting.

POLLUTION AND POWER: THE ENERGY CRISIS

One of the worst air pollutants in the United States is sulfur dioxide. Electric power plants fueled by coal or oil contribute more than half of the total emissions of sulfur dioxide, and other sources include fuel combustion and smelting of metallic ores. Not only the obvious pollutants cause problems. Recent studies suggest that elements present in air and water in only trace amounts may reach toxic concentrations. Lead, nickel, and cadmium, although present only in parts per million in the atmosphere, are already concentrated sufficiently to cause serious concern. Lead and nickel come mainly from gasoline and fuel oil, respectively. Cadmium enters the air mainly from metal refining industries.

The generation of power from fossil fuels is a major source of environmental pollution, and the smelting and refining of metallic ores is another. This makes it a matter of professional as well as personal concern for earth scientists, who are involved in finding, explaining, mining, and exploiting coal, oil, and metallic ore deposits. The situation would be improved considerably if techniques were introduced for removal of sulfur from the fossil fuels before or during combustion or from the effluent gases. Apparently, it will be many years before effective techniques can be in widespread operation, and these will be costly.

World demands for energy are increasing at an exponential rate, and this can only aggravate the problem of pollution. Figure 18-2 is a cumulative diagram schematically showing the world's annual consumption of energy from all sources through the same time interval shown in Figure 18-1. The top curve gives the total of the values for all sources plotted separately in Figure 18-1, and the area under the curve is divided into smaller areas corresponding to the energy from each source.

The annual consumption of energy is high in 1975, but what concerns geologists and economists is the rate of increase. At present, the energy used approximately doubles each 10 years. The top dashed line is an extrapolation of current world demand for energy. This gives minimum values, because the rate is increasing with world population and with the require-

Fig. 18-2 The world annual energy consumption from Figure 18-1 presented in a different way, showing the total annual energy consumption, which is subdivided into different sources. The top dashed line shows a linear extrapolation of requirements for the future. The other dashed lines sketch estimates of future sources. The estimates of availability of energy from fossil fuels (Figure 18-1) fall far below the extrapolated requirements. The requirements will be partly met by the development of nuclear, geothermal, and solar energy sources. The energy gap shown indicates that the world consumption of energy cannot continue to increase at its present rate. Just look ahead a mere 25 years.

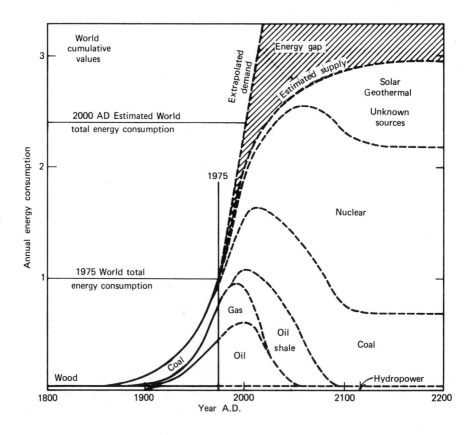

ments of developing countries. Within a few years there will be an enormous gap between world needs and the energy available from fossil fuels.

It has been clear for some time that the world's energy requirements can be supplied largely by oil and natural gas for only a few more decades. An impending energy crisis became evident during 1970. Americans were warned to expect serious winter fuel shortages. During the winter of 1972 to 1973, shortages of heating oil in the Midwest forced closure of schools and other public buildings, and some manufacturing firms had to reduce or stop production altogether.

In early 1974 the potential crisis was brought home forcibly to Americans by the Arab oil embargo. The curtailment and disruption of oil supplies left gasoline stations with insufficient fuel to fill automobile gas tanks. Gas stations rationed supplies, and some went out of business. The American public stopped buying large cars for a few months, and the automobile business went into decline, with thousands of workers being laid off. Airlines reduced the number of scheduled flights because of the shortage and skyrocketing price of fuel, and this put thousands more out of work.

Faced with the energy crisis and its many inconveniences, Americans sought scapegoats. The oil industry was charged with withholding supplies to create shortages and thus obtain higher prices. Reports of large increases in profits encouraged this view. The federal government was charged with poor handling of the problem in many different ways. To a lesser extent, environmentalists were charged with retarding increased energy production for idealistic reasons that ignored the realities of life, and the Arabs were castigated for using oil in political blackmail.

When the oil embargo was lifted and gasoline was again readily available for automobiles, public concern about the crisis diminished. Carefully conducted polls of public reactions to the energy crisis showed that the public believes that fuel shortages are not inevitable in the future, and that the

recent temporary problems will be solved by 1980. A glance at Figure 18-2 shows that the problems will not be solved by oil and gas, and that new sources of energy are urgently required. The energy crisis on a global scale is real and continuing. The Organization of Petroleum Exporting Countries is in an incredibly powerful economic position.

Figures 18-1 and 18-2 show that some of the additional energy requirements can be met by increased coal production, but this falls well below the anticipated demand after the year 2000 A.D., when supplies from oil and gas are expected to decline. Furthermore, although coal is plentiful, it cannot be removed from the ground without paying a high price either for removal of topsoil and the pollution of streams in open cast mining, or for the safety and health of miners. Much of the U.S. coal reserves contain up to 10% sulfur, which is totally unacceptable for use in most urban areas until techniques have been introduced for removing the sulfur. Similar problems face the mining and extraction of oil from oil shales and tar sands during the next 100 years.

Shortages of fossil fuels and rising costs have contributed to the energy shortages, but so has growing concern over the environment and a revulsion against polluting power plants. Throughout the United States, new power plant construction has been blocked by environmentalists for various reasons, involving both pollution and aesthetic factors. Yet, in order to keep pace with the increasing energy demands of our society, including demands for the home appliances of the environmentalists, it has been estimated that the United States must build at least 250 large new power plants during the 1970s.

Utility engineers are seeking new, low-polluting kinds of power to replace the disappearing fossil fuels. It was once anticipated that nuclear power plants would eventually accomplish this, but there is a backlog of orders and costs are increasing rapidly. Environmentalists as well as some scientists are not happy with nuclear power, either, because of potential accidents, long-term radiation hazards, and thermal pollution by the discharge of warm water into lakes or rivers. In the United States, 44 nuclear plants were producing only a few percent of the nation's electricity by 1973 (Figure 18-2).

Legislation has been introduced in several states to prohibit the construction of new nuclear power plants. The main issue is that of potential present and future damage by radiation. Despite many safety precautions and the low chance of accidents, new nuclear plants will probably be located further away from population centers than many of the existing plants, just in case. Liquid wastes left over after the nuclear fuel is used up will remain radioactive and lethal for tens of thousands of years. Safe storage techniques have not yet been developed. These wastes may become a serious problem for future generations. There is another problem. Uranium ore deposits in the United States will probably be used up during the 1980s, leading to dependence on foreign sources, unless the breeder reactors can be developed. Breeder reactors generate more nuclear fuel than they consume.

If nuclear power does not fill the gap between world needs and energy from fossil fuels, what else is there? We saw in Chapter 6 that the dominant energy sources for the geologic cycle are solar radiation and the earth's internal heat, which drives the lithosphere plates. These are the same sources that provided fossil fuels and uranium ore deposits. The fossil fuels are derived from organisms that stored energy from sunshine as they grew. Uranium is one of the elements maintaining the earth's internal energy, and it is brought to the surface to form uranium ore deposits as an end-product of igneous processes. Can we tap these two major sources directly?

Certainly, there will be massive research efforts directed toward harnessing solar energy. The main problem is that in order to be usefully concentrated, the solar energy has to be gathered from very large areas. Experts reach different conclusions about whether or not solar energy can supply a significant proportion of world needs.

The prospect of tapping heat from the earth's interior is now being considered seriously for the first time in the United States. Federal land has been leased for geothermal development since 1971, and exploration for deposits of steam is under way with something of the fervor associated with gold rushes and the opening of oil fields in the early days.

Geothermal resources include energy plus any associated mineral products from steam and hot water that emerges from the earth. Location of a potential geothermal system is difficult unless there is a surface leak, as in areas where there are geysers and hot springs. Our knowledge of geothermal sources is rudimentary, but the general system is illustrated schematically in Figure 18-3. The essential features are a near-surface source of heat such as an igneous intrusion of molten rock, a layer of porous sediment such as a sandstone with the spaces filled with ground water, and an upper layer of impervious sediment such as shale that acts as a trap for the water.

The heat rising from the molten rock heats the water and sets up large convection cells of water and steam, percolating through the sediment. A natural fracture in the upper layer permits steam or hot water or both to escape in a geyser or hot spring. If a well is drilled deep enough it will tap the hot fluids, and these can be drawn off as required, treated, and used to generate electricity. Geothermal energy may be derived from hot water or dry steam, depending on the depth and temperature of the source. Techniques are also being explored for extracting energy from deep, dry hot rock. One method is to explode a nuclear device underground and to circulate water through the cavity so produced.

All methods involve technical problems. The water at depth may contain dissolved salts. Some of these brines are 10 times more salty than seawater. This causes corrosion of the equipment and deposits of minerals in the pipes and valves. On the other hand, some of the dissolved elements

Fig. 18-3 Geothermal power is obtained by tapping a source of deep water and steam that has been heated by an igneous intrusion.

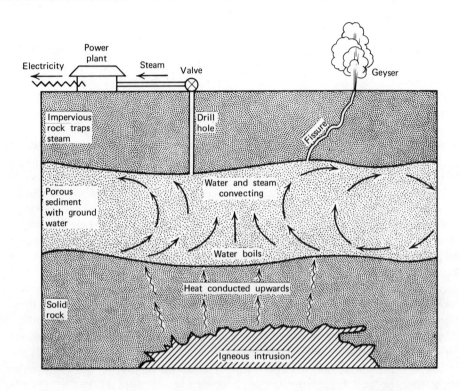

may be extracted and recovered to form subsidiary chemical industries. Sulfurous fumes are given off along with steam, so air pollution is not avoided.

Dry steam fields are in production in Italy, Japan, and The Geysers, just north of San Francisco. The Italian geothermal plant was the first, starting in 1904. The Geysers started producing electricity in 1960 at a modest rate. This had increased by a factor of 8 by 1974, and this should be increased by a factor of 9 by 1976, providing a significant proportion of San Francisco's total requirements. Hot water geothermal fields have been developed in Iceland, New Zealand, Japan, the Soviet Union, and Mexico. Most of these countries have a shortage of fossil fuels and an abundance of volcanoes.

The occurrence of molten rock at high levels in the crust, as depicted in Figure 18-3, is normally restricted to the active volcanic belts of the earth. These are associated with plate boundaries, so once

again the theory of plate tectonics provides a guide for exploration. United Nations technical assistance teams have been sent to many countries to initiate exploration programs for geothermal resources, and a sign of the rising interest is the proliferation of international conferences on geothermal energy.

The Mexicali Valley and the Imperial Valley of California are promising sites for geothermal energy production. Detailed studies by R. W. Rex, formerly of the University of California, indicate that enough electrical generating capacity could be installed to meet most demands of Southern California. The existence of high-level heat sources in this region may be related directly to convection in the mantle beneath the Pacific ocean ridge. This locus of seafloor spreading appears to extend along the Gulf of California and then under the North American continent (Figures 3-7 and 3-12). Apparently, convective uprise of hot mantle material beneath the ridge is causing Baja California to be separated from the continent. The hot springs distributed all over the western United States may be related to the same process, which brings the earth's store of internal heat closer to the surface.

Estimates of potential geothermal energy resources in the United States vary widely. According to various experts, production in 1985 could be sufficient to supply as much as 20% of U. S. power needs or, at the other extreme, less than 1% of 1985 needs. If the largest estimate is correct, and if the production is to be realized, it would be necessary to drill 3800 exploration wells, starting in 1974. This would cost $10 billion per year, about 15% of the risk and development expenditure by the oil industry in the United States in 1969. In fact, the estimated drilling rate for 1974 is only 1/40 of that.

Figure 18-2 shows that the exponential increase in energy consumption will *not* continue because the existing and anticipated energy sources are incapable of supplying world demands. The pattern of energy consumption *has* to change. There is no doubt that the cost of energy will increase considerably, and this will reduce wasteful and nonessential uses. It is difficult to see how the United States can even approach the national aim of energy self-sufficiency by 1980 without reducing consumption. Realistic planning based on harsh facts rather than on the whims of voters is urgently needed.

EARTHQUAKE PREDICTION

Prediction of an earthquake requires the specification of a place, the time, and the magnitude of the event. Until the mid-1960s, this was the domain of seers, fortune-tellers, and religious zealots (Chapter 4). There is a persistent folklore that animals can predict earthquakes. Textbooks as well as popular accounts include stories of dogs barking and pheasants crowing shortly before an earthquake, but no one has managed to develop the animal insights into predictive techniques.

Plate tectonics explains the observation that earthquakes are concentrated in narrow belts, although as yet it gives no satisfactory explanation for the occurrence of major earthquakes within the large plates, such as those of North America, shown in Figure 4-5. World maps of earthquake distribution (Figures 4-6 and 4-7) provide the basis for two kinds of earthquake prediction. Earthquake risk maps are based on the frequency and magnitude of past earthquakes. These show the relative long-term prospects that any particular region will experience an earthquake, and regions within supposedly stable plates are included (Figure 4-5). The second approach involves the location of places in the earthquake belts where only few earthquakes have occurred during the past tens to hundreds of years. These are now regarded as potential sites for large magnitude earthquakes. Neither of these approaches leads to satisfactory prediction, because they do not specify the time very precisely.

Earthquake prediction became an acceptable scientific pastime in 1973, when American geophysicists applied the results of laboratory experiments on rock deformation to explain some peculiarities of earthquake-wave velocities first recorded by Russian scientists in 1969. The dilatancy-fluid diffusion model of earthquakes gained wide acceptance and provided for the first time a physical basis for earthquake prediction and for various phenomena that occur before an earthquake.

The phenomenon of dilatancy was discovered in laboratory studies on how rocks fracture. W. F. Brace of the Massachusetts Institute of Technology and others have demonstrated that microcracks and voids open up in a rock before it breaks. This causes an increase in volume and a change in porosity; the rock dilates. Most crustal rocks contain ground water in the pore spaces between minerals (Chapter 6), and the pore water would be able to move through a rock in this dilated condition. This is the fluid diffusion part of the model.

In Chapter 7 we described and illustrated the *P*-waves and *S*-waves that transmit energy from earthquakes (Figure 7-2). The velocities of these waves, V_P and V_S, vary according to the depths of their paths through the earth (Figure 7-5), but they remain fairly constant for paths through the upper crust in a specific region of limited size. Many active faults generate frequent microearthquakes that can only be recorded by sensitive instruments. Each microearthquake may have a different focus, but they all send waves through the rocks near the fault. Under normal conditions constant values for V_P and V_S are obtained from measurements recorded by networks of instrument stations monitoring a specific source region for earthquake activity. The results obtained are conveniently converted to the ratio V_P/V_S, and this remains constant for most regions.

Figure 18-4 shows results obtained in three different regions, arranged in order of increasing earthquake magnitude. Figure 18-4*b* shows the first discovery from the Garm region of USSR. Each point gives the value V_P/V_S for waves initiated by a specific earthquake at a specific time, somewhere between 1964 and 1966. The magnitude of most of these earthquakes was very small, probably less than 2.0 on the Richter Scale, but there were two larger ones, with Richter magnitudes of 4.2 and 5.4. For most of the two-year period, values of V_P/V_S remained constant but, for an interval before each large earthquake, the ratio first decreased and then increased. The ratio had just regained its normal value when the large earthquakes occurred.

This result was published in 1969, and American geophysicists became aware of it at an international conference in Moscow in 1971. There was a flurry of activity as various groups followed up this observational lead and were led to similar conclusions. The rapid exchange of ideas and data in today's mobile and expansive scientific community makes the problem of who earns scientific priority in such circumstances a sensitive one, as we saw in Chapter 12.

L. R. Sykes of Lamont-Doherty Observatory became acquainted with earthquake prediction research in the Soviet Union in 1971, as a guest of the Soviet Academy in Garm. In August 1972 he and his associates submitted to *Nature* a report of large changes in V_P/V_S in the source region of a series of small to moderate earthquakes in the Adirondack region of New York. A simplified version of some of the results recorded in July 1971 is shown in Figure 18-4*a*. The authors stated that they did not understand the causal mechanism for the changes, but they discussed the processes of dilatancy and changes of pore fluid pressure, which was subsequently formalized in a 1973 paper by C. H. Scholz, L. R. Sykes, and Y. P. Aggarwal.

In the meantime, A. Nur of Stanford published a paper in 1972, in which he quite independently interpreted the Russian results in terms of a model involving dilatancy and fluid diffusion.

At about the same time, J. H. Whitcomb, J. D. Garmany, and D. L. Anderson from the California Institute of Technology checked the records for V_P/V_S from the San Fernando Valley area for the period before the major earthquake of February 9, 1971 (Chapter 4), with results shown in Figure 18-4*c*. There is a similar decrease followed by an increase back to normal when the major earthquake occurred. They also interpreted the results in terms of a dilatancy-fluid diffusion model.

According to the model, the rock near a fault is stressed until dilatancy is caused by the opening of microcracks throughout the region. The pore fluid already present in the rock flows into the new cracks. It is known from laboratory experiments that this increases the resistance of rock to fracture and also reduces the value of V_P. These effects delay the earthquake and cause a decrease in V_P/V_S. Additional pore fluids flow in more slowly from neighboring rocks to fill the empty spaces created by dilatancy. This causes V_P/V_S to increase again, while simultaneously reducing the strength of the rock. At the time the original pore fluid pressure is recovered, or soon after,

Fig. 18-4 Measurements of the velocities of waves from minor earthquakes in earthquake-prone regions may give warning of an impending major earthquake. The three diagrams illustrate the types of results that have been reported from three regions. Each point represents a minor earthquake, and recording instruments permit measurement of the velocities of P-waves and S-waves through the rocks of the region. Under normal conditions, the ratio of the velocities, V_P/V_S, remains remarkably constant. During an interval before the occurrence of a stronger earthquake, with Richter magnitudes as shown in the diagrams, the ratio first decreases and then increases; the stronger earthquake occurs when the ratio reaches its former value. (These diagrams are simplified versions of published data for illustrative purposes; they do not correspond exactly to the actual data.) (Based on data published in Russia by A. N. Semonova, I. L. Nersesov, and I. G. Simbireva, 1969; by Y. P. Aggarwal and others in *Nature*, Vol. 241, January 12, 1973, pp. 101-104; by J. H. Whitcomb, J. D. Garmany, and D. L. Anderson in *Science*, Vol. 180, May 11, 1973, pp. 632-635).

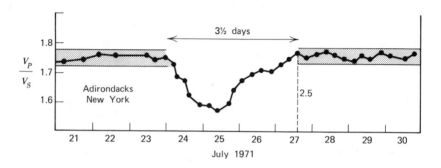

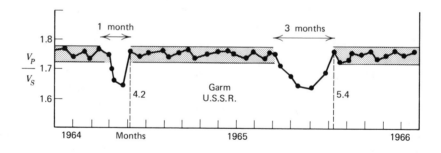

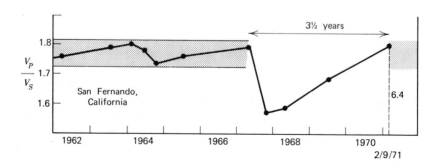

the earthquake is triggered. The time interval between dilatancy and triggering of the earthquake depends on the rate at which water can flow into the region. This depends on the size of the dilatant zone, which is a function of the size of the earthquake that follows.

Figure 18-4 shows that the duration of the period of anomalously low V_P/V_S is a function of the magnitude of the earthquake, as anticipated by the model. The duration of the anomaly is 3.5 days for a magnitude 2.5 event, 30 days for magnitude 4.2, 90 days for magnitude 5.4, and 3.5 years for magnitude 6.4. These data fit a straight line on a graph of logarithm of time in days against the magnitude, and the line indicates an anomaly of about 10 years for a magnitude 7.5 event.

These observations and interpretations therefore include all of the requirements for earthquake prediction. A particular place is monitored with instruments, a decrease in V_P/V_S for earthquake waves passing through the region indicates that an earthquake will occur in that region, the duration of the anomaly is a measure of the predicted earthquake magnitude, and the rate of return of V_P/V_S to normal values permits estimation of the time of the earthquake.

The velocity ratio anomalies may not be symmetrical (Figure 18-4); therefore it is only when V_P/V_S begins to rise again that an initial estimate of the magnitude and the time can be made. The predictions need revision as the return to normal values continues. With the present limited amount of data available, the uncertainty in estimating the time of the event is about 10% of the duration of the anomaly. This is equivalent to about 3 days for an earthquake of magnitude 4.2 and about 4 months for an earthquake of magnitude 6.4.

Some geophysicists believe that dilatancy of very large volumes of rock on the scale necessary for a major earthquake is unlikely, and others dispute details of the dilatancy-diffusion model. Tests include the direct measurement of water flow through the rocks and the measurement of small changes in level and tilt of the surface, because dilatancy and expansion of rock must be reflected by changes in the attitude of the surface. It has been shown for a series of earthquakes in Matsushiro, Japan, between 1965 and 1967, that the directly measured properties agreed closely with theoretical predictions based on the dilatancy-diffusion model. The model also appears to be capable of explaining other phenomena that occur before earthquakes.

Earthquake prediction is on the verge of becoming a reality, at least for some kinds of earthquakes. The faults involved in Figure 18-4 all involve overthrusting (Figure 5-2e). Measurements of V_P/V_S near parts of the San Andreas fault recently indicate changes in V_P/V_S before earthquakes of the lateral slip variety (Figure 5-2a). Some theoretical and laboratory studies of dilatancy, however, suggest that changes in V_P/V_S need not necessarily be followed by an earthquake.

EARTHQUAKE CONTROL

It has been discovered that along parts of some faults in the San Andreas system, relative movement may be accomplished by fault slippage without accompanying earthquakes. This process of fault creep appears to be an important type of deformation, at least locally. Parts of a fault may creep in distinct episodes at rates up to 10 cm per year, while other parts of the same fault may not creep at all. In these locked parts the deformation caused by moving lithosphere plates causes accumulation of strain energy with the prospect that later energy release will produce a large earthquake.

Present knowledge of the processes involved in the generation of earthquakes is not sufficient to guide an engineering program for earthquake control, but hopes lie in the development of methods for releasing locked portions of faults so that accumulation of strain energy is inhibited, and the crustal blocks slide past each other with only small energy-dissipating tremors or, ideally, with none at all. The next best thing to preventing an earthquake is to organize a series of small earthquakes when they are convenient. There is good evidence that earthquakes can be triggered by several means.

The occurrence of earthquakes appears to be linked with the construction of dams and the impounding of large masses of water in reservoirs. They also appear to be associated with the injection of fluids into the ground, and seasonal fluctuations in earthquake activity in some areas show correlation with rainfall. Underground explosions of nuclear devices have also been associated with increased earthquake activity, apparently by releasing natural strain energy within 15 km of the explosion.

The role of pore fluids in earthquake activity received considerable attention before the dilatancy-diffusion model was formulated. The first serious suggestion that earthquakes could be controlled arose from study of the series of earthquakes triggered near Denver by injection of waste fluids into a deep disposal well at the Army's Rocky Mountain Arsenal. Fluids were injected into the fractured rocks below the Arsenal between 1962 and 1966. For six months in 1965, an average of 5.9 million gallons of fluid per month was injected, and earthquake activity reached a peak rate of 45 per month. For the next few months, until the pumping stopped, an average of 2.9 million gallons was injected each month, and earthquake frequency dropped to 18 per month. After the well was closed, earthquakes averaged only 10 per month for 13 months, although 71 were recorded in April 1970.

The case is strong for a cause-and-effect relationship between the injection of fluids and triggering of earthquakes in the Denver region. At the time, questions were raised about the continuation of earthquakes after injection ceased, but it was pointed out that the injected water would migrate only very slowly through the crustal rocks of this region. Until the high fluid pressure at the injection site became equalized throughout the underground reservoir, the possibility of

rock movement and associated earthquakes would have continued.

These observations were followed in 1970 by experiments in the Rangely oil field in northwest Colorado, which confirmed that earthquakes could be turned on and off artificially and at will. In 1973 C. B. Raleigh and J. H. Healy of the U. S. Geological Survey reported that withdrawal of water from a deep faulted zone by pumping from wells reduced the rate of small earthquakes within months. Subsequent injection of water was followed some time later by an increase in the rate of tremors at a depth of nearly 2 km.

It appears that small amounts of water may produce the effect of lubricating a fault, although the process involved is quite different from that of reducing friction on a fault surface. The injection of water may free locked fault systems and release the accumulating strain before it builds up to dangerous levels. This forms the basis for a tentative scheme, illustrated in Figure 18-5, to prevent major earthquakes along the San Andreas fault system in California.

Figure 18-5a represents a series of triplets of bore holes drilled to a depth of about 4 km at intervals of 10 km along the fault zone. The holes in each triplet are 500 m apart, as shown in Figure 18-5b. The first step is to pump water from the outer wells at one triplet. This should switch off the earthquakes in these two regions by increasing the strength of the rocks at depth. The fault becomes locked and strengthened at these points. Then water is pumped into the middle well, as shown in Figure 18-5c, which reduces the strength of the rock at depth making it easier for movement to occur. The hope is that the accumulated strain energy would be released by a series of small earthquakes as the water pressure builds up beneath the middle hole. The process would then be repeated with the next triplet of wells along the fault zone.

There are many technical and theoretical problems to be solved, and there is no certainty that the scheme will ever be put into operation. At present, there is presumably

Fig. 18-5 Control of earthquakes. The method illustrated here has been under careful consideration as a means of controlling and releasing in easy stages the stress associated with the San Andreas fault system in California. A series of small movements and minor earthquakes is safer than one big break along the fracture surfaces, if it can be engineered.

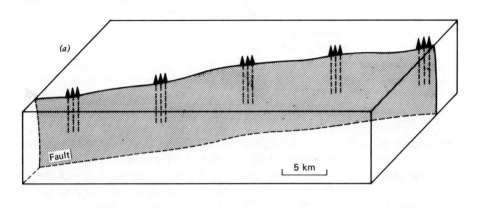

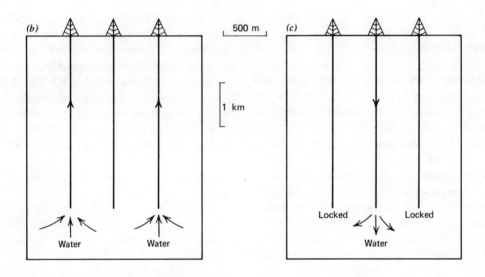

much accumulated strain in parts of the San Andreas fault, and pumping water into a deep well might be sufficient to trigger the disastrous earthquake that the process is designed to prevent by the controlled release of energy. There is no guarantee, as yet, that the locks indicated in Figure 18-5c would successfully limit the fault displacement to the region between them.

About 500 bore holes would be required to tame the San Andreas fault, at a cost of about $1 million per hole. Although $500 million dollars is a lot of money, estimated property damage in the relatively mild San Fernando earthquake of 1971 was $1 billion, and $30 billion in property damage was the estimate for a hypothetical repetition in 1970 of the 1906 San Francisco earthquake (Chapter 4).

Earthquake control may become feasible, but only in regions where the locked fault zones can be reached by drilling. This approach will not work for the earthquakes originating beneath ocean trenches, like the

1970 Peru earthquake. Many engineers emphasize that it is not the earthquakes that kill people, but the failure of buildings that people construct. They maintain that the best approach is not to predict and control earthquakes, but to erect sound buildings, bridges, and dams on relatively safe sites. Then comes the problem of judging the safety of development sites and the effectiveness of building codes and of enforcing the building codes. In California and Japan, where earthquake activity is accepted as a fact of life, building codes include provisions for earthquake-resistant construction. No such provisions are deemed necessary in the central United States where the land is considered to be stable. If the New Madrid earthquakes (Figure 4-5) had occurred this year the wreckage could have been severe.

THE FUTURE

In future years even more than in the past the quality of human life depends on our ability to exploit the world's natural resources without polluting the environment excessively. The general pattern shown by Figures 18-1 and 18-2 for fossil fuels is applicable also to many mineral resources. We are thus approaching a critical stage in the development of the human race, a stage when the demands of an expanding population cannot be met by continued despoliation of the world's known natural resources. Something has to change, otherwise the catastrophic collapse of our industrialized society predicted in *The Limits to Growth* might actually occur.

There have been crises of demand and inadequate supply before, but new developments have always averted disaster. New resources or substitutes were discovered, or something was invented. In Figure 18-2 we have a category of "unknown sources" that might expand to fill the gap between extrapolated demand and estimated supply of energy. Can we afford to assume that something will happen of its own accord to avert the impending indus-

trial and social collapse? I think not. Who, then, will evaluate the situation and make plans and recommendations? Who will make decisions? The influential decisions will be made by politicians and businessmen, guided, we must hope, by advice from scientists and economists. Similarly, while the effective use of earthquake predictions must be planned by social and behavioral scientists, the decisions will be made by government officials. Their considerations must be based on an adequate understanding of the science involved or at least on confident communication with the scientists.

The U.S. Geological Survey began a formal earthquake prediction program in 1974. Japan and the Soviet Union have similar programs. The results in Figure 18-4 may herald success in the development of a reliable earthquake prediction system. Urban dwellers in earthquake belts are accustomed to living with the general threat of some future earthquake, but they choose to remain and enjoy the good life of the city. Success in earthquake prediction could change the situation. A specific prediction that an earthquake of magnitude 7.5 would occur 7 years from today, with an uncertainty of 1 year, could bring with it severe social, economic, and political problems.

Suppose that predictive methods are improved, so that a scientific team was able to predict the occurrence of an earthquake of magnitude 7.5 in a densely populated urban region two days hence. Would the city authorities order evacuation? Imagine the cost of such an evacuation, the social upheaval, and the economic loss through interruption of normal business operations. If the earthquake failed to materialize on schedule, how long would the population stay away, waiting for the shock? Perhaps the authorities would decide that the consequences of a possible earthquake might be less harmful to the city than the definite social and economic disruption of an evacuation. Perhaps they would then issue warnings and advice about taking shelter beneath strong tables or door jambs and mobilize the civil defense. What political recriminations would then result if the prediction was followed by a major earthquake that levelled the city and killed thousands!

If faced with this kind of decision, city leaders would probably regret scientific progress and wish that our ideas on the occurrence of earthquakes had remained in the state of mystery illustrated by the following quotation from Shakespeare's *King Henry IV*.

Oft the teeming earth
Is with a kind of colic pinch'd and vex'd
By the imprisoning of unruly wind
Within her womb; which for enlargement striving
Shakes the old beldam earth and topples down
Steeples and moss-grown towers.

No government could be blamed for an accident arising from Mother Nature's gastrointestinal discomfort. But each year we do learn more about the scientific reasons for what happens in Nature's internal tracts, and this has to be taken into account. Scientific considerations are also involved in decisions about urban and agricultural water supplies, the permissible limits of pollution, the construction and location of nuclear power plants, and many other subjects.

It is quite possible that a decision about whether or not to construct an urgently needed nuclear power plant could be made without adequate appreciation of scientific considerations. A humanistic group of citizens could delay or prevent construction for emotional reasons, however good the scientific reasons for site selection may have been. Similarly, government officials, who remained unconvinced of the significance of a scientific report about a history of microearthquakes in the vicinity some years earlier, could decide to build on the site despite the report.

When we consider the relationship between the human race and the world environment, the existence of intellectual gaps between humanists, physical scientists, and social scientists becomes a matter for serious concern. In today's complex society, with the environment under continual stress, it is more than ever necessary for people in power to have at least an appreciation, and hopefully an understanding, of what physical science is all about.

SUMMARY

Natural resources include water, minerals, rocks, metals, and fossil fuels. Except for water, these materials become concentrated naturally by the rock cycle on time scales of tens to hundreds of millions of years. Plate tectonic theory may provide prospecting guides for minerals, metals, coal, oil, and geothermal energy. Ore deposits with high concentrations of required elements (such as copper) may be mined out before the end of the century. Shortages causing higher prices will make it economically feasible to mine rocks with lower concentrations of the elements. The technology must be developed to recycle the metals and other materials that are concentrated naturally by the rock cycle.

World population and our consumption of natural resources are increasing at exponential rates, and this will bring us to natural limits sooner than most of us realize. Already we experience local pollution at intolerable levels, famine in large areas of the world, and shortages of oil and other resources. Intermittent adjustments in our social and economic systems hopefully will lead to attainment of a harmonious balance between the geological cycles and a world society with stabilized population. Realistic plans on a global scale should be instituted in the near future if unpleasant consequences are to be avoided.

Unless the rate of installation of nuclear power plants increases significantly, there will soon be a large gap between world energy requirements and energy available. Estimates of the future availability of solar energy and geothermal energy are insufficient to fill that gap. Construction of nuclear power plants has been delayed in part through fear of accidental discharge of radioactive materials, especially in earthquake-prone regions.

As a result of recent earthquake studies and laboratory experiments on rock deformation, earthquake prediction is on the verge of becoming a reality. Experiments are also under way on earthquake control by methods that allow crustal blocks to slide past each other instead of jerking.

Success in earthquake prediction could bring severe social, economic, and political problems. Would a city evacuate if an earthquake was predicted?

Leaders in government and industry make decisions affecting thousands or millions of people. With the environment under continual stress, it is more than ever necessary that their decisions are based on adequate understanding of the science involved.

SUGGESTED READINGS

Chapter 1. Introduction: time for a new theory

The chapters cited in this list of comprehensive textbooks deal with the measurement of geological time. These textbooks are cited extensively in the following chapters.

Flint, R. F., and B. J. Skinner, 1974. *Physical geology,* Wiley, New York. Chapter 2.

Gass, I. G., P. J. Smith, and R. C. L. Wilson (editors), 1971. *Understanding the earth,* MIT Press, Cambridge, Mass. Chapter 2.

Geology today, 1973. CRM Books, Del Mar, Cal. Chapter 16.

Hamblin, W. K., 1975. *The earth's dynamic systems.* Burgess, Minneapolis. Chapter 5.

Holmes, A., 1965. *Principles of physical geology,* 2nd ed., Ronald Press Co., New York. Chapter 13.

McAlester, A. L., 1973. *The earth,* Prentice-Hall, N.J. Chapter 10.

Menard, H. W., 1974. *Geology, resources and society,* W. H. Freeman, San Francisco. Chapter 2.

Press, F., and R. Siever, 1974. *Earth,* W. H. Freeman, San Francisco. Chapter 2.

Sawkins, F. J., C. G. Chase, D. G. Darby, and G. Rapp, 1974. *The evolving earth,* Macmillan, New York. Chapter 3.

Strahler, A. N., 1971. *The earth sciences,* 2nd ed., Harper and Row, New York. Chapter 3 and pages 462-464.

Verhoogen, J., F. J. Turner, L. E. Weiss, C. Wahrhaftig, and W. S. Fyfe, 1970. *The earth,* Holt, Rinehart and Winston, New York. Chapter 4.

Collections of articles include the following:

Cox, A. (editor), 1973. *Plate tectonics and geomagnetic reversals,* W. H. Freeman, San Francisco. 49 selected research articles, with introductions by Cox.

Bird, J. M., and B. Isacks (editors), 1972. *Plate tectonics.* Technical papers selected from *Journal of Geophysical Research.*

Press, F., and R. Siever (compilers), 1974. *Planet earth,* readings from *Scientific American,* W. H. Freeman, San Francisco.

Wilson, J. T. (compiler), 1972. *Continents adrift,* readings from *Scientific American,* W. H. Freeman, San Francisco.

Chapter 2. Revolution in the earth sciences

Calder, N., 1972. *The restless earth,* Viking Press, New York. The book of the television program, full of illustrations and vignettes. Pleasant reading.

Cox, A., 1973. *Plate tectonics and geomagnetic reversals,* W. H. Freeman, San Francisco. The nine historical introductions by the editor, with comments and photographs of many scientists, convey something of the excitement experienced by earth scientists during this revolution.

Drake, C., 1970. *The geological revolution.* Condon Lectures, Oregon State System of Higher Education, Eugene.

Glen, W., 1975. *Continental drift and plate tectonics,* Merrill, Columbus, Ohio. Chapter 9 is a historical summary.

Hallam, A., 1973. *A revolution in the earth sciences,* Clarendon, Oxford. A brief historical account overlapping to some extent with the book by Marvin, but with different style.

Kuhn, T. S., 1970. *The structure of scientific revolutions,* 2nd ed., University of Chicago Press, Chicago. First published in 1962, this volume could have been written with the paradigm of plate tectonics as its main example.

Marvin, U. B., 1973. *Continental drift,* Smithsonian Institution Press, City of Washington. A delightful account of the history of an idea, illustrated with many pictures rarely reproduced in other books.

Matthews, S. W., 1973. "This changing earth." *National Geographic, 143,* 1-37. A popular account of sea-floor spreading and continental drift, with good illustrations.

Sanders, H. J., 1967. "Chemistry and the solid earth." Special report by *Chemical and Engineering News,* October 2 issue. This review captures the excitement of scientists sensing that something big is imminent.

Spilhaus, A. F., 1974. "Geophysics—an overview." *Physics Today,* March issue, 23-26.

Sullivan, W., 1974. *Continents in motion,* McGraw-Hill, New York. An account of the revolution by the science editor of the *New York Times.* It is long, entertaining, scientifically correct, and full of insights about scientists at work.

Takeuchi, H., S. Uyeda, and H. Kanamori, 1970. *Debate about the earth,* rev. ed., Freeman, Cooper, San Francisco. A delightful account of the continental drift debate, with concentration on fossil magnetism. This is based on a television program of the Japanese Broadcasting Company.

Wertenbaker, W., 1974. *The floor of the sea: Maurice Ewing and the search to understand the earth,* Little, Brown, New York. Establishment of Lamont Geological Observatory at Columbia University by Maurice Ewing had much to do with the revolution. Parts of the book were presented as a "Profiles" in three issues of the *New Yorker,* November 4, 11, and 18, 1974.

Wilson, J. T. (compiler), 1972. *Continents adrift,* readings from *Scientific American,* W. H. Freeman, San Francisco. Historical developments are outlined by Wilson on pages 1-3, 38-40, 100-101, and 158-160.

Wyllie, P. J., 1971. "Revolution in the earth sciences," *The Great Ideas Today 1971,* 168-237, Encyclopedia Britannica, Inc., Chicago. A general account for the intelligent layman.

Wyllie, P. J., 1971. *The dynamic earth,* Wiley, New York. Historical developments can be traced through Chapters 11 to 15 without tackling the more advanced material in this textbook.

Chapter 3.   The earth's solid surface

Flint, R. F., and B. J. Skinner, 1974. *Physical geology,* Wiley, New York. Chapters 12 and 13 illustrate surface features, and Appendix D includes maps and cross sections.

Hamblin, W. K., 1975. *The earth's dynamic systems,* Burgess, Minneapolis. Chapters 1 and 16 include excellent pictures and diagrams of the earth's surface.

Holmes, A., 1965. *Principles of physical geology,* 2nd ed., Ronald Press Co., New York. This book includes nearly everything. Examine Chapters 2, 12, 24, 29, and 30, and then discover more in the index.

Howell, B. F., 1972. *Earth and universe,* Merrill, Columbus, Ohio. Chapters 9 and 14 include maps and diagrams of interest.

Lowman, P. D., 1972. *The third planet,* Weltflugbild Reinhold A. Miller, Zurich. A magnificent collection of photographs of earth from space.

Menard, H. W., 1964. *Marine geology of the Pacific.* McGraw-Hill, New York. This is full of informative maps and cross sections.

Press, F., and R. Siever, 1974. *Earth,* W. H. Freeman, San Francisco. Chapter 5 illustrates the face of the earth, with numerical data given in Appendix III, and Appendix V describes topographic maps.

Shephard, F. P., 1973. *Submarine geology,* 3rd ed., Harper and Row, New York. The ocean floor described in detail.

Strahler, A. N., 1971. *The earth sciences,* 2nd ed., Harper and Row, New York. This is another book that includes nearly everything. See Chapters 11 and 24 for some surface features, Appendix I for maps and contours, and use the index for whatever you need.

Wertenbaker, W., 1974. *The floor of the sea: Maurice Ewing and the search to understand the earth,* Little, Brown, New York. A fascinating biography of a giant in geophysics. The first part of the book shows how oceanographic ships from Lamont Geological Observatory gathered the data during the 1950s that prepared the way for revolution in the 1960s.

Wyllie, P. J., 1971. *The dynamic earth,* Wiley, New York. Chapters 1 and 2 include informative tables and figures.

Chapter 4.   Earthquakes: their effects and distribution

Canby, T. Y., and J. P. Blair, 1973. "California's San Andreas fault." *National Geographic, 143,* 38-53. An account of the most famous fault in North America.

Cargo, D. N., and B. F. Mallory, 1974. *Man and his geologic environment,* Addison-Wesley, Reading, Mass. Disastrous earthquakes are reviewed in Chapter 10.

Holmes, A., 1965. *Principles of physical geology,* 2nd ed., Ronald Press Co., New York. Chapter 25 describes the Lisbon earthquake that shook Christendom in 1755.

Howell, B. F., 1972. *Earth and universe,* Merrill, Columbus, Ohio. Chapter 13 covers earthquakes.

Menard, H. W., 1974. *Geology, resources, and society*, W. H. Freeman, San Francisco. Earthquakes and their effects are illustrated in Chapters 7 and 8.

Gillette, R., and J. Walsh, 1971. "San Fernando earthquake study: NRC panel sees premonitory lessons," *Science, 172*, 140-143. Lessons from an early warning in urban California.

Press, F., and R. Siever, 1974. *Earth*, W. H. Freeman, San Francisco. Chapter 19 includes much useful information about earthquakes.

Sawkins, F. J., C. G. Chase, D. G. Darby, and G. Rapp, 1974. *The evolving earth*, Macmillan, New York. Chapters 5 and 12 include earthquake distribution and effects. Note Figure 7-1 of page 181 for sinking lithosphere contours.

Smith, P. J., 1973. *Topics in geophysics*, MIT Press, Cambridge, Mass. Many destructive earthquakes are described in Chapter 4.

Vitaliano, D. B., 1973. *Legends of the earth: their geologic origins*, Indiana University Press, Bloomington, Ind. Chapter 5 reviews earthquake lore.

Wilson, J. T. (compiler), 1972. *Continents adrift*, readings from *Scientific American*. W. H. Freeman, San Francisco. Article 15 explains the San Andreas fault and discusses the San Fernando earthquake.

Wilson, J. T., 1972. "Mao's almanac: 3,000 years of killer earthquakes," *Saturday Review*, February 19 issue, 60-64. Earthquakes of the orient.

Yanev, P., 1974. *Peace of mind in earthquake country. How to save your home and life*, Chronicle Books, San Francisco. The title says it all.

Chapter 5. Earthquakes and the theory of plate tectonics

Flint, R. F., and B. J. Skinner, 1974. *Physical geology*, Wiley, New York. Chapters 17 and 18 include plate tectonics.

Gass, I. G., P. J. Smith, and R. C. L. Wilson (editors), 1971. *Understanding the earth*, MIT Press, Cambridge, Mass. Chapter 19 is an account of plate tectonics by E. R. Oxburgh, an innovator in the field.

Geology Today, 1973. CRM Books, Del Mar, Cal. Chapters 7 and 8 include some fine illustrations.

Glen, W., 1975. *Continental drift and plate tectonics*, Merrill, Columbus, Ohio. Chapter 7 puts earthquakes together with plate boundaries.

Hamblin, W. K., 1975. *The earth's dynamic systems*, Burgess, Minneapolis. Pages 55-58 and Chapter 15 cover the basic material.

Menard, H. W., 1974. *Geology, resources, and society*, W. H. Freeman, San Francisco. Plate tectonics and earthquakes are related in Chapter 5, with volcanoes added in Chapter 9.

Press, F., and R. Siever, 1974. *Earth*, W. H. Freeman, San Francisco. Chapters 19 and 21.

Press, F., and R. Siever (compilers), 1974. *Planet earth*, readings from *Scientific American*. W. H. Freeman, San Francisco. Article 11 shows how plates rotate.

Wyllie, P. J., 1971. *The dynamic earth,* Wiley, New York. The first part of Chapter 14 relates the earthquake distributions of Chapter 2 to plates and their movements.

Chapter 6. The geological cycle

This material occupies a large part of most introductory textbooks, and diligent use of indices is therefore recommended.

Flint, R. F., and B. J. Skinner, 1974. *Physical geology,* Wiley, New York. Chapters 2, 4 to 11, 15 to 17.

Garrels, R. M., and F. T. Mackenzie, 1971. *Evolution of sedimentary rocks,* Norton, New York. This is a new approach to the familiar cycles. Chapters 4 to 6 and 9 to 12 are of particular interest.

Geology today, 1973. CRM Books, Del Mar, Cal. Chapters 8 to 13 review the cycle in the context of plate tectonics.

Hamblin, W. K., 1975. *The earth's dynamic systems,* Burgess, Minneapolis. Chapters 4, 6 to 14, and 17.

Holmes, A., 1965. *Principles of physical geology,* 2nd ed., Ronald Press Co., New York. The geological cycle before plate tectonics extends through many chapters in the contents list.

Jacobs, J. A., R. D. Russell, and J. T. Wilson, 1974. *Physics and geology,* 2nd ed., McGraw-Hill, New York. Chapters 15 and 16 explain the life cycle of ocean basins and transported continents in terms of plate tectonics.

Press, F., and R. Siever, 1974. *Earth,* W. H. Freeman, San Francisco. Chapters 4, 6 to 13, and 16 to 18 cover the geological cycles, and Chapter 21 presents plate tectonics as the unifying model.

Press, F., and R. Siever (compilers), 1974. *Planet earth,* readings from *Scientific American.* W. H. Freeman, San Francisco. Articles 12 and 13 illustrate plate tectonic cycles for the evolution of the Andes and the Indian Ocean.

Sawkins, F. J., C. G. Chase, D. G. Darby, and G. Rapp, 1974. *The evolving earth,* Macmillan, New York. Chapters 2 and 6 to 11 cover the cycles, with Chapters 6 to 8 considering in particular the great cycle of plate generation and destruction.

Shelton, J. S., 1966. *Geology illustrated,* W. H. Freeman, San Francisco. The next best thing to going on a field trip.

Strahler, A. N., 1971. *The earth sciences,* 2nd ed., Harper and Row, New York. Parts 3 and 4. Chapters 33 and 34 include statistics for water distribution.

Verhoogen, J., F. J. Turner, L. E. Weiss, C. Wahrhaftig, and W. S. Fyfe, 1970. *The earth,* Holt, Rinehart and Winston, New York. You can find a detailed treatment of most topics in this book. Check the contents list and the index.

Wilson, J. T. (compiler), 1972. *Continents adrift,* readings from *Scientific American.* W. H. Freeman, San Francisco. Article 13.

Wyllie, P. J., 1971. *The dynamic earth,* Wiley, New York. Geological cycles and processes are concisely summarized in Chapter 4.

Chapter 7.   Earthquake waves and the inside of the earth

Clark, S. P., 1971. *Structure of the earth,* Prentice-Hall, N.J.

Geology today, 1973. CRM Books, Del Mar, Cal. Chapter 5 has colorful illustrations.

Holmes, A., 1965. *Principles of physical geology,* 2nd ed., Ronald Press Co., New York. Chapters 25 and 26.

Howell, B. F., 1972. *Earth and universe,* Merrill, Columbus, Ohio. Chapter 10.

Jacobs, J. A., R. D. Russell, and J. T. Wilson, 1974. *Physics and geology,* 2nd ed., McGraw-Hill, New York. Chapters 2 and 10 give an advanced review.

Press, F., and R. Siever, 1974. *Earth,* W. H. Freeman, San Francisco. Chapter 19.

Press, F., and R. Siever (compilers), 1974. *Planet earth,* readings from *Scientific American.* W. H. Freeman, San Francisco. Article 23 describes the fine details that have been detected in recent work.

Strahler, A. N., 1971. *The earth sciences,* 2nd ed., Harper and Row, New York. Chapter 23.

Wilson, T. J. (compiler), 1972. *Continents adrift,* readings from *Scientific American.* W. H. Freeman, San Francisco. Articles 4 and 5 present results from 1955 and 1962 that have since been refined—compare with the references above.

Wyllie, P. J., 1975. "The earth's mantle." *Scientific American, 232*(3), 50-63.

Chapter 8.   The earth's magnetic field and magnetized rocks

Gass, I. G., P. J. Smith, and R. C. L. Wilson (editors), 1971. *Understanding the earth,* MIT Press, Cambridge, Mass. Chapter 4.

Glen, W., 1975. *Continental drift and plate tectonics,* Merrill, Columbus, Ohio. Chapter 6.

Howell, B. F., 1972. *Earth and universe,* Merrill, Columbus, Ohio. Chapter 7.

Phillips, O. M., 1968. *The heart of the earth,* Freeman, Cooper and Co., San Francisco. Chapter 7.

Press, F., and R. Siever, 1974. *Earth,* W. H. Freeman, San Francisco. Chapter 20.

Press, F., and R. Siever (compilers), 1974. *Planet earth,* readings from *Scientific American.* W. H. Freeman, San Francisco. Article 21 reviews the earth as a dynamo.

Strahler, A. N., 1971. *The earth sciences,* 2nd ed., Harper and Row, New York. Chapter 7.

Takeuchi, H., S. Uyeda, and H. Kanamori, 1970. *Debate about the earth,* rev. ed., Freeman, Cooper, San Francisco, Chapters 3 to 5.

Verhoogen, J., F. J. Turner, L. E. Weiss, C. Wahrhaftig, and W. S. Fyfe, 1970. *The earth,* Holt, Rinehart and Winston, New York. Chapter 11.

Chapter 9. The history of the earth's magnetic field: polarity reversals

Cox, A. (editor), 1973. *Plate tectonics and geomagnetic reversals,* W. H. Freeman, San Francisco. Pages 138-147.

Gass, I. G., P. J. Smith, and R. C. L. Wilson (editors), 1971. *Understanding the earth,* MIT Press, Cambridge, Mass. Chapter 17.

Hamblin, W. K., 1975. *The earth's dynamic systems,* Burgess, Minneapolis. Chapter 15.

Menard, H. W., 1974. *Geology, resources, and society,* W. H. Freeman, San Francisco. Chapter 3.

Press, F., and R. Siever, 1974. *Earth,* W. H. Freeman, San Francisco. Chapter 20.

Strahler, A. N., 1971. *The earth sciences,* 2nd ed., Harper and Row, New York. Chapter 25.

Sullivan, W., 1974. *Continents in motion,* McGraw-Hill, New York. Chapter 6.

Takeuchi, H., S. Uyeda, and H. Kanamori, 1970. *Debate about the earth,* rev. ed., Freeman, Cooper, San Francisco. Chapters 4 and 8.

Wyllie, P. J., 1971. *The dynamic earth,* Wiley, New York. Chapter 12.

Chapter 10. Magnetic anomalies and the sea-floor spreading model

Cox, A. (editor), 1973. *Plate tectonics and geomagnetic reversals,* W. H. Freeman, San Francisco. Pages 222-228.

Flint, R. F., and B. J. Skinner, 1974. *Physical geology,* Wiley, New York. Chapter 18.

Geology today, 1973. CRM Books, Del Mar, Cal. Chapter 7.

Gass, I. G., P. J. Smith, and R. C. L. Wilson (editors), 1971. *Understanding the earth,* MIT Press, Cambridge, Mass. Chapter 16.

Glen, W., 1975. *Continental drift and plate tectonics,* Merrill, Columbus. Chapter 7.

Hamblin, W. K., 1975. *The earth's dynamic systems,* Burgess, Minneapolis. Chapter 15.

Press, F., and R. Siever, 1974. *Earth,* W. H. Freeman, San Francisco. Chapters 20 and 21.

Sawkins, F. J., C. G. Chase, D. G. Darby, and G. Rapp, 1974. *The evolving earth,* Macmillan, New York. Chapter 5.

Sullivan, W., 1974. *Continents in motion,* McGraw-Hill, New York. Chapter 6.

Takeuchi, H., S. Uyeda, and H. Kanamori, 1970. *Debate about the earth,* rev. ed., Freeman, Cooper, San Francisco. Chapter 8.

Verhoogen, J., F. J. Turner, L. E. Weiss, C. Wahrhaftig, and W. S. Fyfe, 1970. *The earth,* Holt, Rinehart and Winston, New York. Chapter 13.

Wilson, J. T. (compiler), 1972. *Continents adrift,* readings from *Scientific American.* Articles 8 to 10.

Chapter 11. Continental drift and paleomagnetism

Gass, I. G., P. J. Smith, and R. C. L. Wilson (editors), 1971. *Understanding the earth,* MIT Press, Cambridge, Mass. Chapter 15.

Geology today, 1973. CRM Books, Del Mar, Cal. Chapter 7.

Glen, W., 1975. *Continental drift and plate tectonics,* Merrill, Columbus, Ohio. Chapters 3 and 4.

Hallam, A., 1973. *A revolution in the earth sciences,* Clarendon, Oxford. Chapters 1 to 3.

Hamblin, W. K., 1975. *The earth's dynamic systems,* Burgess, Minneapolis. Chapter 15.

Holmes, A., 1965. *Principles of physical geology,* 2nd ed., Ronald Press Co., New York. Chapter 31.

Jacobs, J. A., R. D. Russell, and J. T. Wilson, 1974. *Physics and geology,* 2nd ed., McGraw-Hill, New York. Chapter 13.

Marvin, U. B., 1973. *Continental drift,* Smithsonian Institution Press, City of Washington.

Menard, H. W., 1974. *Geology, resources and society,* W. H. Freeman, San Francisco. Chapter 6.

Press, F., and R. Siever, 1974. *Earth,* W. H. Freeman, San Francisco. Chapter 21.

Sawkins, F. J., C. G. Chase, D. G. Darby, and G. Rapp, 1974. *The evolving earth,* Macmillan, New York. Chapter 5.

Takeuchi, H., S. Uyeda, and H. Kanamori, 1970. *Debate about the earth,* rev. ed., Freeman and Cooper, San Francisco. Chapters 1, 2, and 5.

Wegener, A., 1966. *The origin of continents and oceans.* Translated by John Biram from the 4th (1929) German edition. Dover, New York.

Wilson, J. T. (compiler), 1972. *Continents adrift,* readings from *Scientific American.* W. H. Freeman, San Francisco. Articles 6 and 7.

Chapter 12.   Paths of polar wandering and migration of the continents

Geology today, 1973. CRM Books, Del Mar, Cal. Chapter 7.

Glen, W., 1975. *Continental drift and plate tectonics,* Merrill, Columbus, Ohio. Chapters 6 and 8.

Hallam, A., 1973. *A revolution in the earth sciences,* Clarendon, Oxford. Chapter 4.

Hamblin, W. K., 1975. *The earth's dynamic systems,* Burgess, Minneapolis. Chapter 15.

Holmes, A., *Principles of physical geology,* Ronald Press Co., New York. Chapter 31.

Jacobs, J. A., R. D. Russell, and J. T. Wilson, 1974. *Physics and geology,* 2nd ed., McGraw-Hill, New York. Chapter 14.

Marvin, U. B., 1973. *Continental drift,* Smithsonian Institution Press, City of Washington.

Menard, H. W., 1974. *Geology, resources and society,* W. H. Freeman, San Francisco. Chapter 6.

Press, F., and R. Siever, 1974. *Earth,* W. H. Freeman, San Francisco. Chapter 21.

Sullivan, W., 1974. *Continents in motion,* McGraw-Hill, New York. Chapters 2 and 6.

Takeuchi, H., S. Uyeda, and H. Kanamori, 1970. *Debate about the earth,* rev. ed., Freeman, Cooper, San Francisco. Chapter 5.

Wilson, J. T. (compiler), 1972. *Continents adrift,* readings from *Scientific American,* W. H. Freeman, San Francisco. Article 11.

Chapter 13.   A revolution proclaimed—the evidence challenged

Adams, F. D., 1954. *The birth and development of the geological sciences.* Reprint of the 1938 edition, Dover, New York.

Calder, N., 1972. *The restless earth,* Viking Press, New York. Chapter 2 covers the conversion of a profession.

Cox, A., 1973. *Plate tectonics and geomagnetic reversals,* W. H. Freeman, San Francisco. The editor's nine introductions to selected groups of research papers are very informative.

Glen, W., 1975. *Continental drift and plate tectonics,* Merrill, Columbus, Ohio. Chapter 9 is a historical summary. See pages 147-148 for some objections.

Hallam, A., 1973. *A revolution in the earth sciences,* Clarendon, Oxford. Chapters 5 and 6 cover the period of rapid change. Chapter 8 discusses dissent.

Marvin, U. B., 1973. *Continental drift,* Smithsonian Institution Press, City of Washington. The revolutionary period is in pages 121-207.

Menard, H. W., 1971. *Science: growth and change,* Harvard University Press, Cambridge, Mass.

Sullivan, W., 1974. *Continents in motion,* McGraw-Hill, New York. Aspects of the revolution are distributed throughout this volume. Chapter 7 discusses "The Doubters."

Wertenbaker, W., 1974. *The floor of the sea: Maurice Ewing and the search to understand the earth.* Much of this book is concerned with the work of scientists at Lamont-Doherty Geological Observatory of Columbia University. The discoveries and reactions of the scientists between 1965 and 1968 make a fascinating story.

Wilson, J. T. (compiler), 1972. *Continents adrift,* readings from *Scientific American.* W. H. Freeman, San Francisco. Historical developments are outlined in a series of reviews by Wilson on pages 1-3, 38-40, 100-101, and 158-160.

Chapter 14.   Voyages of the Glomar Challenger, and deep-sea sediments

Briggs, P., 1971. *200,000,000 years beneath the sea,* Holt, Rinehart and Winston, New York. The story of the *Glomar Challenger* up to 1970, with the discovery of evaporite deposits below the Mediterranean.

Calder, N., 1972. *The restless earth,* Viking Press, New York. Chapters 1 and 2 include Atlantis and the *Glomar Challenger.*

Gass, I. G., P. J. Smith, and R. C. L. Wilson (editors), 1971. *Understanding the earth,* MIT Press, Cambridge, Mass. Chapter 25 describes the Mohole project as a geopolitical fiasco.

Geology today, 1973. CRM Books, Del Mar, Cal. Chapter 13 reviews deep sea sediments and the drilling results.

Matthews, S. W., 1973. This changing earth, *National Geographic, 143,* 1-37. A general article with striking pictures of the Gibraltar Cascades that flooded the Mediterranean desert.

Press, F., and R. Siever (compilers), 1974. *Planet earth,* readings from *Scientific American,* W. H. Freeman, San Francisco. Article 17 is "When the Mediterranean dried up."

Strahler, A. N., 1971. *The earth sciences,* 2nd ed., Harper and Row, New York. Chapter 24 reviews the sediments of the ocean basins that are being explored by the *Glomar Challenger.*

Sullivan, W., 1974. *Continents in motion,* McGraw-Hill, New York. Chapters 8 and 9 review the Mohole and the *Glomar Challenger.* Chapter 10 covers drowned lands and evaporated oceans: Atlantis and the Mediterranean desert.

Vitaliano, D. B., 1973. *Legends of the earth: their geologic origins,* Indiana University Press, Bloomington, Ind. Chapters 8 to 10 discuss Santorin, Atlantis, and ancient Egypt.

Chapter 15.   The diversity and extinction of species

Calder, N., 1972. *The restless earth,* Viking Press, New York. Chapter 5 discusses "Living on rafts."

Gass, I. G., P. J. Smith, and R. C. L. Wilson (editors), 1971. *Understanding the earth,* MIT Press, Cambridge, Mass. Chapters 14 and 18 review evolution, continental drift, and the effect of polarity reversals.

Glen, W., 1975. *Continental drift and plate tectonics,* Merrill, Columbus, Ohio. The effect of continental drift and polarity reversals on species diversity is discussed in Chapter 4.

Hallam, A., 1973. *A revolution in the earth sciences,* Clarendon, Oxford. Ancient faunal distributions occupy pages 100-103.

McAlester, A. L., 1973. *The earth,* Prentice-Hall, N.J. Chapter 12 reviews evolution and the influence of drifting continents.

Press, F., and R. Siever (compilers), 1974. *Planet earth,* readings from *Scientific American,* W. H. Freeman, San Francisco. Article 14 shows how the fossil record contributes to the theory of continental drift.

Sullivan, W., 1974. *Continents in motion,* McGraw-Hill, New York. The diversity of species on drifting continents is reviewed in Chapter 12, and the effect of magnetic polarity reversals is discussed on pages 89-92.

Wilson, J. T. (compiler), 1972. *Continents adrift,* readings from *Scientific American.* W. H. Freeman, San Francisco. Article 12 reviews continental drift and evolution.

Chapter 16.   Causes of plate tectonics

Calder, N., 1972. *The restless earth,* Viking Press, New York. Chapter 4 ends with hotspots and convection.

Hamblin, W. K., 1975. *The earth's dynamic systems,* Burgess, Minneapolis. Convection is discussed on pages 351-352, and hotspots on pages 364-366.

Holmes, A., 1965. *Principles of physical geology,* rev. ed., Ronald Press Co., New York. Chapter 28 includes early views on convection and hotspots.

Jacobs, J. A., R. D. Russell, and J. T. Wilson, 1974. *Physics and geology,* 2nd ed., McGraw-Hill, New York. Convection is reviewed in Chapter 12, and possible causes of plate motion on pages 492-507.

Phillips, O. M., 1968. *The heart of the earth,* Freeman, Cooper and Co., San Francisco. Chapter 6 includes convection.

Press, F., and R. Siever, 1974. *Earth,* W. H. Freeman, San Francisco. Chapter 15 reviews the internal heat of the earth, which drives the convection illustrated in Chapter 21.

Sullivan, W., 1974. *Continents in motion,* McGraw-Hill, New York. Mantle convection and volcanic hotspots are reviewed in Chapters 5 and 14.

Takeuchi, H., S. Uyeda, and H. Kanamori, 1970. *Debate about the earth,* rev. ed., Freeman, Cooper and Co., San Francisco. Chapters 2 and 7 illustrate old and new concepts of convection.

Verhoogen, J., F. J. Turner, L. E. Weiss, C. Wahrhaftig, and W. S. Fyfe, 1970. *The earth,* Holt, Rinehart and Winston, New York. Conditions for convection in the mantle are covered on pages 647-655.

Wyllie, P. J., 1975. The earth's mantle, *Scientific American,* 232(3), 50-63.

Chapter 17. Exploration of the moon

Anderson, D. L., 1974. "The interior of the moon," *Physics Today,* March issue, 44-49.

Apollo 11 Lunar Science Conference, 1970. American Association for the Advancement of Science, Washington, D. C. This is a reprint of the articles published in *Science,* January 30, 1970, following the first Lunar Science Conference.

Baldwin, R. B., 1963. *The measure of the moon,* University of Chicago Press, Chicago. A detailed account of facts and beliefs about the moon before the Apollo program.

Flint, R. F., and B. J. Skinner, 1974. *Physical geology,* Wiley, New York. Chapter 19 includes the moon in an informative account of the earth's neighbours in space.

Geology today, 1973. CRM Books, Del Mar, Cal. Chapters 1 to 3 introduce the earth in space and evolution and present a well-illustrated account of the moon and the other planets.

Hamblin, W. K., 1975. *The earth's dynamic systems,* Burgess, Minneapolis. Chapters 19 to 22 illustrate many lunar features and compare them with features of the other planets.

Levinson, A. A., and S. R. Taylor, 1971. *Moon rocks and minerals.* This book reviews the data resulting from the first study of Apollo 11 rocks. Compare Mason and Melson.

Lowman, P. D., 1969. *Lunar panorama: a photographic guide to the geology of the moon,* Weltflugbild Reinhold A. Muller, Zurich. An excellent selection of photographs organized to provide the reader with a guided tour of the moon, with interpretations added.

Mason, B., and W. G. Melson, 1970. *The lunar rocks,* Wiley-Interscience, New York. A summary of the properties of the Apollo 11 rocks that just beat Levinson and Taylor into print.

McAlester, A. L., 1973. *The earth,* Prentice-Hall, N. J. Chapters 13 to 15 describe the earth in space, with the moon receiving particular attention in Chapter 14.

Press, F., and R. Siever, 1974. *Earth,* W. H. Freeman, San Francisco. Chapter 1 introduces the solar system, and Chapter 23 summarizes current knowledge of the planets with considerable insight.

Press, F., and R. Siever (compilers),1974. *Planet earth,* readings from *Scientific American.* W. H. Freeman, San Francisco. Article 24 reviews "The chemistry of the solar system."

Ruzic, N. P., 1970. *Where the winds sleep: man's future on the moon, a projected history,* Doubleday, New York. Is it science fiction, or will it really become history?

Smith, J. V., and I. M. Steele, 1973. How the Apollo program changed the geology of the moon. *Bulletin of the Atomic Scientists, 29,* 11-15. Many scientists were made to change their minds.

Strahler, A. N., 1971. *The earth sciences,* 2nd ed., Harper and Row, New York. Chapters 5, 8, and 27 describe the solar system and its origin, and Chapter 31 reviews the geology of planetary space including the moon.

Weaver, K. F., 1973. "Have we solved the mysteries of the moon?" *National Geographic, 144,* No. 3, 309-325. A well-illustrated, intelligent review of the discoveries and problems, written for the layman. Three other articles in the same issue "sum up mankind's greatest adventure."

Chapter 18.   Environmental geology and earthquake prediction

Anderson, C. J., 1973. "Animals, earthquakes, and eruptions." *Field Museum of Natural History Bulletin, 44,* May issue, 9-11. Chicago.

Berry, A., 1974. *The next ten thousand years,* Saturday Review Press/Dutton, New York.

Calder, N., 1972. *The restless earth,* Viking Press, New York. Chapter 6 includes earthquake prediction and natural resources.

Cargo, D. N., and B. F. Mallory, 1974. *Man and his geologic environment,* Addison-Wesley, Reading, Mass. One of the recent books bringing geology closer to home. The detailed contents list will guide you to population, water, minerals, energy, earthquakes, and volcanoes, among other topics.

Energy. A special issue of *Science, 184,* April 19, 1974. 25 articles covering people and institutions, policy, economics, fossil fuels and technology, developing technology of sun and earth materials.

Energy and Power. A special issue of *Scientific American,* 1971, 224(3).

Flint, R. F., and B. J. Skinner, 1974. *Physical geology,* Wiley, New York. Chapters 20 and 21 review energy and mineral resources, and describe the effects of man as a geologic agent.

Gass, I. G., P. J. Smith, and R. C. L. Wilson (editors), 1971. *Understanding the earth,* MIT Press, Cambridge, Mass. Chapter 23 advocates research toward earthquake prediction and control.

Geology today, 1973. CRM Books, Del Mar, Cal. Chapter 18 discusses population pressure and the problems of exponential growth and exponential consumption of natural resources.

Kissinger, C., 1974. Earthquake prediction, *Physics Today, 27,* March issue, 36-42.

Hamblin, W. K., 1975. *The earth's dynamic systems,* Burgess, Minneapolis. The environment, natural resources, and the limits to growth are reviewed in Chapter 18.

Maddox, J., 1972. *The doomsday syndrome: an attack on pessimism,* McGraw-Hill, New York.

Meadows, D. H., D. L. Meadows, J. Randers, and W. Behrens, 1972. *The limits to growth,* Universe, New York.

Menard, H. W., 1974. *Geology, resources, and society,* W. H. Freeman, San Francisco. This is my favorite environmental book. Chapters 8, 12, 17, 18, and 19 include earthquakes, climate, and natural resources.

Press, F., and R. Siever, 1974. *Earth,* W. H. Freeman, San Francisco. Chapter 19 includes a section on earthquake risk and prediction by an earthquake expert, and Chapter 24 reviews the future of energy and mineral resources.

Press, F., and R. Siever (compilers), 1974. *Planet earth,* readings from *Scientific American,* W. H. Freeman, San Francisco. Articles 1 to 3 are concerned with energy, Articles 4 to 10 deal with environmental cycles including water, oxygen and carbon, and Article 15 relates mineral resources to plate tectonics.

Sawkins, F. J., C. G. Chase, D. G. Darby, and G. Rapp, 1974. *The evolving earth,* Macmillan, New York. Chapter 12 is environmental, including earthquake hazards and earth resources. Plate tectonics provides a guide to some mineral deposits.

Skinner, B. J., 1969. *Earth resources,* Prentice-Hall, N. J.

Smith, P. J., 1973. *Topics in geophysics,* MIT Press, Cambridge, Mass. Earthquake prediction and modification is covered in some detail in Chapter 4.

Sullivan, W., 1974. *Continents in motion,* McGraw-Hill, New York. Earthquake prediction and prevention, geothermal energy, oil, and metal deposits are reviewed in Chapters 18 to 20.

Watkins, J. S., M. L. Bottino, and M. Morisawa, 1975. *Our geological environment,* Saunders, Philadelphia. The contents list leads to earthquakes and resources. See Chapter 12, "Who is responsible?"

Yanev, P., 1974. *Peace of mind in earthquake country. How to save your home and life.* Chronicle books, San Francisco.

Young, K., 1975. *Geology: the paradox of earth and man,* Houghton Mifflin, Boston. The paradox is man's destruction of the geologic environment that is his provider. People pollution is covered in Chapter 15, and earthquakes are reviewed in Chapter 9. The detailed contents list will lead you to other perils from water, volcanoes, and changing climate.

index